THIS DELTA, THIS LAND

MIKKO SAIKKU

# *This Delta, This Land*

An Environmental History

of the Yazoo-Mississippi Floodplain

*The University of Georgia Press | Athens and London*

© 2005 by the University of Georgia Press
Athens, Georgia 30602
www.ugapress.org
All rights reserved
Designed by Mindy Basinger Hill
Set in Sabon by Bookcomp, Inc.

Printed digitally in the United States of America

Library of Congress Cataloging-in-Publication Data

Saikku, Mikko.
    This delta, this land : an environmental history
of the Yazoo-Mississippi floodplain / Mikko Saikku.
    xvii, 373 p. : ill., maps ; 23 cm.
    Includes bibliographical references (p. 313–357) and index.
    ISBN 0-8203-2534-1 (alk. paper)—
    ISBN 0-8203-2673-9 (pbk. : alk. paper)
    1. Delta (Miss. : Region)—Environmental
conditions—History. I. Title.
    GE155.M57S35 2005
    333.7'09762'4—dc22        2004013923

Paperback ISBN-13: 978-0-8203-2673-3

British Library Cataloging-in-Publication Data available

Parts of chapters 2 and 7 originally appeared as "Down by
the Riverside: The Disappearing Bottomland Hardwood
Forest of Southeastern North America" in *Environment and
History* 2 (February 1996): 77–95; and "'Home in the Big
Forest': Decline of the Ivory-billed Woodpecker and Its
Habitat in the United States," in *Encountering the Past in
Nature: Essays in Environmental History*, edited by Timo
Myllyntaus and Mikko Saikku (Helsinki: Helsinki University
Press, 1999) 87–119, © 1999 by Mikko Saikku and Helsinki
University Press.

*To the memory of Dr. Thomas Wendel*

*(1924–2004)*

*Joka tapauksessa, tietysti kaiken alla,*

*viimeiseen hengenvetoon, kytee typerästi ja*

*epätoivoisesti hennon hento toive, että*

*maailma sittenkin minun kirjoittamisestani*

*nitkahtaisi untuvan verran pehmeämmäksi.*

—PENTTI LINKOLA, "Itke rakastettu maa,"
  *Johdatus 1990-luvun ajatteluun*

# Contents

List of Tables   xi

Acknowledgments   xiii

CHAPTER ONE
Environmental History and the Yazoo-Mississippi Floodplain   1

CHAPTER TWO
A True Ecological Complex   26

CHAPTER THREE
Enter *Homo sapiens*   52

CHAPTER FOUR
The Creation of a Cotton Kingdom   87

CHAPTER FIVE
Taming the Rivers   138

CHAPTER SIX
Bounties of the Bottomland   165

CHAPTER SEVEN
A Transformed Landscape   220

Notes   257

Bibliography   313

Index   359

# Tables

TABLE ONE  Population of the Core Counties of the Yazoo-Mississippi Delta, 1840–1930  107

TABLE TWO  Amount of Cultivated Acreage (Improved Land) in the Core Counties of the Yazoo-Mississippi Delta, 1850–1930  136

TABLE THREE  Land Use in the Yazoo-Mississippi Delta, 1932  224

TABLE FOUR  Distribution of the Forested Area in the Yazoo-Mississippi Delta by Condition and Topographic Location, 1932  225

TABLE FIVE  General Patterns of Human-Induced Environmental Change in the Yazoo-Mississippi Delta, 3000 B.C. to the Present  250–51

# *Acknowledgments*

AFTER GRADUATING from high school some twenty years ago, I was at a loss in which field to continue my studies at the university level. I had always been interested in history, but the study of the natural world also held a fascination for me. A year as an exchange student in the United States had furthermore ignited a serious interest in North American culture and natural history. As a fresh general history major at the University of Helsinki, I was fortunate to attend a lecture series on American environmental history by the visiting Bicentennial Fulbright Professor of American Studies Alfred W. Crosby. Attending his lectures, my academic future suddenly came into focus: it was, after all, possible to combine all those interests in one's work. I wish to thank Al Crosby for that transcendent experience—as well as for other guidance over the years.

My own investigation of American environmental history received an auspicious start at a seminar conducted by another "Bicen," Thomas Wendel. Encouraged by Tom to pursue graduate studies with leading American scholars in the field, I contacted Donald Worster in the Department of History at the University of Kansas. My two years in Don's inspiring program in the early 1990s were crucial for my development as an environmental historian. I hope this book will meet at least some of the uncompromising standards set by my most important academic mentor, who has kindly continued to express interest in my work.

Markku Henriksson has been paramount in establishing North American Studies as an academic discipline in Finland. His contributions to my work over the years are simply too numerous to list here. I also wish to thank Michael Coleman, William Cronon, Robin Doughty, Seikko Eskola, Susan Flader, Ari Helo, Marjatta Hietala, Donald Hughes, Anto Leikola, Ilmo Massa, Neil McMillen, Timo Myllyntaus, Markku Peltonen, Noel Polk, Matti Savolainen, Philip Scarpino, James Sherow, Mart Stewart, Olli Vehviläinen, and Richard White for their comments and encouragement at various stages of my research. My thanks regrettably come too late to reach John G. Clark and Carville Earle.

Over the years, my work has greatly benefited from the insights offered by certain ecologists. Yrjö Haila, Steven Hamburg, Robert Holt, Richard Johnston, the late Olli Järvinen, Torsten Stjernberg, Esa Teräväinen, the late Miklos Udvardy, and Yrjö Vasari made me feel welcome in the scientific community, raising serious doubts about the presumed incompatibility between the arts and sciences.

Fisherman Pentti Linkola's prophetic writings on the natural world and its destruction have been of great importance to me since the 1970s. I wish I could construct a truly valid argument against his utmost pessimism about the future of Earth's biological diversity and the fate of humankind, but my study of environmental history continues to offer precious few tools for such an endeavor.

For the last twelve years, the Renvall Institute for Area and Cultural Studies at the University of Helsinki has been my academic home and primary employer. During this time, I have been privileged to learn from dozens of visiting scholars to our North American Studies program. Of the Renvall "Bicens," the most influential in relation to my own research interests have been Robert Brinkmeyer, Alfred Crosby, the late Robert Crunden, Lynn Dumenil, Harvey Green, Cheryl Greenberg, Frances Karttunen, Jeffrey Meikle, and John Wunder. Robert Bieder, Jay Watson, and Susan Williams are among other accomplished visitors to the Renvall Institute with a keen interest in my work. It is impossible in this short space to name all the other Renvallians who have contributed to the completion of this book, so I shall thank them collectively and name just one, knowing that the others will not be offended for not being singled out. Over the years, Pirkko Hautamäki has read countless variations of the manuscript and offered invaluable advice—and some rather sarcastic comments—regarding my written English.

I have been fortunate to receive considerable funding for my research, most of it ultimately provided by the Finnish taxpayers. A generous grant from the National Council for the Humanities at the Academy of Finland enabled me to begin the project and concentrate on my research for three years. Additional funding for the work at hand has been provided by the Finnish Cultural Foundation, the American Council of Learned Societies, the American-Scandinavian Foundation, the Finnish Graduate School for North American Studies, the Faculty of Arts and the Chancellor's Office at the University of Helsinki, and the Deep South Regional Humanities Center at Tulane University.

The "field work" for this study has been carried out in the United States where I have been fortunate to work as a visiting scholar at four great universities with excellent library and archival resources. The Joyce and Elizabeth Hall Center for the Humanities at the University of Kansas, the American Studies and American Civilization Program at the University of Texas at Austin, the Center for the Study of Southern Culture at the University of Mississippi, and the Deep South Regional Humanities Center at Tulane University each provided a stimulating atmosphere for academic research during my stay. I am especially grateful to Don Worster and the other participants in the Environmental Colloquium of the Program in Nature, Culture, and Technology at the Hall Center for their constructive critique at the time my research questions were beginning to take shape. I furthermore wish to thank the personnel of the following research institutions for their professional help and courtesy: the University Archives at the Roberts Memorial Library, Delta State University, Cleveland, Mississippi; the Louisiana and Lower Mississippi Valley Collections at the Hill Memorial Library, Louisiana State University, Baton Rouge; the Michigan Historical Collections at the Bentley Historical Library, University of Michigan, Ann Arbor; the Official and Special Collections at the Archives and Library Division, Mississippi Department of Archives and History, Jackson; the Archives and Special Collections at the John Davis Williams Library, University of Mississippi, Oxford; the Center for American History at the University of Texas at Austin; and the Special Collections of the Manuscripts Department at the Howard-Tilton Memorial Library, Tulane University, New Orleans. I especially wish to thank Ann Lipscomb Webster of the Mississippi Department of Archives and History, who ensured that new sources pertinent to my study would be uncovered with every visit to Jackson.

Wiley C. Prewitt, Jr., kindly shared his vast knowledge on the Delta's history and natural environment during our unforgettable fishing trips on the Mississippi, while Pekka Hämäläinen provided equally inspiring company in more urbane settings. Alfred Crosby, James E. Fickle, Yrjö Haila, Markku Henriksson, Jack Temple Kirby, John R. McNeill, Thomas Wendel, and Donald Worster each read and commented on the entire manuscript at different stages of its preparation. Their sincere and learned advice and criticism—though often not taken—has certainly improved it, while Derek Krissoff and Jon Davies of the University of Georgia Press and copyeditor Linda Wessels have provided competent editorial support. Magdalena Lindberg was able to prepare the maps on a short notice without compromising quality.

Academic pursuits have been highly esteemed in my family for a few generations, and my parents, sisters, and other relatives have always been most approving of my research. Unfortunately my paternal grandmother, Irene Saikku, will not witness the publication of the book she intently anticipated. My wife Mari Keskinen and our three incredible children, Sara, Otso, and Alli, have undoubtedly borne the heaviest burden from my seemingly never-ending preoccupation with the environmental history of a faraway place but tolerated it admirably. With their kind permission, I dedicate this book to the memory of Tom Wendel, my first graduate teacher and a great friend who died during the copyediting process and never got to see the idea first conceived at his 1988 seminar in its final form.

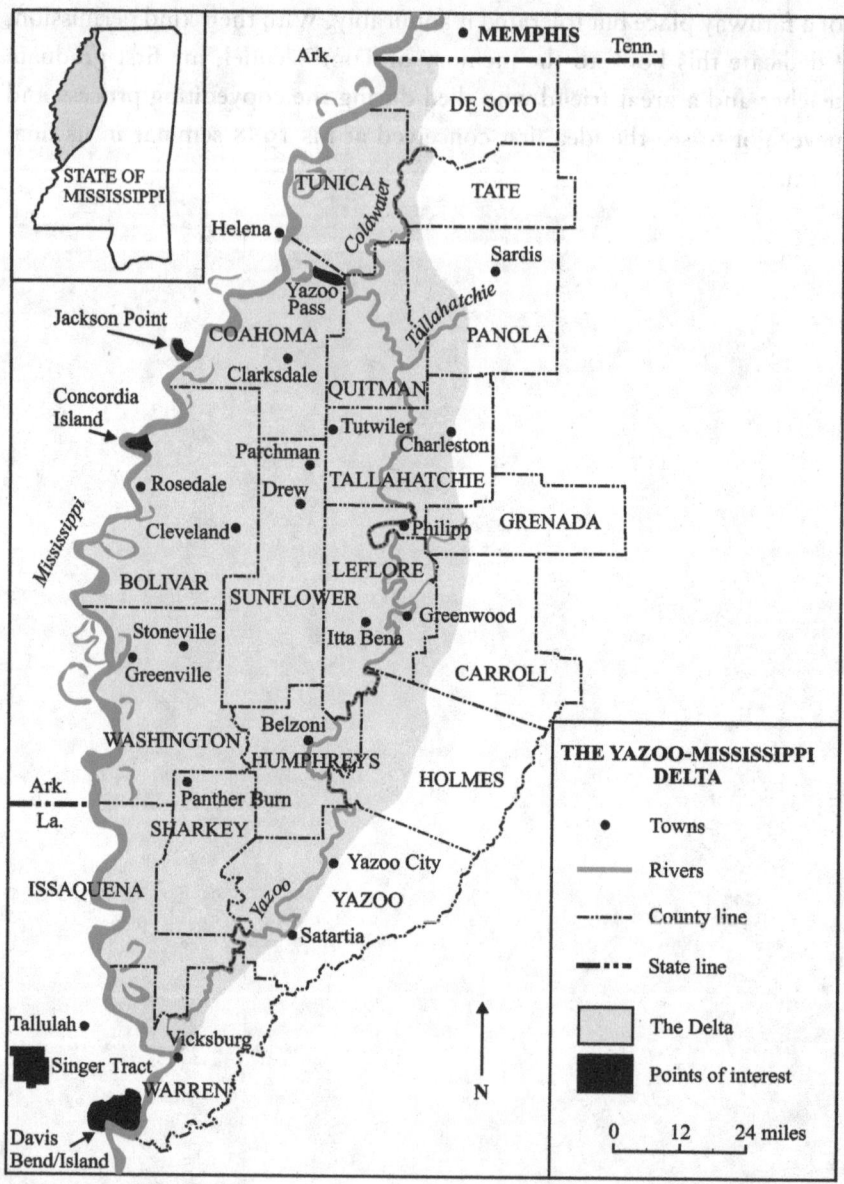

The Yazoo-Mississippi Delta, showing the most important towns and other locations referred to in the text. The county lines and the course of the Mississippi River shown as of today.

THIS DELTA, THIS LAND

This Delta, he thought. This Delta. *This land which man has deswamped and denuded and derivered in two generations. . . .* No wonder the ruined woods I used to know don't cry for retribution! he thought: The people who have destroyed it will accomplish its revenge.

—WILLIAM FAULKNER, "Delta Autumn," *Go Down, Moses*

CHAPTER ONE

# Environmental History and the Yazoo-Mississippi Floodplain

IN 1942, by publishing *Go Down, Moses,* a collection of stories describing the problematic relationships between black and white Mississippians and their natural environment, the then relatively unknown William Faulkner displayed an acute awareness of an immense process that had irreversibly transformed the natural and cultural landscape of his home state.

## The Unknown Delta

European expansion, or, the global dispersion of humans and other organisms of Eurasian origin within the past five hundred years, has resulted in immense environmental change all over the world. Among the most dramatic examples of this phenomenon is the socioecological change in North America during the last four centuries. For example, it has been estimated that between the arrival of the first European colonists in the early seventeenth century and the adaptation of sustained-yield forestry in the first decades of the twentieth century, the original forest cover of the coterminous United States was reduced by more than 80 percent.[1]

The Yazoo-Mississippi Delta,[2] the floodplain between the Yazoo and Mississippi Rivers in the northwest corner of the present-day state of Mississippi, has experienced enormous environmental change since the Civil War. Traditionally, the Delta region is thought to have been transformed between 1865 and the early 1930s from a virgin hardwood forest into an agricultural landscape; the floodplain covered by impenetrable lowland forests had been remade into a thriving New South cotton kingdom where by 1934 only 2 percent of the area could be classified as old-growth forest.[3] While the transformation was undoubtedly real, the long-term history of human-induced environmental change in the Delta emerges as much more complicated.

Considerable human impact on the region's landscapes goes back much further than the late nineteenth century. Already during the antebellum era, the Euro-American settlement had caused significant changes in the ecology of the floodplain. Besides, the Delta settled by people of European and African origin was hardly a true wilderness, as the floodplain had been inhabited and influenced by aboriginal people for millennia. The successive Native American cultures in the Delta had furthermore varied greatly in their population size and utilization of the region's natural resources. Still, for reasons addressed in this study, the early-nineteenth-century Delta landscape encountered by Euro-American settlers and their slaves exhibited more features of untouched nature than would have been the case at almost any other time during the last millennium. In addition, human-induced and natural processes on the floodplain have been intertwined so closely that simple, linear explanations are simply not applicable in describing changes taking place in the natural environment or human societies of the Delta.

Utilizing an interdisciplinary approach within the general framework of environmental history, this study examines the human exploitation of the bottomland hardwood forest habitat in the Yazoo-Mississippi Delta through time, aiming to identify the most significant patterns of environmental change on the floodplain since the arrival of the first humans. Although the region's environmental history during the Native American and early Euro-American periods is an important part of the study, the greatest emphasis will be on the developments of the nineteenth and early twentieth centuries, when the most far-reaching decisions concerning the destiny of the region's natural landscape were made. By narrating the extensive alteration of the natural environment by agricultural clearing, commercial logging, and levee building, and by analyzing its manifold socioecological consequences, I aim to shed light on the following questions: How and why was this so-called last wilderness east of the Mississippi River transformed from forest to field and by whom? How did the human-induced changes in the natural environment affect the original and incoming biota of the region, human and nonhuman, and how did the people involved in the vast environmental change perceive the process? What choices concerning the utilization of natural resources were made deliberately, and to what extent did inhabitants of the Delta act as agents of a more general societal change?

Among topics discussed are the Delta output of agricultural and forest products, human modification of the natural waterways, environmental aspects of the plantation-sharecropper economy, and ethnic and racial perceptions of land ownership and the Delta environment. Special attention is paid to the roles played by different cultural groups in the transformation of the landscape. In addition to chronicling the change and describing the different ecological aspects of the region's changing land use patterns, the study examines connections between environmental and social problems. The theoretical framework for the study attempts to connect the socioecological developments in the Delta to the larger context of American and global environmental history.

Until the 1990s, the American South was largely neglected in environmental historiography when compared to other major regions of the United States, especially the West. While a number of excellent studies have surfaced in recent years, an environmental history of the Yazoo-Mississippi Delta should still be of greater than provincial interest to students of southern history and culture.[4] The history of the Yazoo-Mississippi Delta in the nineteenth and twentieth centuries has attracted scholarly interest in the form of four book-length treatises, namely Robert L. Brandfon's *Cotton Kingdom of the New South*, James C. Cobb's *The Most Southern Place on Earth*, Clyde Woods's *Development Arrested*, and, most recently, John C. Willis's *Forgotten Time*. The Yazoo-Mississippi Delta also plays an important part in a recent examination of the Lower Mississippi Valley's settlement history, John Solomon Otto's *The Final Frontiers, 1880–1930*.[5] These five studies examine different aspects of the region's history, from the political and economic creation of a new plantation empire and the plight of black landowners to the birth of blues music and the growing federal involvement in the twentieth-century Delta economy. None of them, however, devotes more than a few pages to the region's natural conditions, Native American legacy, or environmental issues, and the need for a more comprehensive treatise on the Delta's environmental history remains. Similarly, bottomland hardwood forests receive only fleeting attention in the recent *Mississippi Forests and Forestry* by James E. Fickle, while the examination of human-induced change along the Lower Mississippi is largely restricted to Louisiana in Ann Vileisis's general history of American wetlands, *Discovering the Unknown Landscape*. On the other hand, Robert W. Harrison's authoritative work on the history of flood control and water management in the region concentrates on administrative and technological issues, but largely avoids any examination of the environmental effects of these activities.[6] In their assessment of human exploitation of the Delta's natural resources, previous studies have furthermore largely ignored the problem of change occurring naturally on the Yazoo-Mississippi floodplain. Natural processes are far from being static, and while all human activity results in environmental change, it is against the background of natural change that the human impact of significance must be located and identified.

Because of the interdisciplinary nature and relatively vast geographical area of my study, I rely largely on published material in establishing the extent of social and ecological change in the Delta. Significant literature on the history of southern agriculture, forestry, and engineering is supplemented by early travel literature, naturalists' writings, modern ecological and geographical research, southern fiction, federal and state records, lumber company archives, and private letters and manuscripts as source materials for examining the various social, economic, and ecological processes taking place in the region. The factors that contributed to the successful utilization of the natural resources in the Delta are not unique to Mississippi, southern, or American history: exploitation of disadvantaged people and the natural environment, land speculation, boosterism, immigration, and rapid development of economic infrastructure are familiar themes in modern history. This study of the immense human-induced change that has taken place in the bottomland hardwood forests of the Delta aims to transcend local history and attempts to unveil some general processes and patterns. Such an environmental history of the Yazoo-Mississippi Delta should enable the reader to understand some of today's environmental problems in a broader historical context.

## History, Region, and the Environment

Environmental history established itself in the American historical profession in the 1970s, and a similar development has also taken place in Europe. Although interaction between humans and nature has always played a part in European and American historiography, environmental history in the present sense of the term has been written for only some three decades. What is environmental history and what have the central themes in the environmental historiography of the twentieth century been? In the following pages, I propose to shed light on these questions by trying to define environmental history and making comparisons between different approaches to the field.[7] My intent is furthermore to address some central issues of methodology in environmental history and discuss some conceptual tools that could prove applicable in the identification of factors that influence and modify socioecological change everywhere, including the Yazoo-Mississippi Delta from the arrival of the first humans.

The term "environmental history" may first seem to be composed of incompatible elements. "Nature" and "natural environment" are generally regarded as the antithesis of humanity, while "history" explicitly refers to the (written) story of humanity. However, both "nature" and "environment" are culturally constructed concepts that cannot be examined outside their societal context. What is regarded as nature varies across human societies, and that understanding of nature is not constant but a historically changing set of variables in any given society.

Environmental history may be described as an attempt to study the interaction between humans and nature in the past. Its aim is to deepen our understanding of how humans have been influenced by their natural environment through time and, conversely, how they have affected their surroundings and with what results. This relatively new field of historical study rejects the traditional assumption that human experience has been exempt from natural constraints or that the ecological consequences of past human activity can be ignored. In comparison with traditional historiography, environmental history emphasizes the role of humans as an integral part of their natural surroundings. Modern environmental history strives for a fuller understanding of today's environmental issues and, ideally, provides information for contemporary problem solving. What ecological models does history offer us? What have been the adaptive and maladaptive human societies throughout history and how did they function in their relations with the natural environment? These questions require empirical answers, which environmental history can provide. Even as current environmental problems may differ from former ones, understanding of the past events may prove helpful.[8]

A 1990 *Journal of American History* theme issue devoted to environmental history continues to offer useful definitions and suggest approaches to the core of environmental history.[9] This roundtable discussion of environmental history gathered together some leading environmental historians in the United States. Naturally, the contributing scholars had their own predispositions and preferences on the subject, but in the end a broad agreement was reached. The most important questions within the field seem to be about the different productive strategies of human societies, their ideological backgrounds, their consequences, and making comparisons across culture and place. What kind of human society and natural environment emerge as a result of the interaction between these forces? Environmental history can be of great importance to the general study of human-nature interaction by phenomenologically identifying various past social, economic, and ecological processes and analytically separating relevant patterns from each other. Successful pattern descriptions can identify recurring features of socioecological dynamics and enable enlightened guesses about how they functioned. Detailed description of past events furthermore forces environmental historians to draw analytical distinctions and define criteria for the identification of environmental change.

Finnish environmental historian Timo Myllyntaus has recognized certain distinctive features that characterize current environmental history.[10] In environmental historiography, the study of human-nature interaction often has to focus on long-term change. Although most ecological catastrophes, whether volcanic eruptions or explosions of nuclear power plants, are sudden occurrences that may have long-lasting repercussions, most environmental changes, such as climatic transformations, are slow processes that may take centuries or millennia. Thus environmental history often approaches what the French annalist Fernand Braudel called the history of long duration (*histoire de la longue durée*).

Environmental history is also spatially more flexible than traditional historical research; natural entities, such as drainage basins or other geological formations, are often more important than boundaries created by humans, such as the borders of nation states or economic communities. Thus environmental history regularly applies an international perspective, often extended to a worldwide dimension. With global industrialization and socioeconomic modernization, environmental problems such as industrial pollution have become international problems. Similarly the integration of local economies into the world economy has affected the relationships between human societies and their natural surroundings. The fact that the exploitation of natural resources and the consumption of goods manufactured from these resources often take place in different parts of the world, combined with absentee ownership of production facilities, has caused ecological indifference that small and locally controlled economic systems could hardly accept. Due to their widespread temporal and spatial linkages, human-induced environmental changes of the past continue to affect contemporary life on earth. Soil impoverishment, erosion, deforestation, and pollution of air and water are among current environmental problems that have affected human societies for a long time.

Current research in environmental history displays enormous diversity in its selection of approaches and research subjects; this discussion uses the term "environmental history" as a concept encompassing all these different approaches. It is, however, possible to identify some general orientations within the discipline. The following analysis draws upon Donald Worster's observation that there are three general levels on which environmental history operates: nature itself and the human socioeconomic and intellectual realms. Environmental historians can intertwine these three levels in various ways. Further classifications of environmental historiography have been proposed by others, including Finnish environmental historians Ilmo Massa and Timo Myllyntaus, who recognize four different approaches to or subfields of environmental history.[11]

An approach that can be called historical ecology or ecological history attempts to reconstruct natural environments and their changes in the past, relying heavily on the natural sciences and their methodologies. This subfield aims to comprehend natural systems of the past—how were they organized and how did they function? Environmental historians applying this approach must closely cooperate with biologists and other natural scientists. Ecological studies have often used the term "history" in its broadest sense and made no distinction between human-induced and other changes in natural systems. Environmental historians can therefore add a valuable dimension to the study of natural environments of the past.[12]

Massa claims that ecological history does not comprise only developments in the natural world; it also studies humankind and its culture as a part of a larger ecological system. Defined this way, ecological history shares attributes with the ecological approach to anthropology. Prominent anthropologists from Clark Wissler and Julian Steward to Roy Rappaport and Marvin Harris have studied the relationship between human cultures and their environment using different terminology. These approaches have been called, for example, cultural ecology, ecological anthropology, or cultural materialism.[13]

Probably the most prominent approach in contemporary environmental history is the study of interaction between human modes of production, social conditions, and the environment. This field of study is concerned with connections between the human economy and environmental change in the past. "Mode of production" is a term introduced by Karl Marx for analyzing structural characteristics of societies; it refers to the ways means of subsistence are extracted from the natural environment and divided within a given society. How does this process shape both human society and its natural environment? Mode of production can be used in environmental history as a general description of the basic socioeconomic dynamics that shape the relationship between societies and their natural environment. It need not imply commitment to fixed theoretical positions, such as Marxism. Worster considers this approach the most important subfield of environmental history, where historians "are concerned with tools and work, with the social relations that grow out of that work, with the various modes people have devised of producing goods from natural resources. A community organized to catch fish at sea may have very different institutions, gender roles, or seasonal rhythms than one raising sheep in high mountain pastures."[14]

Yet another subfield of environmental history focuses on environmental change in relation to public control, environmental policies, legislation, and societal decision making in general. As Worster puts it, "[p]ower to make decisions, environmental or other, is seldom distributed through a society with perfect equality, so locating the configurations of power is part of this level of analysis."[15] Environmental historians therefore have to study the political structures of a given society in order to understand its relationship to environment. The values, ideologies, and interests of decision makers greatly influence environmental policies.

Understanding of natural phenomena is formed in human consciousness. The intellectual realm is prominent in the historiography of human ideas on the environment, or the study of how humans have viewed the natural world in their science, religion, arts, and ethics. This subfield focuses on the environment purely from a human perspective, studying the mental and intellectual history of environmental knowledge and consciousness. Different aspects of nature have different meanings in different contexts and for different human beings. Consequently, human ideas about the environment cannot be considered in isolation from their cultural context. Humans throughout history have conceived their environ-

ment in a myriad of ways, and the differences among societies in con-
ceptualization of both nature and the human-made environment are immense. Individual and societal attitudes toward the environment, as displayed in myths, political ideologies, or scientific doctrines, have provided
the basis for human-induced change in nature. Many established cultural
values stem from productive practices and are related to particular elements of nature. Thus they can be connected with particular human societies and their uses of nature. The amply documented differences between
the native and colonial perceptions of nature, for example, support this
assertion.[16]

Environmental history calls for an interdisciplinary approach, as conventional methods of historical research are hardly sufficient and traditional sources cannot provide enough source material on environmental
change. This interdisciplinarity largely results from the extreme diversity
of sources for environmental history. Much of the source materials used
by current environmental historians has been available for generations,
and current research attempts to reorganize the data based on recent theoretical advancements: interdisciplinary synthesis can often be achieved
by combining existing information from diverse disciplines in a new way.

Not surprisingly, the research topic tends to dictate the approach,
source materials, and research methods used. Therefore the source materials used in environmental history vary enormously, from traditional
written documents to data provided by modern science, such as pollen
and sediment studies, dendrochronological findings, and carbon datings.
Methodologies employed by the natural sciences can provide information on past environmental change, whether natural or human induced.
For example, pollen samples found in the bottom sediments of aquatic
systems and floral composition in a given area can provide valuable information on human-induced change in nature. For these reasons, environmental historians have to employ the findings and methodologies of
ecology, biology, zoology, botany, chemistry, geology, geography, meteo-
rology, and many other natural sciences. Environmental historians should
furthermore interpret the history of technology in a new way: the development of technical equipment has had an enormous impact on the way
humans use natural resources.

It can be argued that the skills of an environmental historian are weighed by the researcher's degree of sophistication in interweaving the different approaches and source materials. There is no one accepted paradigm for this task. It must, however, be noted that an environmental historian does not necessarily have to be a specialist in many fields, and not all treatises of environmental history need be grand syntheses embracing all levels. An environmental historian may choose to specialize and supply answers to research questions only within one subfield, maintaining other approaches in the background. Research on as many levels as possible can nevertheless be regarded as the ideal for environmental history as a modern academic discipline.

Conventional anthropocentric historians were reluctant to see nature as anything else than a haphazard natural setting where human activity took place. Environmental historians would strongly reject this assumption; natural environment is seen as an active and often decisive factor in human history, influencing the economic options available and shaping the developmental paths taken. Few would disagree with the claim that "the natural environment is not really passive but is rather a powerful determining force throughout history."[17]

Modern environmental history reassesses commonly accepted views of human history; the argument that humans have by means of religion, science, and technology liberated themselves from the limits set by the natural world seems absurd in the era of nuclear and other industrial accidents, global warming, and ozone depletion. The reappraisal of old truths and the construction of new views of the past have sometimes caused credibility problems for environmental history. Since environmentalism is currently recognized as an influential social movement, environmental history is sometimes regarded as something of an ideologically biased trespasser within academe, an environmentalist offshoot of the so-called new social history, even an academic "fad" that will soon subside. Environmental history has been mechanically coupled with environmental activism, a coupling that overlooks the scholarly tradition of the discipline. It remains true that many, if not most, environmental historians (including myself) are admittedly sympathetic to numerous goals of the environmental movement, such as the preservation of biological

diversity. Furthermore, all approaches to environmental history can be—
and often have been—politicized and ideologized. Such a tendency is not,
however, an inherent element of the discipline. Existing work in the field
proves that it is possible to study environmental history without a political
agenda.[18]

Any integrated study of the history of human-nature interaction should
counterbalance both technocratic fantasies and environmentalist utopias.
Donald Worster has pointed out that one of the aims of environmental history is to "reject naive assumptions about a static, pristine, virgin
world of unspoiled nature."[19] Environmental historians should therefore
aim to clarify environmental issues under debate and place them in their
historical context, rather than provide support for the arguments put
forth by the environmental movement or, for that matter, commercial
interests. William Cronon has even suggested that one of the greatest
contributions of environmental history to the conservation and environmental movements may be in providing counterbalance to ahistorical and
antihistorical impulses within these movements.[20] Deeper understanding
of contemporary environmental problems requires informed knowledge
of past events, and the study of environmental history should provide the
background needed for that quest.

Environmental factors have to be included among the key elements when
historical explanations are constructed. Environmental conditions set the
ultimate boundaries for human societies and their activities—agriculture
is not a viable option for the subsistence of an Inuit community, nor is
whaling for a family of Tuaregs. They do not, however, describe which
one of the possible paths will be taken by humans in a given situation.
This is analogous with thermodynamic principles, defining what is impossible but not indicating which possibility becomes reality. Environment can provide causal explanations in history as nature has always and
everywhere affected socioeconomic development, and changes in the environment can often be cited as causes for societal changes. For example, variations in climatic factors or populations of agricultural pests and
subsequent fluctuations in crop yields could prove crucial to communities with an agricultural subsistence base. Crop failures due to adverse

weather conditions have led to serious famines, and the economic consequences of unsuccessful harvests clearly correspond to numerous social indicators, such as birth, death, and marriage rates.[21]

All human activity results in environmental change, but it is against the background of natural change that the human impact of significance must be located and identified. Human-induced change in nature can never be attributed to a single factor nor the process of change be pictured as a straightforward linear development. In consequence, one of the major problems in the modeling of socioecological complexity is how to differentiate the dynamic boundaries of various subsystems of the human-environment relationship from each other. As Finnish scholar Yrjö Haila points out, "[t]he environmentalist slogan 'everything is affected by everything else' is not literally true, because different processes are dynamically independent of each other to variable degrees."[22] In any case, usable criteria for a meaningful analysis of the relationship between humans and their natural environment are difficult to come by.

Our understanding of the character and implications of environmental problems, both past and present, has changed in recent decades due to conclusions originating from political and sociological studies, as well as from an increasingly nuanced view of natural sciences. Environmental problems stem from various dynamic processes, both spatial and temporal, in which culture and nature are inseparable. The dynamism and complexity of this relationship results in unanticipated consequences even for seemingly simple interactions. Many environmental problems, such as global warming and ozone depletion, illustrate that human-induced and natural change are not easily distinguishable from each other. Furthermore, human understanding of environmental problems does not stem directly from nature itself, but is always more or less constructed through a social process. A problem is a social construct, and all problems are defined within culture and articulated through language. This implies that not only is finding solutions to environmental problems socially mediated, but also recognizing them in the first place. The interplay between sociocultural and political processes in problem construction is dynamic: the identification and evaluation of environmental problems is constrained by social and political factors in addition to physical ones.[23]

Useful concepts in any socioecological analysis are those of the "first" and "second" natures, the former referring to natural entities and processes that occur without human influence and the latter (popularized by Michael Pollan's 1992 book of the same title) to environmental elements produced by human activity. It is important to notice that second nature does not consist only of physical entities and structures but also of various sociocultural factors, such as institutions, rules, customs, and habits, which can confine human activities in a mode similar to a physical barrier. Nature, whether "virgin" or modified by humans, constrains human activities. "Barge transportation," writes Haila, "has to keep to waterways; whether these are canals or rivers is of secondary importance. Besides, natural and human-induced processes are mixed as far as the maintenance of waterways goes: both canals and rivers used for barge transportation require draining and regulation."[24]

The history of "wilderness" offers a good example of the difficulty of making distinctions between first and second natures, and between nature and culture in general. Regard for wilderness, or, nature supposedly untouched by human activities, increased in the United States since the late nineteenth century as the American landscape was profoundly altered by growing human exploitation of natural resources. However, the notion of wilderness, embraced by generations of nature conservationists, has proven problematic of late, as the term cannot be given any simple empirical meaning. Assumed wilderness areas have generally been inhabited or influenced by aboriginal people for millennia, and therefore do not qualify as real wilderness defined as above.[25]

Numerous factors indicate that the traditional image of science, built upon a strict subject-object dualism, has become outdated and that "environment" as a scientific concept simply cannot be "externalized" from the social sphere—the recognition and evaluation of environmental problems is a thoroughly social process. Historians of ideas on the environment have proved beyond doubt that scientists often promote, both consciously and unwittingly, their own role through sociological mechanisms, and as a type of social practice, science is predetermined by a whole variety of social and philosophical conditions. Thus it can be difficult to conceptualize the relationship between humans and environmental problems in a fruitful way, since traditional ecological research is committed to an objectivistic model of knowledge.[26]

Concepts such as "ecohistorical period" and "ecohistorical formation" have been used as tools to recognize past and present environmental problems and to constitute environmental policies. These concepts seem applicable also for this study, as adequate spatial and temporal scaling of human activities and natural processes are of critical importance for environmental history.

An intriguing classification of different time layers in human history has been proposed in the monumental work of the French annalist, Fernand Braudel. Historical change occurs on several time scales, and previous developments act as constraints for what can happen at present. Braudel recognized three temporal scales for historical analysis, geographical time, social time, and individual time, and implied corresponding spatial scales. Braudel considered geographical time as the slowest time scale, referring to a long-term temporal and continent-wide change, and used civilizations (which are more or less independent of each other) as its units. Social time incorporates socioeconomic structures, such as economic and political institutions, fluctuating on shorter temporal scales and also on spatial scales that vary according to the nature of each conjuncture. Braudel's individual time refers to events affecting the lives of individual persons. It consists of the temporal life cycles of human beings and the spatial scale of their immediate social surroundings, both constrained by conjunctures. Geographical time is almost undetectable for individuals because its effects can hardly be experienced in their lifetime. Societal development is not uniform, as the evolution of various socioeconomic spheres causes differentiation. The dominating spheres in a particular society also dictate its dynamics of exploitation of natural resources. Truly significant changes in human ways of exploiting nature relate to larger socioeconomic structures, which change more slowly than political conjunctures. Various multilayered processes of socioecological change occur on the time scale of conjuncture, often triggered by events occurring in individual time. On smaller scales, the differences between and across societies become more pronounced.[27]

Yrjö Haila and Richard Levins have elaborated on Braudel's ideas and applied the concept of "ecohistorical period" to describe socioecological change in a long historical perspective. The notion refers "to such periods in history in which human activities have led to (relatively) uniform changes in nature over vast areas."[28] The concept of different ecohistorical periods is thus a very broad but convenient way to conceive turning points in environmental history. Such loose classification does not exclude the construction of more detailed analyses or the use of regional approaches.

Ecohistorical periodization can be closely related to changes in the mode of production of human societies. Haila and Levins accept Eric Wolf's grouping of modes of production into three basic categories: kin ordered, tributary, and capitalist. Kinship—which is not necessarily a biological category—as the foundation of society ultimately limits the size and complexity of kin-ordered groups: formal administrative structures are poorly developed, and the division of labor is based upon age and gender. Such societies typically obtain their means of subsistence in fairly direct interaction with local natural conditions, but can also maintain exchange networks with other, unrelated kinship groups. The purpose of trade—and ultimately all commerce—is to exchange locally available items for locally unavailable ones. In the tributary mode of production, social labor is largely mobilized by exercising power: political or military coercion by the central power extracts surplus wealth from primary producers. The direct influence of the central power remodels the relationship of the tribute-providing communities with their natural environment. This development, however, is not uniform, and local economies can remain almost unchanged. It is thus the emergence of the capitalist mode of production that marks a real turning point in the human relationship with the natural world. In the idealized capitalist marketplace, all elements of nature can be brought within the sphere of economic activities. Because products of nature are now measured solely by their exchange value—compared with each other using the same economic standard—they lose their specificities and the exploitation of nature can be universalized.[29]

Some 100,000 years ago, *Homo sapiens* established a previously unknown mode of existence.[30] The species' intelligence and behavioral flexibility, manifested in the development of human language, had resulted in the formation of permanent social groups. Ecohistorical periodization can be applied from this turning point forward. Over time, various social groups became increasingly less subject to the constraints of their natural environment. Evaluating a group's degree of ecological independence is a basic task in any ecohistorical periodization.

Irreversible human influence on earth's ecological systems is not new: preagricultural societies were—and still are—strongly influenced by their natural surroundings but able to modify their environment. For example, numerous extinctions of large herbivores (and many other species depending on them for survival) in different parts of the world thirty thousand to ten thousand years ago correlate with the arrival of human populations to these regions. The emergence of agriculture, followed by drastic growth in human populations and eventual urbanization, began a new phase in global environmental history some ten millennia ago. The shift from hunting and gathering to agriculture was a complex process that generally led to systematic and extensive growth of human influence on the natural environment.[31] In some parts of the world, it resulted in the permanent settlements, domestication of animals, clear division of labor, centralized administration, and written languages of the so-called ancient civilizations. Trade in agricultural and other products gradually emancipated humans from the dependence for subsistence on local conditions and eased the transmission of social and technological innovations across cultures. Trade thereby created new socioeconomic structures that can be used in ecohistorical periodization. The development of such structures was complex and, on closer inspection, displays immense temporal and spatial variation.[32]

Most environmental historians would agree with the claim that the greatest shift in the human-nature relationship since the beginning of agriculture was the rise of industrial capitalism connected to the economic and ecological expansion of Europe. The process gained momentum with the great explorations in the fifteenth century and, eventually, resulted in global colonization and a tremendous increase in the volume of trade and the importance of the marketplace. Monoculture, or, the extensive cultivation of certain agricultural products over vast areas solely for the emerging world market, began as early as the fifteenth century in the newly colonized areas.[33] From the beginning, the new, interconnected network induced vast environmental changes in Europe and its economic peripheries. Harold Innis's classic staple theory is often applicable for environmental history in the context of local and supralocal control of commodity production. Industrialization and economic development in the new peripheries of the European economy were built upon the export of natural resources in the form of raw or semirefined products, or, staples. Local economic development soon became dictated by the supralocal needs of the world economy.[34]

Recent environmental historiography has proved beyond doubt that the extraordinarily rapid downfall of many aboriginal nations in the areas of European colonization was largely caused by smallpox and other germs imported by the European conquerors and not by the natives' military or political inferiority. Aided by their biological allies, such as crops, weeds, domesticated animals, pests, and Eurasian microorganisms that created virgin soil epidemics in the Americas and elsewhere in the New World, the Europeans were able to create ecological representations of their homelands everywhere in the temperate zone of the globe. The useful concept of the so-called Neo-Europes, referring to the non-European temperate regions around the world that were subjected to successful colonization by European life-forms since the Middle Ages, was coined by Alfred W. Crosby in his classic *Ecological Imperialism*. The European colonists were also able to adopt from the conquered lands the economically viable crops and other products of the aboriginals—such as the maize and the potato cultivated by Native Americans—and use these to their own advantage in creating new commercial empires.[35]

A series of transformations in the European economy and society, combined with technological innovation, resulted in the final breakthrough of industrial capitalism in the early nineteenth century and created the modern world with its environmental problems of unprecedented scale. In capitalist modernization, self-sustained economic growth was achieved for the first time in history. The old patterns of colonial trade were fortified, and the production of raw materials for export increased all over the world. Global economic trends and increasing integration of the world market contributed to escalating uniformity in the human use of natural resources. Human exploitation of nature under modern capitalism, however, is not omnipotent, since during recession phases production pressure on economic peripheries typically diminishes and natural systems may recover. Although maximization of profit, the basic aim of capitalist production, typically conflicts with the sustainable yield of natural resources, it can also offer incentives for rational management. It must furthermore be noted that traditional subsistence methods, as practiced under heavy population pressure, have often resulted in even more dramatic change in the natural environment.[36]

Donald Worster claims that the many "revolutions"—the Scientific, the Capitalist, and the Industrial—in Europe after the late fifteenth century were only surface manifestations of a more fundamental change of thought: the emerging secular, progressive, and rational culture of economic and scientific materialism. Worster nominates Adam Smith, author of *The Wealth of Nations* (1776), as the representative modern man and the embodiment of the cultural shift. For Smith and his followers, nature had become instrumental, measured by whatever human uses it could serve, and possessed value only to the extent that it had been "improved" by human labor. Unlike the practitioners of Christianity, Islam, or other traditional religions, the adherents of rational greed did not conceive that the natural world lays any obligations on humans.[37]

A new period in the history of the human-nature relationship, characterized by a dramatically increased human capacity to cause irreversible change in nature, commenced around the mid–twentieth century. Haila and Levins call this stage "the ecological crisis," as human activities for the first time reached the potential of modifying the basic boundary conditions of the very existence of human life on earth, and deterioration of the natural environment threatened its life-supporting potential everywhere.[38]

Ecohistorical periods, as discussed above, refer to patterned changes in the human use of nature. Haila and Levins have introduced the complementary notion of "ecohistorical formation" for recognizing distinct ways in which different societies relate to nature. Such ecohistorical formations can be characterized "by features which will help distinguish different kinds of dynamics both in the mode of production and in the related ecology."[39] A given mode of production can be practiced under different conditions and in different natural habitats. Thus comparative analyses of ecohistorical formations should strive both to recognize similarities in the relations to nature of a particular mode of production across different environments and to detect relevant differences between different modes of production in their relations to nature in similar environments. Haila and Levins have prepared the following questions for the identification of ecohistorical formations:

> What is the habitat, vegetation, array of soils and climate?
> What is produced?
> With what raw materials and by what tools?
> Who does the work, and in what way (individually or socially)?
> Who owns or controls the means of production, and who disposes of the
>    product?
> What becomes of the surplus?
> Of what does wealth consist?
> What is the relation between production and reproduction?
> Why are things produced?
> What keeps the ecohistorical formation as it is?
> What forces lead to change in the system?[40]

Some environmental historians would discredit the above list as biased toward an overly materialistic interpretation of history and for neglecting the intellectual and spiritual realm of human experience. That the list displays a materialist emphasis is true to some extent. Nevertheless, these fundamental questions should be addressed to some degree in any analysis of the relationship between a given society and its natural environment.

Construction of pattern descriptions in environmental history can be difficult because of the problems connected to scaling, a fairly recent but critical issue in ecology. Problems of scaling arise from the widely differing characteristic spatial and temporal extensions of ecological processes and human activities. The understanding of past and present environmental problems, however, calls for adequate scaling of socioecological change.

Socioecological change in a historical perspective is without doubt real and largely irreversible. It is not, however, uniform, although it may initially appear so. Mechanistic projections of long-term change such as Ernst Friedrich's *Raubwirtschaft* or the classic model of population growth by Thomas Malthus are therefore misperceptions.[41]

Environmental historians of the 1970s and 1980s customarily saw contemporary scientific ecology and its idealized mechanical systems as a model that could easily be applied to environmental history. The celebrated concept of "ecosystem," fully developed in the work of Eugene P. Odum since the 1940s, refers to "[a]ny entity or natural unit that includes living and nonliving parts interacting to produce a stable system in which the exchange of materials between the living and nonliving parts follows circular paths."[42] Odum's cybernetic approach expanded the ecosystem concept beyond its narrow fieldwork origins and made it a comprehensive definition with great theoretical and applied significance. The classic ecosystem approach had adapted from population and community ecology the notion that the entities studied possess unambiguous spatial extensions. Such closed systems, however, do not exist in nature, as both energy and matter flow through them. Furthermore, the process of ecological succession in Odumian ecology was directional and culminated in stability. Recent ecological research has, however, convincingly illustrated that nature is not static, and the idealized concept of the discipline as the study of orderly flows of energy in and between stable systems is no longer valid.[43]

Space and time in today's ecology are no longer uniform Newtonian concepts; rather, they are constituted by the multitude of ecological processes themselves. Cause-effect chains across dynamically separated domains are difficult to deduce from linear systems, and relationships between different actors in the models have become much more complicated. Modern environmental historians have to adapt to the view of a dynamic natural world. There has been some resistance to this view among scholars who long for a harmonious and stable nature, which is easier to examine—and to protect. I have to agree with Dan Flores's view that acknowledging the fact of ongoing, natural disturbance in nature does not prevent environmental historians "from critiquing human disturbances that were foolish in an anthropocentric sense, or reprehensible from the perspective of the diversity of life."[44] In any case, the idea of built-in equilibrium in the natural world has lost much of its scientific credibility, and most ecologists today would agree with the claim that the proper concern in assessing the harmfulness of environmental change is whether ecological systems in question retain their resilience, i.e., does the change remain within bounds that enable the recovery of the systems after disturbance.[45]

Change in nature is constant and ubiquitous, and life itself is a process that produces and reproduces regular patterns from unstructured flows of energy. Many external factors, such as flows of energy and matter, are essential for the survival of ecological systems. More or less constant, they can be looked upon as perimeters of the system in question. Inconstant but frequent external factors can be classified as "disturbances," and rare but drastic ones as "catastrophes."[46]

"Ecological complex" has been formulated by Haila and Levins as "any ecological entity that is separated from the surroundings by differences in the rate intensity of processes inside versus processes outside."[47] Examples of such entities obviously include individual organisms, but also larger structures: the chemical and ecological processes and interactions on an island differ greatly from those in the surrounding ocean. All entities are, however, relative, as their boundaries are not fixed and allow an influx of energy from outside sources. The recognition of different ecological complexes is furthermore complicated by the bias of human perspective—it is easier for humans to recognize organisms similar to themselves as entities in comparison to significantly larger (such as biomes) or smaller (such as individual bacteria) complexes. Each level of ecological organization is distinguished by processes that have their specific spatial and temporal scales: molecules within a cell reproduce faster than the cell, cells faster than the organism, organisms faster than the population in which they belong, and so on.

Haila and Levins have applied the division of temporal and spatial dimensions of ecological change into "micro-," "meso-," "macro-," and "megascale" processes. Microscale is constituted by processes characteristic of individual organisms, while mesoscale refers to time span of years, decades, and centuries, corresponding with demographic changes within the population level. Macroscale processes cover large areas, demand hundreds of thousands of years, and largely determine the species pool from which ecological communities are assembled. The largest, or, megascale processes, cover entire continents and may take tens of millions of years.

Ecological units characteristically maintain their structure not only despite but also because of external disturbances. An important class of such events is often called a "disturbance regime," referring to a certain type of landscape created and maintained by a disturbance that occurs with characteristic frequency. The formation of such disturbance regimes is typically a mesoscale process. Wildfires in northern coniferous forests or tallgrass prairies, hurricanes in tropical forests, and flooding in alluvial plains are examples of disturbances that are triggered by factors more or less external to the landscape itself, but essential to the maintenance of the whole complex in longer perspective. Disturbance regimes can provide a general norm for acceptable human intervention in nature: in order to preserve biological diversity and productivity, human modification should remain comparable to that caused by natural occurrences within disturbance regimes.

Natural catastrophes, as compared to disturbances, are unique events with spatial and temporal dimensions dramatically exceeding the resistance and resilience of the ecological complexes affected. All time-space scales are subject to catastrophes, but the definition of the term varies according to the size of the ecological complex in question. Mass extinctions of species, such as that of the dinosaurs at the end of the Cretaceous period, are catastrophes on the macrolevel of evolution, whereas a normal forest fire affects an individual tree in the same way on the microlevel of organisms. A most significant difference between natural and human-induced catastrophes is that pertaining to their pace: the latest natural mass extinction some 70 million years ago took hundreds of thousands of years, while the current wave of extinctions proceeds on the temporal scale of centuries, if not decades.[48]

It is the correspondence of natural and human scales that determines what kind of human-induced change is of consequence to a given ecologi-

cal system. Human-induced environmental change results from the interaction between human-induced and natural dynamics on a shared scale.
This correspondence also explains why intensive change on a limited scale
is possibly insignificant (except for the individual organisms subjected to
it) whereas a hardly observable change occurring uniformly on a large
scale can be truly disastrous, although its consequences might not be visi-
ble in the lifetime of individual organisms. Therefore, the most significant
human-induced changes in history are results of systematic activities over
vast regions and long time periods. Furthermore, as no single human being can through his/her activities cause relevant change on a larger scale,
some sort of social formation is needed for creating human-induced environmental change of consequence. Natural resources are exploited and
transformed into commodities within a social matrix.[49]

An important notion, related to the ecohistorical formation, has been
championed by Dan Flores. He claims that "bioregion" should be recognized as a precise and useful term for environmental historians. Flores asserts that "the first step in writing environmental histories of place
is recognition that natural geographic systems—ecoregions, biotic provinces, physiographic provinces, biomes, ecosystems, in short, larger and
smaller representations of what we probably ought to call *bioregions*—
are the appropriate settings for insightful environmental history."[50] Bioregional histories should thus commence with geology, landform, and climate history. The second basis for bioregional history, beyond ecological
parameters, is constituted by the diversity of human cultures across both
space and time.

   An element that distinguishes Flores's bioregional history from the traditional approach to history of places is a precise spatial application of
Braudel's *histoire de la longue durée:* instead of making wide geographic
generalizations in shallow time, deep time should be analyzed in a single
locality. Bioregional history is therefore the story of different but successive cultures occupying the same space. As space plus culture equals
place, these cultures create their own succession of places. The structure
of dialogue between nature and human culture resembles the relationship
between habitat and species. Ecological ideas in human cultures can be

seen as adaptive packages of "captured knowledge" about living at a certain locality, and, consequently, human cultural adaptation and knowledge transmission may be understood as essentially analogous to natural selection.[51]

Flores has criticized environmental historians for too often starting out with a single interpretive framework and forcing their studies of topographically, ecologically, and culturally diverse and geographically vast territories into some facsimile of that model. A better approach is to aim for "deep time, cross-cultural, environmental histories of places—and after a sufficient number of such case studies have been done, *then* to look for patterns."[52] In the following chapters, I attempt to construct a bioregional history of the Yazoo-Mississippi Delta. The study approaches its subject, the history of human-induced change in the region's natural environment, on many different levels and scales. I try to intertwine the three basic levels of environmental history—the natural, socioeconomic, and intellectual—in my examination of that change.

I fully acknowledge the manifold problems associated with successful identification and bounding of ecosocial processes in history. Employing some of the central notions from the preceding methodological discussion, I still venture to attempt that in one geographical locality, the Yazoo-Mississippi Delta. I aim to find out whether the Delta as a geographical construction is even a valid entity for examination in this context. Can the notions of ecological complex and bioregion be readily applied in reference to the Delta? What were the natural preconditions for particular productive practices in the region, and how were those practices merged into their larger socioeconomic environment? What kinds of ecohistorical formations were identifiable in the Delta from the arrival of the first humans? Also, can an ecohistorical periodization be proposed for the region, and in what detail? And finally, has the human-induced environmental change in the Yazoo-Mississippi Delta remained within bounds that would enable the recovery of the original landscape in the future, and if not, when and how was the resilience of the bottomland hardwood forest complex destroyed?

The rich deep black alluvial soil which would grow cotton taller than the head
of a man on a horse, already one jungle one brake one impassable density of
brier and cane and vine interlocking the soar of gum and cypress and hickory
and pinoak and ash, printed now by tracks of unalien shapes—bear and deer and
panthers and bison and wolves and alligators and the myriad smaller beasts . . .

—WILLIAM FAULKNER, *Big Woods*

# CHAPTER TWO

# *A True Ecological Complex*

PROBABLY THE BEST-KNOWN definition on the geographi-
cal extent of the Yazoo-Mississippi Delta has been provided by Delta author David L. Cohn—"[t]he Mississippi Delta begins in the lobby of the
Peabody Hotel in Memphis and ends on Catfish Row in Vicksburg."[1]
This culturally apt description of the social extremes of a New South
plantation empire does not, however, offer enough help in defining the
Delta as an ecological complex or as a bioregion; that endeavor calls for
classifications provided by the natural sciences.

## Defining the Delta

By the mid-1970s, U.S. public land management agencies had realized
how important land classification, on the basis of variations in climate,
vegetation, and landform, could be in finding solutions to management

problems. The differences were officially recognized in a 1976 map, "Eco-regions of the United States," which attempted to systematically divide the country into "ecosystem regions." The map was later supplemented by a book by Robert E. Bailey of the U.S. Forest Service, explaining the principles of the classification system and creating a new taxonomy for American nature. Bailey's work divided the United States into sixty eco-regions, grouped under four "Domains"—"Polar," "Humid Temperate," "Dry," and "Humid Tropical"—arranged in twelve "Divisions," thirty "Provinces," and forty-five "Sections."[2]

Section 2312 in this classification, "Southern Floodplain Forest," is a vast alluvial bed of some twenty-five million acres along the Lower Mississippi from the mouth of the Ohio River to the Gulf of Mexico, ranging from 30 to 40 miles in breadth at its northern end to 150 miles at the river's mouth in present-day Louisiana. It was created by the meandering Mississippi River and attained its current form during the final cycle of the latest glaciation. Misi Sipi, the "Great River" of the Ojibwa nation, drains the world's third largest basin. The drainage area of the greater Mississippi covers some 1.2 million square miles, or more than 40 percent of the coterminous United States, from New York to Wyoming. The main channel of the river, from Lake Itasca in northern Minnesota to the Gulf of Mexico, flows southward over a path almost 2,500 miles long. Near its midreach, the Mississippi is joined by its two largest tributaries, the Missouri and the Ohio, and, along its lower reaches, by several other major streams. Below the confluence of the Mississippi and the Ohio, the greatly increased flow results in the formation of the immense stream that supports the vast alluvial plain. In the northern section of the Lower Mississippi Valley, extending from the present-day Cairo, Illinois, to Helena, Arkansas, the river channel is broad and shallow, with numerous islands and sandbars. Below Helena, the channel becomes narrower and deeper and the islands and sandbars less common.[3]

Among the many floodplains constituting the Lower Mississippi Valley is the Yazoo-Mississippi Delta, which is actually more oval than deltoid in form. The channel of the Mississippi, from Memphis to Vicksburg, forms the western boundary of the Yazoo-Mississippi floodplain. The eastern boundary is defined by a series of bluffs that begin just below Memphis and run south to Greenwood and thence southwesterly along the Yazoo River, which meets the Mississippi just above Vicksburg. The enclosed area is approximately two hundred miles long and seventy miles across at its widest point, encompassing about 4,415,000 acres or some 7,000 square miles of alluvial floodplain from 32.4 to 35.0 degrees north latitude and 90.2 to 91.5 degrees west longitude. The floodplain is relatively flat with a gentle slope to the south, averaging from one-quarter to one-half foot per mile.[4]

In addition to the Delta, the 13,400 square mile drainage basin of the Yazoo River and its tributaries covers much of the rolling loess hills to the northeast of the alluvial plain. Flow characteristics between these two topographic regions vary distinctively, but all of the drainage eventually passes through the Coldwater-Tallahatchie-Yazoo river system to the Mississippi. Rainfall from the eastern portion of the basin tends to run off rapidly, carrying heavy loads of sand silt. The main stem of the Yazoo is formed at the confluence of the Tallahatchie and Yalobusha Rivers above Greenwood, discharging into the Mississippi after some 170 miles. The backwater area of the Yazoo extends some 60 miles northward from Vicksburg to near Belzoni and contains close to 2,500 square miles of alluvium. Some 6,200 square miles of alluvial land west of the Yazoo is drained through the Big Sunflower River, Deer Creek, and Steele Bayou. The area above Belzoni includes almost 3,700 square miles of alluvium and over 10,000 square miles of hill country and is subject to headwater flooding. The primary hill tributaries are the Little Tallahatchie, Yocona, and Yalobusha Rivers, while the Coldwater, Tallahatchie, and Yazoo serve as drainage outlets.[5]

Altogether, some 4,790,000 acres in the present-day state of Mississippi are located within the Mississippi Alluvial Plain, comprising just under 20 percent of the total extent of the Lower Mississippi Valley.[6] The Delta's "core" counties of Bolivar, Coahoma, Humphreys, Issaquena, Leflore, Quitman, Sharkey, Sunflower, Tunica, and Washington lie entirely within the Mississippi-Yazoo floodplain. In addition, varying amounts of land in Carroll, DeSoto, Grenada, Holmes, Panola, Tallahatchie, Tate, Warren, and Yazoo counties are of alluvial origin and belong to the Mississippi-Yazoo floodplain; these counties can be called "boundary" or "border" counties between the Delta and the loess hills to its east. The Mississippi River counties of Claiborne, Jefferson, Adams, and Wilkinson, as well as part of Warren County, are south of the Delta proper but contain narrow strips of alluvium, supporting forest vegetation similar to the Yazoo-Mississippi floodplain.[7]

The whole Delta in Bailey's classification belongs to Section 2312, which is located in the Outer Coastal Plain Forest Province (2310) within the Subtropical Division (2300) of the Humid Temperate Domain (2000). The Southern Floodplain Forest Section covers approximately 42,600 square miles or 1.4 percent of the United States. The eastern edge of the Delta contributes to the section's eastern boundary with Section 2215, Oak-Hickory Forest, and Province 2320, Southeastern Mixed Forest.[8]

The climate of the Southern Floodplain Forest Section shows a small to moderate annual temperature range and is characterized by the absence of really cold winters and the presence of high humidity. Abundant rainfall is well distributed throughout the year, ranging from forty to sixty inches. Average annual temperature ranges between 60° and 70°F. Forest typically forms the natural vegetation for areas with a humid subtropical climate, and—as the name of Section 2312 implies—botanical classifications can offer further possibilities in defining the Delta as an ecological complex.[9]

At the time of the arrival of the first European colonists in North America, around 1600, the eastern half of the continent, from the Mississippi River to the Atlantic Ocean, and from the forty-seventh parallel in southern Canada southward to the coastal plains of the Carolinas, except for the northern extension of the Mississippi prairie, was generally covered with forest. The types of forests included coniferous woodlands in the north, deciduous woodlands further south, and tropical savanna in Florida, with local variations everywhere to add complexity.[10] Among these were the bottomland hardwood forests typical to southeastern river valleys and covering the entire floodplain between the Yazoo and Mississippi rivers.

Eastern North America carries one of the most complicated and variable aggregations of vegetation in the temperate regions of the world. It has literally hundreds of species of trees, most of them hardwoods. Not surprisingly, the eastern forest as a whole is usually known as the Eastern Deciduous Forest. Authors have, in turn, subdivided the major types of natural forest vegetation found within the Eastern Deciduous Forest. The type of forest found in the Delta and on other southern floodplains is among the most distinctive of those subdivisions because of its unusual geographic characteristics and sharply bounded topography. These vast bottomland and alluvial swamp forests, which occur naturally on river floodplains of the southern United States, are often grouped together as the Southeastern Bottomland Hardwood Forest. Such bottomland forests are found wherever streams or rivers more or less regularly swell beyond their channels. Consequently, they are integral parts of delicate hydrological systems. The floodplain forests are dominated by woody species that show morphological and physiological adaptations enabling them to thrive in an environment where the soils within the root zone may be inundated or saturated for varying periods during the growing season.[11]

The bottomland hardwood forest type goes by many other names as well. For example, in the classic 1924 map of the natural vegetation of North America by H. L. Shantz and R. Zon, the bottomland hardwood forests are called River Bottom Forests. A. W. Küchler applied yet another term in his 1964 map, which has become a standard for discussion of the potential natural vegetation types of the United States. "Southern Floodplain Forest" (type 113), in Küchler's classification, is a dense, medium-tall to tall forest of broadleaf deciduous and evergreen trees and shrubs and needleleaf deciduous trees that occurs along the river and stream floodplains of the southeastern United States. Because of the necessarily large scale used, these two maps do not delineate the total extent of southern floodplain forests in detail. Furthermore, vegetation classification systems rarely recognize the dynamic aspects of landscapes and communities and tend to classify the natural world as if human influence did not exist. Thus such classifications, while reflecting the broad pre-Columbian vegetation types of North America, do not identify the way Native American cultures used and modified the landscapes.[12]

A more elaborate map of the major southeastern river floodplains and the associated bottomland hardwood communities was prepared by the U.S. Forest Service in 1960. It illustrates how the coastal boundaries of the floodplain forests are defined at the point where the rivers and streams enter the sea and the vegetation shifts to coastal marshes. In areas where there is little or no intrusion of seawater, the deltas of southern rivers support vegetation similar to bottomland swamp forests. Inland boundaries can be more difficult to establish, as the decrease in the amount of floodplain and the shift from the dominance of depositional to erosional processes along headwaters may be gradual. As a rule, compositionally distinct floodplain and upland forests can be juxtaposed over short longitudinal distances by differences in elevation.[13]

The bottomland hardwood forests are found in the general area of 28 to 38 degrees north latitude and 75 to 95 degrees west longitude. These forests occur mainly along the major rivers and tributaries that extend into upland pine sites. The bottomland hardwood forests characterize the alluvial lands of the Mississippi Valley as far north as southern Illinois, as well as the river valleys of the Coastal Plain from the Mississippi River eastward along the coasts of the Gulf of Mexico, on the Florida Peninsula, and northward through the Carolinas. The largest single area of bottomland hardwood forest occurs on the alluvium of the southern Mississippi River Valley. In addition to the Mississippi, its tributaries, such as the Yazoo, White, Arkansas, Red, Quachita, Tensas, and St. Francis Rivers, naturally support extensive areas of bottomland hardwood forest, as does the Atchafalaya, today a distributary. Other large rivers that drain into the Gulf of Mexico and have supported considerable areas of bottomland forests include the Pearl, Tombigbee, Alabama, Pascagoula, Chattahoochee, Apalachicola, Suwannee, Sabine, Neches, and Trinity. Comparable rivers flowing into the Atlantic include the Altamaha, Ogeechee, Santee-Cooper, Pee Dee, Cape Fear, Neuse, and Roanoke. The extensive swamps of southeastern North America, such as the Dismal Swamp, between modern-day Virginia and North Carolina; the Pamlico Swamp of North Carolina; the Santee River swamp of South Carolina; the Okefenokee Swamp, between Georgia and Florida; the swamps of the St. Francis Basin in Arkansas and southern Missouri; and the numerous large swamps of the Lower Mississippi Valley similarly support bottomland hardwood forest vegetation.[14]

The bottomland hardwood forest region of the southeastern United

States is noted for its long growing seasons, hot summers, and high humidity. The climate of the region may be classed as humid to subhumid. Annual precipitation averages fifty inches and is distributed throughout the year. Late summer to early fall is usually the driest part of the year. Moderate droughts occur irregularly every few years, but severe ones only once every two or three decades. The bottomland hardwoods region is characterized by a relatively long frost-free period, averaging about 240 days. Prolonged periods of below-freezing temperatures are untypical to the area. Mean January temperatures range from about 40°F in the north to about 55°F along the Gulf Coast, while mean July temperatures average about 80°F throughout the region. The summers are not only hot but very humid. During the warmest months violent thunderstorms with considerable displays of lightning constitute a typical climatic aspect which, in the form of fires, assumes significant importance to the forest composition in the region. Ice storms, tornadoes, and also hurricanes in coastal areas are occasional but characteristic climatic phenomena that may furthermore affect local conditions and the forest composition. These can cause considerable damage but have limited impact on a regional scale.[15] On May 6, 1858, Rowland Chambers of Yazoo County was awed by a storm that produced hail "as large [as] a hen[']s egg," weighing an ounce.[16] An early-nineteenth-century traveler described a tornado's effect on a bottomland hardwood forest in the following way:

> [T]he tempest came on, sweeping every thing in its way. Its current was generally about a quarter of a mile wide, running an easterly course; and bearing down all opposition. Providentially, we lay on the opposite side from whence it came, or we would have suffered materially; for the next morning we could discover the opposite shore to be full of trees of enormous size, laying prostrate in the river, and opening a prospect for miles on both sides, exhibiting to view the resistless power and might of the Author of nature.[17]

The Lower Mississippi Valley has also witnessed major seismic activity in recent times. Beginning in December 1811, a series of violent earthquakes near New Madrid, Missouri, destroyed an estimated 150,000 acres of timberland, most of it bottomland hardwood forest, created new lakes, and transformed thousands of acres of farmland into swamps.

Modern weather patterns in the Lower Mississippi Valley and the Yazoo-Mississippi Delta cannot be sufficiently explained by latitude (the region's proximity to the equator). Warm season patterns in the Delta are controlled by the Bermuda High, a semipermanent tropical high-pressure area, whereas cold season weather is controlled by cold fronts originating in the northwestern and north-central parts of the continent. Weather statistics for the Yazoo-Mississippi Delta reveal a climate representative of the southeastern bottomland forest region. Annual precipitation in the Delta averages fifty-two inches, while recorded yearly extremes for the city of Greenwood in the east-central Delta range from thirty-five to seventy-four inches. As a rule, more than 50 percent of the Delta rainfall occurs between November and April. The average monthly temperature for August is 83°F throughout the Delta, while January averages range from 42°F in the north to 48°F in the south. Occurrence of killing frosts correlates with latitude, and, consequently, the growing season in the Delta ranges from 220 to 240 days.[18]

Current weather patterns can ultimately be traced to volcanism in South America during the Late Tertiary period (5 million to 2.5 million B.P.) that created a continuous land bridge between North and South America and diverted warm ocean currents northeastward along the eastern coast of North America.[19] The tectonic closing of the Isthmus of Panama partly caused the global climate system to be highly sensitive to changes in incoming solar radiation and, consequently, susceptible to oscillations between glacial and interglacial climatic regimes. More than twenty 100,000-year glacial/interglacial cycles have occurred during the Quaternary period, or the last 2.5 million years; each transition between peak glacial and interglacial extremes has represented an immense change that has triggered processes of extinction, migration, and speciation. The present-day Delta has not been directly glaciated during the Quaternary period but the changes in global atmospheric circulation patterns that caused the glacial advances and retreats have affected the region intensely.

During the last full-glacial interval, some 20,000 to 16,500 years ago, the sea level was probably more than one hundred yards below the present-day level, while the Southeast provided a refuge for both boreal and temperate biotic communities of eastern North America. The climate was considerably cooler than today, and species of the spruce family (*Picea*) dominated the southern floodplains fed by glacial meltwater. Late-glacial climatic changes commenced by 16,500 years ago; spruce populations began to decline with the warming climate and were replaced by invading hardwoods, especially members of the oak family (*Quercus*). Analysis of fossil pollen records in sediment cores reveals the cypress family (*Taxodium*) present in the Southeast between twelve and ten thousand years ago.

Pollen data from the lower Yazoo-Mississippi Delta during the interval from sixteen to ten thousand years ago indicate climate somewhat cooler and moister than today. The forests of the region seem to have formed a relatively open woodland populated by various deciduous trees. Ten to five thousand years ago the climate became warmer and drier during a hypsithermal interval, and pollen data for the Delta during this period suggest full establishment of the deciduous forest and considerable closure of the forest canopy. The sea level attained its modern position between 5,000 and 3,500 years ago, producing swamps and marshes on lower elevations of the Southeast. The warming climate had also brought increased storm activity and lightning strikes with ensuing forest fires, and by four thousand years ago, fire-resistant southern pines (*Pinus*) had come to dominate the drier uplands of the Southeast. During the last five thousand years, modern vegetation patterns and somewhat cooler climatic conditions have taken shape in the Southeast and the Yazoo-Mississippi Delta. Pollen from various species of oak and hickory (*Carya*) dominate the Delta palynological record, while the climate has remained approximately the same. However, there have been fluctuations in climatic conditions during the last five millennia. For example, a warm span lasting from the eleventh to the early fourteenth centuries A.D. was followed by the Little Ice Age from mid-fifteenth to mid-nineteenth centuries, with a general decrease in temperatures and an increase in precipitation.[20]

The physical character of the present landscape of the Yazoo-Mississippi Delta and the fluctuating hydrologic regime that dictates its ecological functioning formed during the late Pleistocene and Holocene epochs. A sequence of marine deposition of sands, silts, and clays, combined with intermittent rises in the sea level and subsequent uplifting of the land, resulted in a series of terraces making up the Coastal Plain of the present-day southeastern United States. As the land surface was raised, an undulating topography was formed by the developing seaward drainage patterns. Continual uplands erosion and eroded materials deposition along the streams and at the sea outlets formed the alluvial bottomlands. A relatively abrupt decline in the load of glacial outwash carried by the Mississippi occurred roughly nine thousand years ago and transformed the former regimen of multichannel braided streams into a meandering one. Much of the present landform in the Delta seems to have resulted from the meandering of the Mississippi River between 4,800 and 2,600 years ago. Although the Mississippi River was undoubtedly the major factor in the creation of the floodplain, the Yazoo and its tributaries drain substantial portions of the neighboring upland and have deposited considerable amounts of alluvium along their courses. The role of the smaller rivers in the continuous formation of the floodplain undoubtedly grew more pronounced after the Mississippi stabilized in its present meander belt.[21]

Because of the gentle angle of seaward dip, the Mississippi and its tributaries meander across their floodplains, transporting, eroding, and depositing alluvial sediments. Floodplains and their rivers are in a continual dynamic balance between the building and removal of structure. The southeastern rivers constantly modify their valleys by cutting channel banks and forming new land, building natural levees and creating oxbow lakes—and therefore develop a very complex array of geomorphological features. The extremely low topographic relief of the floodplain can be deceptive, as an elevational change of a couple of inches creates distinctively different hydrologic conditions, soils, and biotic communities.

The Mississippi and its tributaries drain extensive basins, many with high precipitation levels. As the rainfall runoff rapidly drains to the flat Coastal Plain, the rivers spill out of their channels. Flooding in the Lower Mississippi Valley can be divided into three major types: headwater, backwater, and tidal flooding. Headwater and backwater floods occur throughout the Mississippi Alluvial Plain, including the Yazoo-Mississippi Delta, whereas tidal flooding results from tidal surges caused by tropical storms and affects only the southernmost portion of the Lower Mississippi Valley, close to its mouth. Headwater flooding typically results from rainstorm activity over drainage basins of the Mississippi tributaries and affects lands adjacent to the streams. Such flooding along the Mississippi commonly occurs as a result of excessive rainfall in the central parts of North America. Combined with snowmelt in the upper tributary areas, the rain produces the great spring floods of the Mississippi. High water stages on the Mississippi can cause backwater flooding along its tributaries, caused by the damming effect of the main stem waters. Excess water in the main channel holds or slows runoff from the tributaries and can even create a reverse flow of the tributary stream some distance upstream from its mouth. This typically results in a long inundation of the adjacent floodplain. Despite the magnitude of its floods, the Mississippi is not as unique as an eighteenth-century traveler wanted to believe, stating "that no part of the waters which overflow its banks, ever return to their former channel: this is a circumstance, which I believe is not to be met with any other river in the world."[22]

Bottomland hardwood forests are by nature adapted to flood regimes. Floodplains in their natural state absorb and delay some of the flood's energy. The dense concentrations of tree trunks in the bottomland hardwood forests provide enough friction to slow flow and eliminate many vortices. At the same time, floodplains often provide enough cross section to discharge the waters and prevent backwater flooding. Consequently, there is a great difference between water courses with or without forested wetlands in the way flood waters pass downstream.[23]

Once there is reduction in the runoff, the levels of the rivers will drop rapidly, except in the low-lying swampy areas, where sizable areas may remain inundated for several weeks. Consequently, much of the floodplain acreage is—in its natural state—subject to rapid inundation during any season of the year. Because of the relatively small drainage area and the subsequently modest water volume carried by the stream, the extent of the floodplain is limited along smaller watercourses. However, several species of bottomland broadleaf trees, especially those typical of short submergence habitats, can be found on the banks of such streams. In these moist creek bottoms, areas of bottomland hardwood forest dissect the pine-dominated higher lands of the Southeast.[24]

Bottomland hardwood forests are found generally on the so-called first bottoms and terraces. First bottoms are the floodplain created by the present drainage system and are of comparatively recent origin. Under natural conditions, they are subject to frequent flooding. Terraces (sometimes also known as second bottoms) were formed by earlier drainage systems; they represent former floodplain and are not hydrologically connected with the present river system, except during superflood stages. The transition from a first bottom to a terrace may be distinct or gradual. Alluvial landforms of the Yazoo-Mississippi Delta formed rather recently and belong to the first bottoms category. Over time, the lateral movement of the river across its floodplain has created various depressions and rises in the form of meander scrolls. "Ridge and swale" is a term often used for this terrain built by the earlier and present drainage systems. In addition, new areas of land, or fronts, are always being formed by the river along its banks.[25]

Floodplains have typically been created by two major aggradation processes: the constant deposition of alluvial sediments on the inside curves of rivers and overbank flooding. Natural levees (sometimes called high ridges) adjacent to the river channel have resulted from the deposition of coarser materials when water has overflowed the channel bank. The portions of natural levees fronting actively flowing streams are often called batture lands. Natural levees typically slope sharply toward the river and more gradually away from it and constitute the highest terrain on the floodplain. The finer sediments are deposited farther away, forming broad areas of "slackwater" or "backswamp" with poorly drained clay soils with high organic content often described as "buckshot." Variation in elevation within the slackwater area results in low ridges, flats, sloughs, and swamps. Low ridges are banks or fronts of former stream courses, typically one to ten feet above the adjoining flats. They are covered by water only during floods. The flats constitute the general terrain between ridges and have little topographical relief. The sloughs are the remains of nearly filled stream courses or present drainageways; water collects only seasonally in these shallow depressions. The swamps are distinct depressions that contain water except during extreme drought. Large swamps are the result of later terraces blocking the seaward drainage.

Sedimentation occurs also as the soil from a caving bank is deposited on the opposite (convex) side of a river curve, resulting in the formation of so-called point bars. Thomas Nuttall offered a sound description of the process in 1819:

[T]he river appears singularly meandering, sweeping along in vast elliptic curves, some of them six to eight miles round, and constantly presenting themselves in opposite directions. The principal current pressing against the center of the bend, at the rate of about five miles per hour, gradually diminishes in force as it approaches the extremity of the curve. Having attained the point of promontory, the current proceeds with accumulating velocity to the opposite bank, leaving, consequently, to the eddy water, an extensive deposition in the form of a vast bed of sand, nearly destitute of vegetation, but flanked commonly by an island or peninsula of willows. These beds of sand, for the most part of the year under water, are what the boatmen term bars. The river, as it sweeps along the curve, according to its force and magnitude, produces excavations in the banks; which, consisting of friable materials, are perpetually washing away and leaving broken and perpendicular ledges, often lined with fallen trees, so as to be very dangerous to the approach of boats, which would be dashed to pieces by the velocity of the current.[26]

A nineteenth-century traveler on the Mississippi could grow tired of the monotonous scene of "always seeing a caving bank on one side, and an advancing sand-bar, covered with willows and poplar, on the other," for hundreds of miles.[27] Over time, point bars grow above the ordinary high-water stages, begin to support an increasing amount of vegetation, and become stabilized as a part of the floodplain as fronts. Eventually, as these projections become longer and the meander curves sharper, the river shortens its course by making a "cut-off" across the land. Continuing sedimentation then transforms the abandoned section of river channel into an oxbow lake and, eventually, a swamp. Oxbow lakes are the only natural lacustrine features in an alluvial valley.

Hundreds of different soils are recognized for the southeastern United States. The nature of the soils in an alluvial river valley is determined largely by the materials present in the drainage area that serves as the erosional source of the deposits of mineral soil. Bottomland soils may be clayey, silty, or sandy, depending on the origin of the deposits and their location in relation to the river. Alluvial soils derived from the materials of the Atlantic Coastal Plain are sandier and coarser than alluvial soils derived from erosion of the grassland prairie soils, which, as in the Lower Mississippi Valley, are high in expanding clays. As a result, differences between general forest cover types of southeastern bottomlands largely arise from differences in the soils' geological origin. As explained above, flood waters deposit coarse sands on channel banks and fine clays farther away from stream channels, thus modifying the sites constantly by creating natural levees and point bars. As silt builds up, rivers and streams change directions, further altering sites and the forest composition.

Because the bottomland soils are mostly composed of materials with a particle size small enough to inhibit nutrient leaching and to retain moisture, these soils are generally more fertile and moist than the adjacent upland soils with predominantly sandy composition. The fertility is furthermore enhanced by the continual replenishment by flooding and the high organic matter content of bottomland hardwood forests soils. The thickness of alluvial sediment in the Lower Mississippi Valley ranges from 15 to 260 feet, with an average thickness of 125 feet in the northern part of the valley.[28]

Soils of the Yazoo-Mississippi Delta belong to the Inceptisol order, suborder Aquepts. Called Haplaquepts, they are naturally slowly drained soils typical of the broad, nearly level floodplain of the Lower Mississippi Valley. Removal of original vegetation and adequate drainage can transform these soils with their ample available nutrients into highly productive agricultural lands, "the richest land this side of the valley Nile" according to Big Daddy, the Delta planter in Tennessee Williams's Cat on a Hot Tin Roof.[29] The nineteenth-century settlers generally divided the Delta soils into two classes, loam and clay. While the sandy loam soils found along the riverbanks were fertile enough, they could not match the nutrient content of clay-dominated soils of the backswamps. These poorly drained areas were popularly known as "buckshot lands," because the soil dried "into angular bits the size of a buckshot, and [was] of a lead color."[30]

There is a marked difference between the soil composition in the Delta and the adjoining bluffs to its east. The soils of the hills bordering the Mississippi-Yazoo floodplain belong to the suborder Udalfs of the Alfisol order. They are eolian (wind-blown) loess deposits underlain by marine sediments. Udalfs erode easily and, when intensively cultivated, loess hills soon lose their productivity. An early-nineteenth-century traveler found "[t]he river bottoms generally, and some of the cane brake hills, not being exceeded for richness in the world, while some ridges and tracts of country after being cultivated for a few years, are so exhausted, as to become almost barren."[31]

## The Nonhuman Biota of the Delta

Floral and faunal uniqueness is at least partly a matter of scale, and any area becomes "unique" if it is large but not too large. It is relatively easy to characterize different areas on the basis of dominant floral and faunal elements. A classic example of this is the 1858 division of the world into six zoogeographical regions on the basis of bird distributions by the English ornithologist P. L. Sclater. On a considerably smaller scale, the bottomland hardwood forests of southeastern North America have their own areas of unique flora and fauna. This results not as much from the overall level of endemism evident in the floodplain forests, but from the forests' evolutionary history and unique combination of different species. In response to glacial cycles, the southeastern river valleys have served as important dispersal corridors for many species to and from higher latitudes.[32]

The diversity and species richness of plant communities in the southeastern bottomland hardwood forests can be immense. Moist sites typically contain a more diverse mix of species than wet ones. In the better-drained areas of the floodplain, forests become very dense. Trees at these sites can reach a very large size, often over three feet in diameter. In 1808, Henry Ker explored the bottomland forests along the Lower Mississippi and marveled how "after making our way through a thick cane brake, we entered a beautiful wood of large timber, lifting their heads to an immense height, and seeming to envy each other's glory. The sycamore tree is the monarch of the forest; its size is incredible to all but those who have seen it. In this place I measured one which was thirty-five feet in circumference, making it about eleven feet six inches in diameter."[33] Small trees, shrubs, and lianas are abundant in the bottomland hardwood forests. The forests of the nineteenth-century Yazoo-Mississippi Delta were among the most luxurious in their species richness and biological diversity:

Along the elevated ridges fronting the streams the white oak, the willow oak, the shell-bark and mocker-nut hickories, the black walnut in great numbers, the yellow poplar and the sassafras large enough to furnish canoes of great size, the mulberry, the Spanish oak, the sweet and black gums are the principal forest trees, with an undergrowth in the openings of dogwood, various haws, crab apples, wild grapes, buckthorns, etc. In the forests covering the lower lands, which slope back to the swamps and reservoirs, the cow oak takes the place of the white oak, while the over-cup white oak occurs everywhere in the more or less saturated soil. Here the sweet gum reaches its greatest size, and here grow also in great perfection the bitter-nut, the elms, hornbeams, white ash, box-elder, and red maples of enormous size. The honey locust, water oaks, and red and Spanish oaks are equally common. Here, among the smaller trees, the holly attains its greatest development, with hornbeams and wahoo elms, while papaws, haws, and privets form the mass of the dense undergrowth, which, interspersed with dense cane-brakes, covers the ground under the large trees.[34]

As with other types of forests, tolerant species gradually replace intolerant ones in the process of natural succession. The most important environmental condition in the bottomland hardwood forests is the length of the hydroperiod, which varies in time and space across the floodplain. The zonation of bottomland tree species reflects their relative flooding tolerance. In 1896, the interior of Concordia Island in Bolivar County supported "the thickest timber I had ever seen. Oak, gum, ash, hackberry, and poplar stood so thick, with no underbrush, only big blue cane growing rank and tall, almost to the limbs of the trees." On the riverfront, however, "the woods were thinner; there were fewer trees but larger—big old cottonwoods and sycamores that seemed to me when I looked up like their tops were lost in the sky."[35] Many plant species of the bottomland hardwood forests typically require or tolerate recurring or long inundation, and diameter growth of trees typically correlates with the amount of flooding. Flooding during the growing season has the greatest effect on species survival, and extended flooding, while beneficial to many species of trees, raises mortality among others. Consequently snags are abundant in mature bottomland hardwood forests.[36] In his famous account of a cougar hunt in the Delta bottoms, John James Audubon marveled at "the singular effects produced by the phosphorescent qualities of the large decayed trunks which lay in all directions around" him.[37]

The main pioneer woody species of recent alluvium are cottonwoods (*Populus deltoides* and *P. heterophylla*) and willows (*Salix* spp., especially black willow, *S. nigra*).[38] These species are intolerant of shade and cannot succeed themselves. Cottonwoods grow on ridges having coarse-textured soil, while black willow establishes itself on fine-textured soils of flats. An early-nineteenth-century traveler marveled at the "enormous cotton-wood trees" bordering the Mississippi, "more than six feet in diameter, and occasionally festooned with the largest vines which I had ever beheld."[39] A slow succession of plant communities occurs also as sloughs and swamps fill with sediment. Usually, the pioneering species at these sites is black willow, followed by baldcypress (*Taxodium distichum*) and water tupelo (*Nyssa aquatica*) in swamps and overcup oak (*Quercus lyrata*) and water hickory (*Carya aquatica*) in sloughs. In areas where the stream channel has undergone fairly recent changes, levee deposits typically support mixtures of river birch (*Betula nigra*) and silver maple (*Acer saccharinum*) in addition to willow and cottonwood.

During later successional stages, communities with shorter hydroperiods are typified by laurel oak (*Quercus laurifolia*), red maple (*Acer rubrum*), American elm (*Ulmus americana*), green ash (*Fraxinus pennsylvanica*), and hackberries (*Celtis* spp.). Low ridges may be dominated by sweetgum (redgum, *Liquidambar styraciflua*), or black tupelo (blackgum, *Nyssa sylvatica*), pecan (*Carya illinoensis*), shagbark hickory (*Carya ovata*), white oak (*Quercus alba*), and black walnut (*Juglans nigra*), depending on elevation and, consequently, the length of the hydroperiod. Dense stands of cane (*Arundinaria gigantea*) furthermore characterize ridges in a natural state. Because of poor drainage the more elevated riverine flats (terraces) may support species similar to the low ridges of the first bottoms. Swamps, sloughs, and other areas of deeper water and long hydroperiods are characterized in later succession also by planertree (water elm, *Planera aquatica*).

Floodplain sites with short hydroperiods also support swamp chestnut oak (*Quercus michauxii*) and water oak (*Quercus nigra*). In the driest ridges along the Yazoo, beech (*Fagus grandifolia*) and magnolia (*Magnolia* spp.) may have occurred naturally with the more typical bottomland flora. Other species of woody plants commonly found in the southeastern floodplain forests include honeylocust (*Gleditsia triacanthos*), boxelder (*Acer negundo*), red mulberry (*Morus rubra*), pawpaw (*Asimina triloba*), possumhaw (*Ilex decidua*), persimmon (*Diospyros virginiana*), and swamp privet (*Forestiera acuminata*), along with different vines (e.g., *Ampelopsis arborea, Berchemia scandens,* and *Campsis radicans*) and several species of grapes (*Vitis* spp.). American sycamore (*Platanus occidentalis*) is another typical tree of the bottomlands, as noted by Thomas Nuttall in 1819: "On the river lands, as usual, grows platanus or buttonwood, upon the seeds of which flocks of screaming parrots [Carolina parakeet, *Conuropsis carolinensis*] were greedily feeding."[40] Spanish moss (*Tillandsia usneoides*), a common epiphyte of the moist southern woods, creates an eerie forest landscape that early European travelers and settlers to the southernmost Delta commented on: "The first thing I observed in the woods, as new to me, was the long moss hanging in dingy gray streamers from the limbs of the trees. The whole tree-top, like the head of Medusa, before Minerva changed her beautiful locks, hung thick with long, flowing tresses. These streamers are five or six feet long. To see all the trees draped with them, it gives the woods a lovely mournfulness—a beautiful gloom."[41] The observer furthermore noted that "[t]here is not, that I know, an object of nature that produces such a number of sepulchral images as the view of the cypress forest, all shagged, dark, and enveloped in the festoons of moss."[42]

At least one area in the Yazoo-Mississippi Delta has seemingly not been inundated in recent history, even during most severe flooding. Dogwood Ridge extends through the floodplain from what is today Coahoma County to Holmes County and ranges from two to eight miles in width. It separates the drainage areas of the Sunflower and Yazoo Rivers and seems to have supported a floral assemblage slightly different from the rest of the floodplain. The species for which the ridge was named—dogwood (*Cornus florida*)—is intolerant of flooding and prefers well-drained soils, as are sassafras (*Sassafras albidum*) and prickly pear (*Opuntia* spp.), also found on this apparently flood-free area within the Delta. It must further-more be noted that of the 3,563 tree specimens present in the early Delta land reports from the 1830s, over 30 percent are intolerant of flooding (i.e., can survive only one growing season or less of inundation). Given that there was a major Mississippi flood in 1828, only a couple of years before the beginning of the land surveys along the main stem of the Yazoo, the presence of such tree species indicates that portions of the Delta have remained relatively dry even during serious flooding.[43]

Stream migration, accompanied by the constant erosion and deposition of sediments, creates a continuous pattern of disturbance in the southeast-ern bottomland hardwood forests. The dynamic floodplain environment therefore seldom develops a stable "climax" community. Ice storms, tor-nadoes, and wildfires are examples of other, less constant but character-istic disturbances. Mary Hamilton vividly described successional patterns in the so-called Cyclone tract in Sunflower County. In 1897, some years after a tornado had carved a path two to four miles wide and eighteen miles long through the forest, "[c]ane, undergrowth, blackberry briars, grape and poison oak and muscadine vines, all growing to enormous size out of that rich Delta ground, interlaced through that fallen timber so it was one great tangled mat."[44] Insects, plant diseases, and wildlife preda-tion on seed and seedlings further affect the composition of bottomland hardwood stands. Almost all tree species of bottomland hardwoods are vulnerable to fire. During average or wet years, natural fires are very infre-quent; occasional prolonged droughts, however, can result in devastating wildfires that kill or severely damage younger vegetation and provide en-try for decay in mature trees.[45]

The southeastern bottomland hardwood forests present an extremely heterogeneous mixture of tree species with changes so subtle that classify-ing specific forest cover types within the complex is a difficult task.[46] The U.S. Forest Service today classifies forest land according to the presence or absence of certain species groups, dividing the bottomland hardwoods into oak-gum-cypress and elm-ash-cottonwood types, while the U.S. Fish and Wildlife Service includes them within the class of Palustrine Forested Wetlands under three different headings. Other, more detailed classifica-tions of bottomland hardwood forest have been proposed; the Society of

American Foresters recognizes some twenty different forest cover types for bottomland hardwood forests, with at least twelve of these found in the South.

Yet another description was developed by John A. Putnam, the trailblazer in southern hardwood forestry. Putnam's classification divides the bottomland flora into eight primary types with several variations. The widely distributed sweetgum–water oak type is usually found on terrace flats and in first bottoms, while the white oak–red oak–other hardwoods type occurs mainly on first bottoms and terrace ridges. The overcup oak–bitter pecan type is characteristic of low, poorly drained flats, sloughs, and backwater basins. The cottonwood type pioneers fronts and well-drained flats, while hackberry-elm-ash is normally a successional type following cottonwood on low ridges, flats, and sloughs in first bottoms and on terrace flats and sloughs. The willow type invades fronts in sloughs and low flats. The riverfront hardwoods type, dominated by sycamore, pecan and green ash, is transitional between cottonwood or willow and the sweetgum–water oak type and occurs on all front lands except deep sloughs and swamps. Finally, the cypress-tupelo-gum type characterizes low and poorly drained flats, deep sloughs, and swamps.[47] However, only the more recent data on the bottomland hardwood forest and its inhabitants separate these different plant communities. Consequently, the broad terms "bottomland hardwood forest" and "floodplain forest" are used interchangeably throughout this historical study, referring to all of the different plant communities found within this major forest type.

Southeastern floodplains provide a diverse habitat for a variety of animals.[48] The location of bottomland hardwood forests at the interface between aquatic and terrestrial systems results in the so-called edge effect: the diversity and species richness of animals tend to be greater at the ecotone, or the edge between two distinct ecological complexes. The edge effect is pronounced in the bottomland hardwood forests of the Yazoo-Mississippi Delta, as they are bounded by both aquatic systems and upland forests. This effect, coupled with the diversity of habitats found within the floodplain, provides for an abundance of fauna.[49] The linear distribution of the floodplain forests along rivers and streams also offers protective pathways for many animals to migrate between habitats. Phenology and mast production in the bottomland hardwood forests are nonsynchronous with upland communities and further contribute to their value as animal habitat.

Animals of the bottomland hardwood forests, like the plants, have adapted to the distinctive hydrological cycles of the floodplain and tend to occupy certain areas within it. Some animals maximize their utilization of the floodplain during its inundation, while others prefer partial inundation or drydown conditions. The hydrological cycle typically favors species that are highly mobile (e.g., spiders), arboreal (e.g., tree frogs), can survive inundation as adults (e.g., salamanders), or can cope with the hazard of drydown by going underground (e.g., crawfish). During the flood season, the floodplain is important as a breeding and feeding area for many aquatic species, especially fish. Aquatic systems also provide homes and important sources of food for many semiaquatic and terrestrial species. Surface water furthermore moderates temperature extremes in the floodplain and serves as an escape habitat for many animals.

The major autotrophs of bottomlands are the trees that provide food directly or via detrital pathways to animals associated with the bottomland hardwood forests. Whether during the flooded or unflooded state of the floodplain, most animals characteristic of the bottomland forests are detritivores that support a host of predators. A wide diversity of invertebrates live on the floodplain, often in enormous numbers. In fact, the density of many invertebrate species seems to be considerably higher in the permanently flooded portions of the bottomlands than in the nearby rivers.

A variety of fish species inhabit the floodplain both temporarily and permanently, and the value of bottomland sloughs and swamps for their spawning and feeding during the flooding season is great. As a rule, most larger fish remain temporary residents in the floodplain. Amphibians' ability to adapt to fluctuating water levels, combined with their semiaquatic habits, enables many species to thrive in a range of bottomlands habitats, from the permanently inundated to the relatively dry. Characteristic species include the two-toed amphiuma (*Amphiuma means*), the lesser siren (*Siren intermedia*), and various salamanders (*Desmognatus* spp., *Ambystoma* spp.) and frogs (*Rana* spp., *Acris* spp., *Hyla* spp.). Reptiles are not as numerous but include snakes (*Regina* spp., *Akgistrodon* spp.), the snapping turtle (*Chelydra serpentina*), the box turtle (*Terrapene carolina*), and various turtles of the genus *Kinosternon*. The animal that best characterizes the South in the minds of many, the American alligator (*Alligator mississippiensis*), is also present in the southern bottomlands. The fascination of early European travelers on the Mississippi with this large reptile is captured in the 1797 description by Francis Baily:

> But what attracted my attention the most was the enormous alligators which we saw basking in the sun on logs near the shore. Here they would expose themselves, lying with their monstrous jaws wide open, and apparently asleep; but not absolutely so, for on our approach to them they would flounce suddenly in the water, and scatter the foam to a considerable distance. . . . Their appearance is enough to terrify the eyes of any beholder; and dreadful is their rage when attacked *in the water,* which appears to be their natural element; but on the land an escape from them is easily made, as their motion is very slow.[50]

Food resources in the bottomland hardwood forests attract mammals as diverse as the opossum (*Didelphis virginiana*), the gray squirrel (*Sciurus carolinensis*), the flying squirrel (*Glaucomys volans*), rabbits (*Sylvilagus* spp.), the raccoon (*Procyon lotor*), and the bobcat (*Felis rufus*). At the time of the European conquest, canebrakes of the bottomland ridges provided food and cover for large herds of white-tailed deer (*Odocoileus virginianus*).[51] Also, the black bear (*Ursus americanus*), the cougar (*Felis concolor*), and the red wolf (*Canis rufus*) were among the original fauna of the bottomland hardwood forests of the Yazoo-Mississippi Delta. Even the American bison (buffalo, *Bison bison*) and the elk (wapiti, *Cervus canadensis*) have been encountered in the habitat.

Extensive and long-lasting flooding, however, limits mammals' access to many food resources. Some species, such as the opossum, cannot survive under flood conditions, while others, such as the raccoon, are likely to remain even on a largely submerged floodplain. Wetland mammals of the region, such as the beaver (*Castor canadensis*), the muskrat (*Ondatra zibethica*), and the river otter (*Lutra canadensis*), are even better adapted to the inundated floodplain. Beavers can affect local conditions by building dams that change hydrologic conditions and result in flooding and increased mortality of certain plant species. The additional disturbance is not restricted to the killing of plants not adapted to prolonged flooding, as beavers also modify successional patterns by creating open habitat. The cutting of trees along the water favors the invasion of early successional shrubs and small trees that form the mainstay of the beaver diet.[52]

The suitability of a forest area for birds depends on many factors, including stand structure, foliage volume and the number of vertical foliage layers, habitat patchiness, and moisture gradient. Most of these are quite favorable for many bird species in bottomland hardwood forests of natural state; for example, the supplemental nutrient and soil input provided by flooding enhances productivity (e.g., foliage development) in bottomlands. Ecologists have claimed that a more productive and diverse community of plants (such as a bottomland hardwood forest as compared to a longleaf pine forest) is likely to support a greater richness of herbivorous species, and so on throughout the food web. Numerous studies have demonstrated that especially the bird density in the bottomland hardwood forests is remarkably higher than in other types of southeastern forests. Accordingly, the forests of the Yazoo-Mississippi Delta have supported an abundant and diverse avifauna.[53] Christian Schultz, writing in 1808, insisted that the forests along the lower Mississippi supported songbirds in an abundance unsurpassed elsewhere in the United States:

> [A]t the moment the gilded foliage of the lofty trees acknowledges [the] appearance [of the rising sun], the whole feathered creation, as if with one accord, pour forth their gratitude in one general hymn. The woods on both sides of the river, ever since we passed the Arkansas, appeared to be literally alive with its numerous feathered inhabitants; and although we generally kept the middle of the river, which is one mile in breadth, yet we could hear the general chorus much better than on shore. I do not recollect ever to have heard any thing to equal this charming natural concert.[54]

The most important natural variable in bottomland hardwood forests, the extent and duration of flooding, has profound and predictable effects on the bird communities. For example, birds that use the understory vegetation, such as the ground-nesting Kentucky warbler (*Oporornis for-*

*mosus*), are not found on flooded sites. Conversely, species such as the prothonotary warbler (*Protonotaria citrea*) are confined to these wetter sites. Flooding creates habitat for aquatic birds and affords protection from mammalian predators to colonial nesters such as the snowy egret (*Egretta thula*), the yellow-crowned night heron (*Nycticorax violaceus*), and the red-winged blackbird (*Agelaius phoeniceus*). The wild turkey (*Meleagris gallopavo*), regularly encountered in the floodplain forests, finds old stands of gum or cypress growing in water ideal for roosting. Extended flooding raises mortality among trees, creating nesting sites in the form of snags for various woodpeckers and the cavity-nesting wood duck (*Aix sponsa*). Dead trees also serve as foraging sites for woodpeckers.

Many species of birds have been associated with the mature bottomland hardwood habitat. Breeding bird censuses have shown that, for example, the pileated woodpecker (*Dryocopus pileatus*), the red-bellied woodpecker (*Melanerpes carolinus*), the yellow-billed cuckoo (*Coccyzus americanus*), the Acadian flycatcher (*Empidonax virescens*), and the red-eyed vireo (*Vireo olivaceus*) are consistently abundant in the remaining mature bottomland hardwood forests. The presently endangered wood stork (*Mycteria americana*) nests in tall hardwood stands and feeds in the associated aquatic systems inhabited by the purple gallinule (*Porphyrula martinica*) and the common moorhen (*Gallinula chloropus*). Many species of raptors can be encountered at the bottomlands, the most characteristic of them being the Mississippi kite (*Ictinia mississippiensis*), the swallow-tailed kite (*Elanoides forficatus*), the red-shouldered hawk (*Buteo lineatus*), and the barred owl (*Strix varia*).[55]

Numerous species of wood warblers (Parulinae), some with special affinities, are found in the bottomland hardwood forest habitat. The northern parula (*Parula americana*) often builds its nest in Spanish moss. The Swainson's warbler (*Limnothlypis swainsonii*), the hooded warbler (*Wilsonia citrina*), and the Kentucky warbler frequent the understory of bottomland hardwoods, while the prothonotary warbler fancies cavities near or over water for nesting. The critically endangered, if not extinct, Bachman's warbler (*Vermivora bachmanii*) possibly nested in the Delta bottomlands, as did two other disappeared species, the Carolina parakeet and the ivory-billed woodpecker (*Campephilus principalis*).[56] Audubon encountered both species during his 1820 journey to the region, and two of the six known Mississippi specimens of the ivory-billed woodpecker were "secured from a great wooded area, virgin timber, of cottonwood, cypress, red gum, American elm, hackberry, ash-leaf maple, etc." near Rosedale, Bolivar County, in March 1893.[57]

In wintertime the bird community of the bottomland hardwoods is dominated by visitors. During this critical period, species that occupy other habitats during the breeding season, such as the yellow-bellied sapsucker (*Sphyrapicus varius*), the ruby-crowned kinglet (*Regulus calendula*), and the white-throated sparrow (*Zonotrichia albicollis*), find the foraging opportunities and/or the climate in the southeastern bottomland hardwood forests more suitable. Many species of ducks and other birds associated with freshwater habitats occupy adjacent aquatic systems during the winter months, as noticed by Audubon in 1820: "The *Yazoo River* flowed a Beautifull Stream of transparent Watter, Covered with 1000ds of Geese & Ducks and filled With Fish."[58] The bottomland hardwood forests of southeastern North America have undoubtedly supported an exceptionally high diversity and density of permanent resident, breeding, and wintering birds, and the value of remaining habitat for several species of birds is hard to overestimate.

Although there have been only a few studies comparing the fauna of bottomland forests with the fauna of upland forests, the bottomland hardwood forests appear to have a distinctive fauna. Bottomland hardwood areas are not as rich in species diversity in certain animals, such as lizards, snakes, and some insects, as upland forests. However, the diversity of bottomland fauna is higher in amphibia, turtles, birds, and mammals. In frogs, salamanders, shrews, and birds, the bottomland hardwood forests show markedly higher densities.[59]

The vast bottomland hardwood forests of the pre-Columbian Delta were supported by the floodplain of the Mississippi and Yazoo rivers. The soils of the forests were alluvial, derived from the deposits of sand, silt, clay, and calcareous sediments left by the shifting courses of meandering rivers. Within the Yazoo-Mississippi Basin, shallow water frequently covered a sizable portion of the floodplain for varying periods during the year.

Although the region was generally flat, numerous physical features of relief created notable differences in drainage and hydroperiod. The result was a very complex arrangement of soils and biotic communities. Flooding was a vital feature in the maintenance of these systems. The seasonal abundance of water and rich alluvial soil contributed to the formation of vegetation that clearly distinguished the bottomland hardwoods of the Delta from the upland forests of the hills to its east. The diverse vegetation of the bottomland hardwood forests stabilized soils and supplied organic matter supporting animal communities. Water, soils, flora, and fauna were closely interdependent in the bottomland hardwood forests of the Yazoo-Mississippi Delta, together forming a valid ecological complex.

The evolution of distinctive, species-rich flora and fauna in the Yazoo-Mississippi Delta was structured by processes operative on both evolutionary and ecological time scales. Biotic communities as known during historical times assembled only since the last major glaciation, or during the past twenty thousand years. Like other southeastern vegetational assemblages, the bottomland hardwood forests of the Delta have not been stable either in composition or in location over time. Still, pollen data suggests that elements of an oak-hickory forest have been present at the Delta for the past sixteen thousand years, providing a relatively established zone for faunal—and human—habitation.[60] The pre-Columbian biota of the Yazoo-Mississippi Delta was a mosaic of plant and animal communities pursuing compositional equilibrium, but still responding dynamically to impacts of climate, geomorphic processes, fire, and Native American populations.

[T]he (themselves) nameless though recorded predecessors who built the mounds to escape the spring floods and left their meagre artifacts: the obsolete and the dispossessed, dispossessed by those who were dispossessed in turn because they too were obsolete: the wild Algonquian, Chickasaw and Choctaw and Natchez and Pascagoula, peering in virgin astonishment down from the tall bluffs at a Chippeway canoe bearing three Frenchmen—and had barely time to whirl and look behind him at ten and then a hundred and then a thousand Spaniards come overland from the Atlantic Ocean . . . [and] then came the Anglo-Saxon, the pioneer, the tall man, roaring with Protestant scripture and boiled whisky . . .

—WILLIAM FAULKNER, *Big Woods*

CHAPTER THREE

# *Enter* Homo sapiens

THE ABORIGINAL HUMAN INHABITANTS of North America, usually known as Indians or Native Americans, have often been portrayed as a group of modern environmentalists in their land-use practices. Native Americans, according to such accounts, lived off nature's bounty and had left no mark upon the land at the time of European conquest. This romantic assertion, however, has little to do with the actual life of Indians and greatly ignores their influence on the natural environment of pre-Columbian North America. Furthermore, attempts to portray Native Americans as innocents living on "virgin land" ultimately make them a part of the natural world, thus denying both their history and cultural heritage.[1]

Indian land-use practices, of course, differed significantly from those of the European colonists and were, by contemporary Western standards, considerably less advanced. The notion that Native Americans had passively adapted to their natural environment, however, is false; Indians modified vegetational assemblages, manipulated animal populations, and created habitats best suited to human settlements. The natural environment offers many ways for humans to live in a given area but gives no clues as to what the optimum land use for human purposes might be; only culture can provide the values for defining such use. Indian uses of regional environments varied considerably by tribe, and those uses greatly influenced the pre-Columbian landscape.[2]

Given the great diversity of Native American cultures, making generalizations about their land-use practices is difficult, at best. Certain common features, however, emerge when these practices are contrasted with those of Europeans. Livestock domestication was almost nonexistent, and crop-raising aboriginals of North America almost always integrated their horticultural subsistence practices into hunting, fishing, and gathering economies. Usually there was a clear sexual division within these economies: men concentrated on hunting and fishing, women on horticulture and gathering. Many Native American economies protected themselves from environmental fluctuations by moving about in ecologically and seasonally defined cycles to take advantage of the seasonal abundance and variety offered by nature. Throughout the North American continent, including the Delta, security of subsistence was usually obtained by incorporating a wide variety of environmental resources into these cycles; when one resource failed to appear during a given season, others were still available.[3]

## The Native Delta

Similarities between different cultural artifacts found in the present-day state of Mississippi seem to justify the designation of the Yazoo-Mississippi Delta and the adjoining loess hills into a definable archaeological province.[4] Earlier archaeological approaches provided only a very general description of the Native American use of natural resources, but during the last decades archaeologists, anthropologists, and ethnohistorians have made significant contributions to our understanding of the relationship between aboriginal people and their physical environment in southeastern

North America. For example, Indian agricultural practices in the Southeast do not appear to have been as uniform as previously portrayed.[5]

The presence of modern humans (*Homo s. sapiens*) in North America at the time of the European conquest has traditionally been explained by the so-called land-bridge theory, which assumes that the ancestors of present-day Native Americans were Asian hunters of the late Pleistocene epoch. These people seem to have entered the continent between 14,000 and 13,000 years ago via the then-dry Bering Strait and a newly opened corridor in the Mackenzie River Valley between the Cordillerian and Laurentide Ice Sheets, increased in population, and eventually dispersed over North and South America. Attempts to route and date the arrival of the first Paleo-Indians are, however, subject to great controversy, and it has even been proposed that the event could have taken place by sea and up to 50,000 years ago—maybe even earlier. In any case, these people seemed to have developed many basic techniques of stoneworking and environmental exploitation—except for livestock husbandry—by the time of their arrival in North America.[6]

Traditionally, archaeologically defined Indian cultures of the Southeast have been discussed in terms of their chronological and stylistical relationships in time and space. Numerous attempts have been made to develop a descriptive sequence of time units on the basis of archaeological criteria. As noted by Bruce D. Smith, this has often resulted in great confusion in a form of competing "cultural" chronologies. Accordingly, he proposes a "natural" approach to the subject, based on the temporal boundaries of the three Holocene time units of early Holocene (12,500 to 6,000 years before the present), middle Holocene (8,000 to 5,000 B.P.), and late Holocene (the last 5,000 years).[7] This tripartition of pre-Columbian cultural development correlates with the major climatic trends during the Holocene and facilitates comparison between general environmental trends and human subsistence. As noted above, the general climatic amelioration following the meltdown of the Laurentide Ice Sheet was punctuated by the onset and termination of a warm and dry episode, known as the hypsithermal. It is, however, impossible to escape the various cultural chronologies when familiarizing oneself with the existing research on the archaeology of the Yazoo-Mississippi Delta, and, consequently, the following account employs a general division of pre-Columbian history into certain distinctive cultural phases. [8]

Distinctive artifacts, such as specialized bifaced fluted stone points that were used as spear tips, exemplify Native American cultural development during the Paleo-Indian period (circa 12,000 to 10,500 years ago). These spear tips, commonly found along many southeastern rivers, indicate the presence of substantial human populations hunting large mammals for their subsistence. Paleo-Indian projectile points have also been found along the eastern edge of the Delta. Paleo-Indians could have been oriented toward the utilization of essentially modern faunal and floral assemblages, but probably also pursued now-extinct Pleistocene megafauna. Various floral elements constituting the understory regime of the bottomland hardwood forests, along with mushrooms and fungi, probably provided much of the potential subsistence base for Native Americans in the Yazoo-Mississippi Delta ever since the Paleo-Indian period. Also, mast of different oaks and hickories, trees that have been present in the area for at least ten millennia, presumably contributed greatly to the Native American diet. [9]

The cultural transition period between latest Paleo-Indian and earliest Archaic times (approximately 10,500 to 9,500 years ago) is usually called the Dalton period. During this epoch, Native Americans were probably organized into close-knit bands of some twenty to thirty individuals bound together socially by family ties created and reinforced through marriage and subsisting on hunting, fishing, and gathering of wild plants. Due to environmental variability across the Southeast, such groups could either have operated out of a settlement area, with camps initiated for specialized resource needs, or shifted seasonally in response to environmental opportunities.

Increasing regional differentiation of the natural environment and human culture seems closely interrelated in Archaic North America. In the Mississippi Basin, the Early and Middle Archaic periods, the eras extending from circa 9,500 to 5,000 years before the present, coincided with drier climate patterns that resulted in erosion of slope soils and reduction in the ground cover of the uplands. Increasing sedimentation along the rivers, on the other hand, led to the further stabilization of vegetation on the Mississippi floodplain (which switched into a meandering regime some 9,000 years ago), with a subsequent increase in productive habitats such as oxbow lakes. At the same time, the relative impairment of upland habitats for human subsistence probably encouraged an increase in the utilization of bottomland resources by Native Americans.

The Early Archaic and Middle Archaic periods still lack an absolute chronology for cultural invention and development of hunting and fishing implements. In any case, spear throwers (atlatls), fish hooks made of bone, grooved stone axes, and earth ovens seem to have been in use by the end of the Middle Archaic period. The domestication of the dog had also occurred during or before this cultural period. Between 5,000 and 2,500 years before the present, the establishment of modern climatic patterns coincided with intensifying use of riverine resources, growth in interregional exchange, and the initial appearance of cultigens and ceramic containers.

By at least the advent of the Late Archaic period (5,000 to 3,000 years ago), a significant transformation had occurred in the political and social life of Native Americans in parts of the Southeast. Studies of archaeological sites, especially the famous Poverty Point, a settlement some 3,600 to 2,600 years ago in what is today northern Louisiana, indicate that substantial human populations were now living in more or less permanent concentrations on southeastern floodplains. Large villages may even have served as geographical centers for larger "chiefdoms" and, for the first time, hint at a possible shift from the kin-ordered to the tributary mode of production in some parts of the Southeast. Significantly, the existence of such highly organized populous settlements without a fixed agricultural base would have no parallel in the Old World. Wider trade interactions seem to have been established in the Southeast and beyond, facilitating the transfer of new technologies and other cultural innovations. These are evident, for example, in the new stoneworking and cooking techniques adapted by Native Americans of the region. Archaeological sites of the period in the Delta have yielded typical Poverty Point artifacts, including some of the earliest pottery in the Southeast. Steatite and ceramic vessels presumably facilitated the use of nuts and acorns by Native Americans, allowing for more efficient removal of tannic acids from acorns and extraction of oils from nuts by boiling.

Archaeologists have traditionally claimed that the agricultural practices of southeastern Indians originated from Mesoamerica. After the disappearance of the North American megafauna some 10,000 years ago,
southeastern Indians had turned to a mix of small-game hunting, fishing,
and collecting, but the cultivation of crops also began to diffuse northward from Mexico.[10] It is, however, quite possible that the cultural accep-
tance and cultivation of different species of squash (*Cucurbita* spp.) began
with indigenous rather than imported varieties already 7,000 to 5,000
years ago. Developing slowly, agriculture in the Southeast commenced
at the latest during the Late Archaic period. Deliberate plant food production is confirmed at southeastern archaeological sites by seeds found
outside their natural range of distribution or differing from their wild
ancestors. At least three plants, the sumpweed (marsh elder, *Iva annua*),
the sunflower (*Helianthus annuus*), and the goosefoot (lamb's quarter,
*Chenopodium* spp.) seem to have been domesticated 4,000 to 3,000 years
ago. The bottle gourd (*Lagenaria*), a tropical cultigen of Mexican origin,
was introduced to the region over 4,000 years ago and was widely grown
by 2,500 years before the present. For example, the Poverty Point people
seem to have practiced small-scale cultivation of bottle gourd and squash,
presumably both for containers and the edible seeds. Developments in
hunting technology are also evident in archaeological finds from Late Archaic sites along the Central and Lower Mississippi River Basin. Characteristically included are stone weights of the ingenious bola weapon, used
for catching various species of birds in wetland habitats along their migration route, the Mississippi Flyway. Further south, along the Gulf Coast,
wintering waterfowl seems to have been trapped with nets specially made
for this purpose.[11]

The Woodland cultural period commenced approximately 3,000 years
ago, stylistically dated by innovations in pottery vessels. The so-called
Lake Cormorant culture was present in the upper Yazoo-Mississippi
Delta since approximately 2,500 years ago. Small villages were situated
to utilize both riverine and upland environments, while simple exchange
networks continued to move materials such as copper and steatite across
the Southeast. During the Middle Woodland period (or, Hopewell period,
after an archaeological site in present-day Ohio), between about 2,000 to

1,600 years before the present, southeastern Indians had begun to design large ceremonial centers with pyramidal mounds. The Lake Cormorant culture of the Delta adapted many Hopewellian traits from the north during the so-called Marksville period. Indian mounds of the period could attain the size of over one hundred feet in diameter and more than thirty feet in height. The mounds acted as burial sites for socially important individuals and contained elaborate ceremonial artifacts in addition to tombs constructed of logs. Cypress logs nearly three feet in diameter were sometimes used in the burial chambers. Permanent village sites with their extensive burial mounds indicate a greater degree of ecological independence attained by southeastern Indians. At the same time, they implicate widespread cultural uniformity and shared belief systems, accompanied by major sociopolitical changes.

The evolution of specialized burial sites is often explained by the emergence of a new type of social leader, the so-called Big Man, whose position of authority is not inherited but created by his personal efforts. An emerging Big Man builds a following within his own community and then harnesses the surplus production of the whole community to his own ambition. Overcoming the tendency of small autonomous communities to remain isolated and independent, the Big Man is also responsible for establishing and maintaining reciprocal exchange and communication networks with other communities.[12]

During the late Woodland period (approximately A.D. 200 to 800[13]), a new culture emerged in the Yazoo-Mississippi Delta: the Baytown culture shows increasing reliance on horticulture, but remains of wild plants such as different hickories and oaks, as well as walnut and persimmon, are still commonly found at archaeological sites. Permanent villages of up to fifty inhabitants seem to have been common on the Mississippi floodplain, and typical Baytown pottery has been found on almost every Delta natural levee. The widespread growth and dispersal of Native American populations during this period seems to have occurred without strong cultural integration. Changes in burial customs indicate that Big Men were no longer present, as the large mounds reserved for them are replaced by smaller communal cemeteries.

Grooveless stone axes (celts) and special tools used in the construction of dugout canoes (adzes) had been developed during the late Archaic period. They became more common during the Woodland period and imply increasing exploitation of forest resources by Native Americans. Stone tools and fire were used to clear patches for agriculture within riverine woodlands. The Hopewellian culture had developed without the cultivation of corn (maize, *Zea mays*), which was introduced along the Mississippi and Ohio rivers only around the year 400 via the Southwest. The long-term utilization of plant species belonging to the starchy seed complex, such as maygrass, knotweed, little barley, and goosefoot, presumably made it relatively easy for the southeastern Indians to switch over to an agricultural system dominated by corn. By the beginning of the next millennium, this crop had become the basis for Native American agriculture east of the Mississippi. Two general types of corn were grown by the eastern Indians: Tropical Flint and Northern (Eastern) Flint. The former was a small-cobbed, many-rowed race, while the latter had a large cob with eight to ten rows. Tropical Flint was initially the variety most often grown in the Lower Mississippi Valley.[14] Utilization of a revolutionary technological innovation for hunting and warfare, the bow and arrow, emerged in the Southeast more or less simultaneously with corn agriculture.

Most archaeological attention has been paid to the subsistence technology of the last precolonial period, the "Mississippian period," which commenced during the ninth century and ended with the arrival of Europeans in the Southeast.[15] This cultural tradition rose along the Mississippi floodplain between modern-day St. Louis, Missouri, and Vicksburg, Mississippi, and spread to river valleys throughout much of the region, producing societies that were vastly different from those that had existed before. At any given time during the Mississippian period, this cultural phase was represented by a dozen or so clusters of settlements along different segments of the Mississippi and Ohio River Valleys. Many such clusters survived only for decades, while some persisted for most of the Mississippian period. The designation "Mississippian" encompasses numerous tribes with varying sociopolitical complexity, but a common denominator, in addition to corn agriculture, was the concentration of human settlements around ceremonial centers, recognizable by large and conspicuous temple mounds. Hence the traditional name for these cultures given by Europeans was "Mound Builders."[16] In the lower Yazoo-Mississippi Delta, the Mississippian culture merged with the Plaquemine culture of more southern origin, and was later of great influence on the Natchez and other tribes of the Lower Mississippi Valley.

The Mississippian culture in the interior Southeast, including the Delta, was characterized by its sociopolitical organization, sedentary villages on riverine locations, and relatively firm agricultural base. Convincing documentary and archaeological evidence exists of extensive areas devoted to rain-fed agriculture, especially maize cultivation, and the use of hoes and spades in the Mississippian Delta. Innovations in agricultural technology seem to have resulted in a significant surplus of corn, and the emergence of communal storage pits suggests increasing social planning. Between the twelfth and fourteenth centuries, simple chiefdoms directed by local councils seem to have transformed into more complex systems characterized by paramount (regional) chiefs and social stratification. Commission of great public works, such as bigger temple mounds, had become possible. While most of the Mississippian villages were small and unfortified settlements surrounded by farmsteads, larger towns were also being built, especially at confluences of waterways. They seem to have acted as ceremonial centers with temple mounds grouped around a central plaza, and may have functioned as regional commerce centers.

The population of a ceremonial town often exceeded two thousand inhabitants, but even greater human concentrations emerged during the Mississippian period. The town of Cahokia, near present-day East St. Louis, Illinois, was occupied from at least 1100 to 1350 and had probably over thirty thousand inhabitants during its heyday around 1250. It remained the largest settlement in the history of what is today the United States until Philadelphia surpassed it only in 1800. After the year 1200, there is strong evidence of direct contact between Cahokia and the Yazoo-Mississippi Delta, and a settlement by northern migrants seems to have been established along the upper Yazoo.[17]

Stronger and lighter pottery, tempered with mussel shell, was being manufactured in the Southeast during the Mississippian period. It is possible that the advent of durable shell-tempered pottery was related to new ways of processing corn; lime additive was now required to prevent pellagra in a diet progressively dominated by maize.[18] The bow attained a new application in drilling, and shell beads became prestigious trade items, providing a regionally accepted form of commerce exchange. Trading monopolies on limited resources, such as basalt and salt, also developed during the Mississippian period. Other important trade items included slaves who could be captured in intertribal wars and used for agricultural and household labor. Mutilation of slaves' feet was common, preventing escape attempts but still enabling them to work the fields.

The elusive belief systems of Mississippian people are only rarely reflected in their material culture. The sun was typically the most important component of the deity force for the Mississippian Indians, developed to the greatest extent among the nation later known as the Natchez. Native American conceptions of nature and culture presumably intersected in two ways: humans were seen to be a part of nature by virtue of being biological beings within a material world, while other natural beings, such as plants and animals, were also parts of a human cultural world. Consequently the boundaries between human and nonhuman communities in Indian thought are difficult to define. In many Native American belief systems, spirits of nature had the power to transmit to humans the qualities of the animals affected by human action. It is hard to say whether Native Americans understood their place in nature in terms of different social relationships between humans, nonhumans, and spirits, or in ritual classifications. In any case, Indian religions and rituals typically saw human beings' role in facilitating the movement of natural cycles as crucial, and this enabled them to make an effort to control the natural environment in different ways. Native Americans gave serious consideration to the relationships between living things and to the dynamics of their natural environment. Their understanding of many natural processes, however, differed sharply from that of the contemporary science of ecology. While Native Americans did not regard nature as a mere "thing" to be exploited, some of their actions—such as the wasteful hunting of many game species—were not rational by the standards of modern wildlife management.[19]

Native American farmers of the Mississippian period showed a preference for floodplain locations because of their need to grow corn without the use of plows and fertilizers; silt deposited by floodwaters enriched the soil and enabled continuous cultivation and sedentary populations. Garciliaso de la Vega's late-sixteenth-century, secondhand account of the de Soto expedition described Mississippian adjustment to flooding in the absence of levees and other flood control devices. He stated that because of the frequent flooding, "the people built their houses on the high land, and where there is none, they raise mounds by hand, especially for the houses of the chiefs; . . . with galleries around the four sides of the house where they store their food and other supplies, and here they take refuge from the great floods."[20] Occasional severe floods were endured for the sake of the naturally productive alluvial soil. Besides, the bottomland soils were, unlike the harder upland soils, easy to till with Indian tools. The most favored landforms for Mississippian agriculture in the Yazoo-Mississippi Delta were the older natural levees on recently abandoned channels. Because of the nature of alluvial deposition, these lands consisted of sandy loams, which were better drained and more easily worked than the clays of the backswamp areas. The soils still held enough moisture to sustain agriculture without the use of irrigation works and were the last to be inundated during floods.

Native Americans established fields by cutting and burning existing vegetation, paying attention to soil quality. To clear land for agriculture from the bottomland hardwood forests, Mississippian farmers presumably first cut the canebrakes and girdled the trunks or gashed the roots until the trees were dead. The stumps and dead trees, as well as brush, were then burned.[21] After the land had been cleared, the soil was broken for cultivation with simple wooden implements and the crops were planted. So-called swidden agriculture (i.e., periodic rotation of agricultural area, meaning the sequential abandonment of established fields and subsequent regrowth of natural vegetation, reclearing, and recultivation of the same fields) does not seem to have been practiced along the lower Mississippi, so the use of slash-and-burn techniques was probably restricted to the establishment of permanent agricultural acreage.[22] Tillage as practiced by Indians differed significantly from the contemporary European practices: broadcast seeding was virtually nonexistent, and crops were planted in rows with each stalk or plant hoed to keep down the weeds. Native Americans commonly used the method of hill planting: as an individual plant grew, loose dirt was scraped around it in order to suppress the weeds. The hills, which were twelve to twenty inches in diameter and about three feet apart, were used over and over again in successive seasons and could grow to sizable mounds of earth. As the soil between the hills remained unbroken, there was little danger of soil erosion in the Indian fields, and this contributed to the sustenance of soil fertility.

Mississippian farmers were probably the first eastern Indians to cultivate beans (*Phaseolus* spp.), which added to the corn-dominated diet the essential amino acids lysine and trytophan needed for protein synthesis. Despite their obvious nutritional contribution, beans seem to have been cultivated only occasionally. In addition to corn, beans, and previously adapted cultigens, tobacco (*Nicotiana rustica*) was also grown. All of these American products were unknown to Europeans until the conquest of the New World. In order to achieve maximum yields from their fields, Indian farmers of the Southeast could have used two agricultural methods: intercropping and multiple cropping. By intercropping (planting several cultigens together in the same fields) Native Americans efficiently utilized the limited area of productive land and enabled the crops to complement each other. The long growing season of the Southeast also made multiple plantings possible: early and late varieties of corn often yielded two annual crops from the same fields. Mississippian corn yields may well have averaged around forty bushels per acre.[23]

Native American crops complemented each other in a number of ways. In an ideal case, the cornstalks would have provided a place for the beans to climb, while the squash vines covered the ground and efficiently suppressed competing weeds. Beans, when cultivated, replaced some of the nitrogen in the soil that corn depleted; this partly enabled Native Americans to cultivate the same fields over long periods of time without the use of fertilizer. Periodic flooding fertilized the fields, but intercropping was probably even more important for the sustenance of established agricultural areas. Only after the fields had markedly declined in fertility were they abandoned and new areas cleared.[24]

Native American corn agriculture in the Mississippi Valley was presumably very productive and, combined with hunting, fishing, and gathering, usually provided the tribes in the region with a secure subsistence. Gardens may well have been cultivated by Mississippian people, and there is some archaeological evidence suggesting that they may have protected aggregations of the edible butternut (*Juglans cinerea*), which in the Delta occurs near the southern limits of its natural range. Despite agricultural innovations, gathering and hunting were still important activities for Mississippian farmers. Mississippian tribes of the interior Southeast typically combined their regular horticultural cycle with annual cycles of hunting and gathering: crops were planted in spring, and other economies sustained communities until the fall harvest. Various wild fruit, berries, nuts, and acorns were commonly gathered by women and children. Bottomland locations presented seasonal opportunities for fishing and hunting, and protein from animal meats, especially deer, waterfowl, and fish, supplemented the diet dominated by corn. Various species of fish and migrating waterfowl may have contributed over half of the animal protein consumed by the Mississippian people occupying southern floodplains. After the annual floods had receded, the floodplains were dotted with ponds and sloughs filled with a variety of fish that were often extremely easy to catch in the changed conditions. Fish were caught mainly with nets, but they could also have been drugged with various plant products. Most of the fishing and turtle collecting was presumably carried out during the summer months, while an intensive period of deer hunting probably took place during the winter. The ideal Mississippian subsistence system was consequently a highly diversified one.[25]

The changes in the natural environment wrought by Europeans since the seventeenth century have been so immense that it is very difficult to tell what impact the many tribes of Native Americans had on the native fauna of North America. Bison probably contributed to the Native American diet in the Delta, as La Salle reported their occurrence along the Mississippi in 1682, and one was killed in the vicinity of Memphis, just north of the Delta, as late as 1723.[26] The white-tailed deer, however, was the likely candidate for constituting the single largest source of terrestrial animal protein for Native Americans in the region. Raccoon, rabbits, squirrel, and opossum seem to have been of lesser importance. Much of the bottomland hardwood forest habitat is ideally suited for deer; in addition to varied browse and herbage supplied by different plant communities, mast provides large quantities of acorns and nuts. Consequently, the carrying capacity range for bottomland hardwood forests has been estimated at one deer per five to fifteen acres.

Under natural conditions, the greatest amount and diversity of understory vegetation on a southern floodplain—and, consequently, the largest deer populations—are to be found in edge habitats, which are created by recent stream course changes, windthrow, or fire. It was therefore typical for many Native American tribes, in addition to clearing land for agricultural reasons, to burn sections of the forests to ease the hunter's task and open areas for the grass and other growth that game, especially deer, could feed on. By retarding normal successional patterns and maintaining edge habitats, this burning of the woodlands by Native Americans benefited their main game and consequently themselves. Analysis of faunal remains at Native American settlements in the lower Delta indicate a progressively expanding amount of edge and open habitat: the amount of arboreal species such as squirrels, raccoons, and opossums decreases through time, while edge species such as deer and rabbits become more common as prey.[27]

The influence of anthropogenic fire in creating game habitats nevertheless varied from region to region in the Southeast. In abandoned Indian fields of the Delta and the adjoining loess hills, fire could have maintained optimal deer habitats, such as meadows with dense and shrubby margins, but along the Gulf Plain the use of fire, to some extent, degraded these habitats. Fire gives a competitive edge to a certain fire-resistant conifer, the longleaf pine (*Pinus palustris*), and anthropogenic fires probably contributed to the formation of vast coniferous forests in the region. Without fires, whether natural or human set, the region would have consisted mostly of oak-hickory forests, which have a higher average carrying capacity for deer and many other game animals.[28]

The timing of deer hunting season in late fall and winter was advantageous for the hunters, but also for their prey. Large-scale hunts were easier to conduct after the leaves had fallen and the weather become cooler. The deer had also developed winter coats and were in full flesh after feeding on the mast, available since the fall. While mast utilization led to greater deer concentrations, their mating season furthermore made bucks unwary and easier to hunt. Significantly, the fawn were by early winter old enough to survive the death of the doe, so doe hunting did not constitute a double blow to the population.[29]

Both the Mississippian and the later historic Indian nations encountered by Europeans seem to have divided their landscapes into two categories: the core region of permanent settlements and cultivated fields, and the borderlands utilized mainly for hunting purposes. The utilization of hunting grounds by different Native American groups presumably correlated with the size of the deer populations affected. As long as the deer populations remained high, hunts took place peacefully, but low deer populations increased competition and prompted warfare between Native American groups. This in turn made the disputed borderlands dangerous for all Indian hunters and decreased pressure on deer populations, allowing them to recover.[30]

Deer meat could be cooked fresh or dried and smoked for use during the winter months. In addition to nutritional use, deer was important for other uses. Deerskin, tanned with deer brains, was the principal material for clothing. Antler tips were utilized as arrow tips, and dried sinew and entrails could be used for bow strings. Bow strings were made also from the dried gut of the black bear, another important mammal for the Mississippian Indians. It provided for heavy winter robes and bed coverings, and its tough skin was commonly used for footwear. Pierced bear claws were commonly used as necklaces. In addition to edible flesh, the bear yielded large amounts of fat, which was processed into bear oil.[31] Slabs of fat were rendered over fire, spiced with sassafras, stored in jars, and used for various purposes, from cooking to dressing the hair. Fish provided variety to the Native American diet the year round; the large catfish of the family Ictaluridae presumably yielded large amounts of high-quality protein. Freshwater molluscs (especially *Villosa lienosa, Pleurobema cordatum,* and *Viviparus subpurpureus*) could have been used by the Delta Indians as a reserve source of animal protein during the low water season.[32]

Our knowledge of the Native American impact on the birdlife of the Delta remains scanty. They certainly hunted migratory waterfowl and turkey for food and killed some other species for their feathers, but it is doubtful that this had any significant long-term effect on the populations of species affected. Concentrated efforts to hunt ducks and geese during migration were common, while turkeys and other species of birds were probably taken mainly during exploitation of other resources. For example, turkeys utilize acorn crop simultaneously with the deer, and the concentration of these two species in specific habitats at certain times of the year would have made combination hunting relatively easy. The enormous flocks of the passenger pigeon were frequently utilized by eastern Indians, both at nesting sites and on migration, but there are no explicit records concerning hunting of this species by Native Americans within the bottomland hardwood forest complex. Similarly, there is little information about Native American exploitation of the nesting colonies of herons, egrets, and cormorants. Such rookeries located in the cypress swamps of the Delta would have contained thousands of birds and constituted a valuable supply of nutrition.[33]

There is interesting evidence about aboriginal pressure of a different kind on the populations of the now-extinct ivory-billed woodpecker, a species confined to the bottomland hardwood forest habitat. The English naturalist Mark Catesby gave the best-known account of the subject in 1731: "The bills of these Birds are much valued by the Canada Indians, who make Coronets of 'em for their Princes and Great Warriors, by fixing them round a Wreath, with their points outward. The Northern Indians having none of these Birds in their cold country, purchase them of the Southern People at the price of two, and sometimes three, Buckskins a Bill."[34] Audubon also claimed to "have seen entire belts of Indian chiefs closely ornamented with the tufts and bills of this species," and that "its rich scalp attached to the upper mandible forms an ornament for the war-dress of most of our Indians."[35] Consequently it is not entirely astonishing to note that a beak of the species, along with one of a pileated woodpecker, has been found in an Indian (probably Arapaho or Cheyenne) grave in Colorado. As no direct commercial route is known to have existed between the Indians of the Plains with those of the Southeast, it seems possible that the huge bill of an ivory-billed woodpecker was a valued item that could change hands several times and travel for hundreds of miles.[36]

Ethnobotanical and palynological evidence suggests that considerable forest clearing took place on southeastern floodplains during the Mississippian period, as Native American technological innovation and environmental manipulation were expanding. Growing Native American impact on forests is indicated by large-scale manufacture and trade of increasingly sophisticated woodworking tools, such as large stone-hoe blades, in the Mississippi River Valley and along its immediate tributaries. Wood from different hickories was used for heavy containers, pestle and mortar sets, bows and arrows, and as siding for winter houses. Cane was widely utilized by Mississippian Indians: it was used as siding for summer houses and for the construction of fish traps, seines, and fences. Baskets and mats were woven from it, and hollowed stalks could be turned into blow guns. Hollowed trunks of large sycamore and cottonwood trees were fashioned into river boats. Native Americans seem to have overexploited baldcypress in certain areas, as the tree's northern range limit adjacent to the greatest human concentrations in the Lower Illinois and Central Mississippi River Valleys was truncated during the Mississippian period.

For reasons still unclear, the Mississippian pattern of settlement seems to have been in marked decline well before the arrival of Europeans. Examination of skeletal remains suggests poor nutrition and widespread infectious disease. In the absence of irrigation systems, a climatic downturn in the form of prolonged droughts could have been devastating on a corn-based subsistence economy. Corn as a crop is very sensitive to moisture stress; summer droughts of only a few weeks severely reduce yields and longer droughts can destroy the crop completely. The implicit subsistence crisis at a time when many potentially productive areas remained unsettled furthermore indicates unequal access to resources and severe social constraints on dispersal. During the fifteenth century, Native American populations in the Mississippi Valley seem to have concentrated in the larger fortified towns in response to systematic and increasing warfare between powerful chiefdoms. Expansion in Native American construction of fortifications, such as log palisades, temporarily increased their impact on the bottomland hardwood forests, already affected by slash and burn agriculture. On the other hand, much peripheral agricultural acreage seems to have been abandoned and used only for hunting purposes during the late Mississippian era.[37]

Extensive archaeological excavations along the lower Yazoo have significantly expanded our knowledge of pre-Columbian land use in the Delta. The Lake George site at the western edge of present-day Yazoo County has yielded an enormous amount of information indicating that during the Holocene interglacial interval, or, the last 12,500 years, Native Americans had a progressively greater impact on the faunal and floral assemblages of the Yazoo-Mississippi Delta.[38] From the Archaic period onward there is substantial evidence of an increasingly sophisticated exploitation of natural resources by Indians, which allowed for the birth and increase of sedentary populations and the evolution of a more centralized form of government. These developments in turn seem to have led to increased complexity in other aspects of culture, as evidenced by the growth of trade and ceremonialism. The pre-Columbian history of the Yazoo-Mississippi Delta is complex, as the region served as a frontier between northern and southern cultural developments. The southern half of the Delta evidently participated more consistently in these developments, and cultural climaxes such as Poverty Point seem to have evolved largely in the more densely populated southern Delta. On the other hand, the most dramatic example of cultural change in the region, the Mississippian tradition, entered the Delta from the north.[39]

In the pre-Columbian Southeast, portions of major river valleys came to be particularly influenced by sedentary human populations. Native Americans maintained villages and extensive agricultural fields at the best locations along all the major rivers for long periods. Native Americans certainly exerted additional disturbance on the ecological complex of the Delta bottomland hardwood forest. They converted portions of the Delta from forest to agricultural land, exploited many tree species for fuel and housing, and modified biotic assemblages in various ways. They also introduced new species, such as maize and other cultigens, to the region.

In the Yazoo-Mississippi Delta, Indian settlements and agricultural fields were concentrated on natural levees in the proximity of rivers. Fields in the vicinity of permanent villages could become quite extensive and, combined with other land use practices, resulted in localized deforestation. Still, the typical Native American use of the bottomland hardwood forest apparently did not endanger the natural variety of the land or its capacity for self-renewal. Furthermore, much of the Native American use of timber remained insignificant in relation to deforestation in the bottomlands. It was mostly the vicinity of scattered Indian villages on natural levees that became a patchy landscape of fields, grasslands, and young woods, and actions by the Indian generally resulted in local modification of the bottomland hardwood forest rather than its large-scale eradication. In the absence of flood control devices, human-induced change of the Delta environment was largely restricted to the highest elevations within the floodplain, affecting directly only a very small percentage of the total land area between the Mississippi and the Yazoo Rivers. It is interesting to note that John Lawson, writing in 1714, remarked that the agricultural Indians of North Carolina did not necessarily clear and cultivate the most fertile lands along the rivers, as the removal of huge trees in the bottomlands presented too great an inconvenience.[40] On the other hand, James Adair pointed out that the eighteenth-century Choctaw of central Mississippi—whose tribal legends and customs clearly indicate a Mississippian ancestry—customarily cleared only the best riparian lands within their settled areas and planted them densely, aiming to minimize the amount of tedious land clearing.[41] Such accounts make it tempting to speculate that much of the old-growth bottomland hardwood forest in the Delta may have been excluded from conversion to fields by Native Americans, even when extensive corn agriculture was being practiced. In any case, vast areas of the bottomland hardwood forest remained untouched by agricultural activities, especially in moister locations.

The total number of aboriginal people in America at the time of the European conquest has been a subject of great controversy among historians and archaeologists and remains only a matter of speculation, though an estimate of four to seven million inhabitants in precontact North America seems quite sensible. Accordingly, estimates of the percentage of land under cultivation within the range of the southeastern bottomland hardwood forest complex during the Mississippian period vary considerably.[42] Whatever the size of the Native American populations of the Delta, commercial intercourse remained restricted despite wide trading networks. Native American technology was adapted to the needs of personal and communal rather than market-oriented production. The Native American goal of production was security, not the maximization of resource use. Land and labor were not commodities to be sold, but natural and social elements for communal well-being.

The Indian had neither the need nor the means to remake the landscape as thoroughly as the colonists were to do. For example, the white settlers were to use fire as a tool to clear the way for plants and animals introduced from Eurasia and Africa, whereas the aborigines had used it to benefit certain favored American species. Native Americans furthermore lacked domesticated grazing animals. Early European accounts customarily claimed that the Indians encountered in the South lived in harmony with their natural surroundings. Whether this was or had been true in regard to the utilization of natural resources by the Indian populations, the escalating clearing of forest for timber harvest and cultivation by the European colonists, combined with expanding alteration of the floodplain hydrological regime, represented a major shift in the degree to which humans affected their environment in the southeastern bottomlands and the Yazoo-Mississippi Delta.

## Economic and Ecological Imperialism

Hundreds of Native American tribes and several distinctive culture areas existed by the time of European intrusion in what is now the contiguous United States. Accordingly, dozens of tribes inhabited the Southeast, each with different cultural adaptations and subsistence economies. Some well-known tribes and nations of the region include the Cherokee and the Catawba of the southern Appalachians and the Creek of the interior of present-day Alabama and Georgia. The Guale lived on the coast of Georgia, while the Apalachee, the Calusa, and the Timucua inhabited coastal Florida. The Gulf Coast of Mississippi was populated by the Acolapissa, Bayogoula, Biloxi, and Pascagoula. Tribes resident on the Yazoo River

and its tributaries at the time of European contact included the Chakchiuma, Choula, Houma, Ibitoupa, Koroa, Taposa, Tiou, and Yazoo, while
the Tunica occupied a portion of the northernmost Delta. The Natchez
lived along the Mississippi south of the Delta, while the more populous
Choctaw and Chickasaw nations resided in the interior of the present-day
state of Mississippi.[43]

The first Europeans to enter the Yazoo-Mississippi Delta were Spanish
conquistadors led by Hernando de Soto, who explored parts of the South
during the sixteenth century. Veterans of the Spanish conquest of Central
America, de Soto and his party of close to six hundred men had sailed
from the Spanish colony in Cuba on May 18, 1539, for Tampa Bay. From
Florida, they then departed for an arduous three-year plundering excursion of interior North America. After exploring the present-day states
of Florida, Georgia, North Carolina, Tennessee, Alabama, and Mississippi, de Soto and his men became the first Europeans to travel across
the Yazoo-Mississippi Delta and the Mississippi River. The de Soto party
crossed the river in 1541 and journeyed across Arkansas into Oklahoma,
then traveled down the Arkansas River to its mouth where de Soto died
on May 21, 1542. Survivors of the journey finally reached the Spanish
colony of Rio Pánuco in Mexico in the fall of 1543.[44]

For over a century after that, the Mississippi Valley remained without
European presence, and when a new claim of ownership was made, it
came by way of Canada, not the Gulf of Mexico. During their travels in
1673, the Jolliet-Marquette expedition met with native inhabitants of the
Delta. However, the first European to realize the size and importance of
the Mississippi River was the French explorer René Robert Cavelier, Sieur
de La Salle. In 1682 La Salle descended the Mississippi to its mouth and
claimed possession of the river and all the land draining into it for Louis
XIV of France. La Salle envisioned French sovereignty over the heartland
of North America in the form of a chain of settlements extending from
Quebec via the Great Lakes and the Mississippi to the river's mouth. His
aim was furthermore to thwart the English colonization effort that was
already approaching the Lower Mississippi Valley from the east. La Salle
first conceived the Mississippi Valley as a geographical entity and named
it Louisiana in honor of his king.

The task of establishing a permanent French colony on the Mississippi fell to the Le Moyne brothers. Jean Baptiste Le Moyne, Sieur de Bienville, was commissioned to reclaim the Mississippi Valley for France. Bienville assisted his brother, Pierre Le Moyne, Sieur d'Iberville, in establishing France's presence on the Gulf of Mexico. In 1699 Iberville made the first known entry into the Mississippi River from the Gulf of Mexico. Together the brothers founded the cities of Biloxi and Mobile and both served as governors of Louisiana, maintaining the struggling colony despite France's neglect. In 1702 Iberville secured an alliance with the Choctaw against the English. When the royal colony was transferred to the Company of the West in 1717, Bienville was appointed commandant general and subsequently founded the city of New Orleans in 1718. When the company abandoned Louisiana because of aboriginal pressure and declining profits, he was again appointed governor of the royal colony and served until his retirement in 1743. Despite the efforts by La Salle and his successors, France ultimately failed to realize its control over interior North America. The English were more tenacious in their attempts to conquer the New World, and their colonization of eastern North America was rapid: the first permanent British settlement, Jamestown, was founded in Virginia in 1607, and by the middle of the eighteenth century there were already more than ten British colonies on the Atlantic coast.[45]

The French had been able to monopolize the commerce with the coastal tribes close to their expanding settlements on the Gulf. French commerce along the Mississippi was, however, threatened by competing English traders, who arrived within the present-day state of Mississippi in the late 1690s and began wide commerce with the natives. Textiles, firearms, metal tools, and alcohol were typically exchanged for pelts, especially deer skins. The ongoing trade war between the French and English had manifold consequences for Native American societies in the Southeast. The Choctaw remained ambivalent in their loyalties to Europeans, but the Chickasaw, residing farther away from the French, allied with the English. In addition to pelts, the English were eager to purchase Indian captives who could be transported to Carolina and the West Indies to be sold and used as slaves on plantations. Chickasaw war parties consequently entered the Delta to raid the Chakchiuma, Choula, Houma, Koroa, and

Yazoo and turned their catch over to English traders, virtually extermi-
nating many of the smaller tribes. Chickasaw slave raiders ventured far
and wide, even crossing the Mississippi and the Ohio in pursuit of new
trading stock.[46]

European intrusion into the interior Yazoo-Mississippi Delta during
the seventeenth and eighteenth century was largely restricted to the lower
portions of the floodplain. The French settlement Fort St. Pierre was established on the Yazoo in 1719 but declined rapidly. British influence was
similarly confined to the southern part of the Delta, where they, as allies of
the Chickasaw, traded with the Yazoo and other Native American tribes.
Early descriptive accounts of the Delta by Europeans therefore relate to
the southernmost part of the region and to the lands adjacent to the Mississippi River. Early French accounts provide some information on the
natural environment of the Delta; especially interesting are the references
made to the existence of vast grasslands.[47]

French agricultural expansion along the lower Mississippi and into the
area populated by the Natchez, and their attempts to force the Chickasaw to abandon the English as trading partners, led to a series of wars
during the first half of the eighteenth century. Natchez attempts to drive
out the French settlers came to an end in 1729 when French troops, aided
by Choctaw mercenaries, captured hundreds of Natchez, including the
last paramount chief, the "Great Sun." The captives were transported
to the French West Indies as plantation slaves, while those who escaped
capture dispersed across the South and were integrated into other Indian
nations, especially the Chickasaw, Creek, and Cherokee. The French also
used Choctaw mercenaries in their bloody but largely unsuccessful warfare against the Chickasaw between 1720 and 1752, while the Quapaw
were employed in the annihilation of the rebellious Yazoo.[48]

By the mid–eighteenth century, geopolitical changes in North America created new pressures for native societies. The conflict of interests
between the British and the French in the continent climaxed, and the
empires' commercial and military contest for allies effectively involved
Native Americans in the Seven Years' War between 1756 and 1763. After the withdrawal of the French from the continent in 1763, political

leverage held by Indians in international affairs diminished considerably. According to the Peace of Paris, the British received most of French North America: Canada, the Illinois country, and the eastern half of the Mississippi Valley. The French had, however, in 1762 ceded western Louisiana to the Spanish, who were still in possession of Florida. Therefore, many southeastern Indian nations still occupied the strategically important position of being situated between the English and the Spanish. By 1765, the British had taken possession of most of the present-day state of Mississippi, including all of the Yazoo-Mississippi Delta. The land and its inhabitants were now administered by the governor of the new British colony of West Florida and the superintendent of Indian affairs. British mercantile depots in many Choctaw and Chickasaw towns became important exporting centers, as these nations were used as middlemen in trade with tribes in Spanish Louisiana west of the Mississippi.

In 1776, thirteen British colonies on the Atlantic seaboard seceded from the empire and established an independent nation. The expansion of the United States quickly proceeded as the young nation acquired vast areas of land from Britain, France, and Spain between 1783 and 1819.[49] In 1803 all French holdings west of the Mississippi River were obtained by the United States via the Louisiana Purchase, while Florida was acquired from Spain in 1819. The Choctaw and Chickasaw nations had been heavily involved in the warfare between the British and Americans during the American Revolution. Originally allies of the British, the nations signed the Treaty of Hopewell after the British surrender in 1786 and came under the protection of the United States. In the contest during the 1790s between the Spanish and Americans for control of the Old Southwest, the Choctaw and Chickasaw then generally allied with the United States. The alliance with the United States, later evidenced by loyalty to that nation during the War of 1812, did not prove very rewarding for the Indians: the history of the tribes in the nineteenth century is a chronicle of retreat, land loss, and relocation.

During the 1780s, the United States aggressively aimed to further reduce the military and political power of Native American nations, and the native societies became ever more vulnerable.[50] President Thomas

Jefferson's dream of yeoman America had no room for traditional Indian societies and strove to open more native land to white settlers. By 1800, American settlers had crossed the Appalachians and the frontier (technically defined as a region of more than two and less than six people per square mile) had already advanced to modern-day Ohio (admitted to the Union as a state in 1803), Kentucky (1792), Tennessee (1796), and westernmost Georgia. Only thirty years later, by 1830, the frontier line cut across northern Missouri (1821) to its western boundary and similarly had advanced in Arkansas (1836) and Louisiana (1812) to their western bounds.[51]

The territory of Mississippi, created on April 8, 1798, was admitted to the Union in 1817, and the pressure to reduce tribal domains intensified rapidly, as white Mississippians could not tolerate the idea of sovereign Native American nations within the state's borders. Besides, as the quality of much of the native-owned acreage had already been proven by generations of use, the greed of the European settler for the core of Indian territory is easily understandable. Before long, the settlers began to arrive. An early traveler on the Mississippi commented that "[i]n such a wild part of the country . . . , where as yet there had been no regular survey of the land, of course these squatters are just as free to perch on the banks of the river, as the buzzards or vultures are to take possession of the cotton-wood trees growing above them." Still, "[i]n process of time, many of them become useful citizens of thickly peopled territories, of which but a few years before they were the only inhabitants."[52]

Meanwhile, "Yazoo"—the name of an obscure frontier river—had become widely recognized in the United States in connection with a series of schemes designed to absorb large parts of the present states of Alabama and Mississippi for speculative purposes. In 1789, the state of Georgia passed an act selling vast tracts of lands (including a large portion of the Yazoo-Mississippi Delta) to three companies (the Virginia Yazoo Company, the Tennessee Yazoo Company, and the South Carolina Yazoo Company) without a clear title to the lands in question. Opposed fiercely by the Spanish governor of Florida and several Indian tribes, the three companies unsuccessfully proposed several settlements, and in the end the sales had to be abandoned; contrary to the boosters' claims, "this

Stupendous Business" did not "render the Admiration of the World."[53] In 1795, the sales were revived as the Georgia legislature again granted the same tracts with some additional land to new speculators. Legislative corruption in the process was clearly evident, and in 1796 the new legislature repealed the act. In the meantime, the companies had succeeded in selling most of the land to new speculators. They in turn undertook to defend their land titles in the federal courts. After a strenuous legal battle, the U.S. Supreme Court in 1810 declared the 1796 Georgia statute unconstitutional and Congress later compensated the claimants with over four million dollars. This was the first time the U.S. Supreme Court voided a state statute, effectively weakening the powers of state governments by expanding the scope of federal court review of state laws. The Delta had for the first of several times become involved in the tug-of-war between federal and states' rights. In the process, federal protection of land speculators was established in the United States.[54]

The conquistadors' visits to the interior in the mid–sixteenth century were brief and unsuccessful in terms of plunder and did not seem to leave any mark of lasting consequence. However, the Delta was densely populated when de Soto and his party crossed it in 1541, but almost deserted by the time the La Salle expedition arrived in 1682. Traveling across the Southeast, de Soto and other conquistadors, directly or indirectly, influenced almost all Native American cultures in the region. In addition to the transmittal of various Eurasian microorganisms, the horse (*Equus caballus*) was reintroduced in North America by the Spanish, and it later became important for many Indian tribes. The de Soto party started out with more than two hundred horses and some seven hundred pigs (*Sus scrofa*), and many of these undoubtedly escaped or became the property of Native Americans.[55]

The introduction of Old World diseases to which Native Americans had no immunity, such as smallpox, measles, scarlet fever, and whooping cough, decimated Native American populations across the Southeast. It is possible that over 80 percent of the aboriginal population in the Central and Lower Mississippi Valley, including the Yazoo-Mississippi Delta, was extirpated during the sixteenth and seventeenth centuries, mostly by virgin soil epidemics.[56] The Mississippian populations, seemingly already in decline, were especially vulnerable because of their dense population concentration. Wave after wave of disease often reached and killed off Indian populations well ahead of the European explorers. Except for a couple of greatly weakened centers, the classic Mississippian culture had been destroyed and forgotten by the beginning of the eighteenth century.

The effects of the diseases introduced by Europeans were augmented by continuous warfare among natives. The cumulative effect seems to have been the large-scale relocation of the survivors, and, at the same time, the opening of the Delta to long-distance migrations. For example, the Tunica, encountered by de Soto in 1541 at the large settlement of Quizquiz in the northern Delta, seem to have been newcomers to the region. Still, the Tunica were to endure further migrations during the seventeenth and eighteenth centuries before settling in central Louisiana. Thus many of the smaller tribes encountered in the southern Delta by the French at the beginning of the eighteenth century, such as the Tunica and Ofo, were recent migrants to the region, and had but vague cultural affiliation with the indigenous natives. The only occupation of the Delta by that time probably consisted of a few remnant populations along the Lower Yazoo composed of many different ethnic groups.[57] Many survivors were still in the process of moving farther downriver. In 1699, French missionaries estimated the Indian population of the Lower Yazoo to be approximately two thousand, of which the Tunica made up more than half. Some of the other inhabitants of the region, such as the Yazoo and Chakchiuma, may well have been descendants of the great Mississippian nations.

The effective Native American occupation of the Yazoo-Mississippi Delta came to an end by the 1730s. The native tribes had disappeared or were dispersed through disease, warfare, and relocation, and the Delta no longer supported permanent human populations of significance. Instead, it had become an occasional hunting ground for deer and other products for European trade by the remaining larger Native American nations of the hill country. The gradual depopulation of the Delta by disease and slave raiding during the sixteenth and seventeenth centuries opened vast amounts of prime habitat to deer. Given the high reproductive potential of the species, it seems certain that deer populations in the Delta rose significantly. For many descendants of Mississippian peoples who had left the Delta and settled in the hill region to the east, the move resulted in a paradoxical situation by the eighteenth century: former hunting grounds had turned into areas of settlement and cultivation and vice versa.[58] In addition to succumbing to disease, many smaller tribes had been diminished by slave raiders and warfare to a point where they were forced to join stronger groups and lose their identity. By 1750, only three to four groups of Native Americans, representatives of the Choctaw, Chickasaw, Tunica, and possibly the Quapaw nations, were to be found in what is today Mississippi. By the beginning of the nineteenth century, only the Chickasaw and Choctaw remained. The Choctaw and Chickasaw, whose populations numbered about twenty-two thousand and five thousand, had been the most successful nations in maintaining their subsistence systems during epidemics. In addition to surviving the initial virgin soil epidemics in disproportionate numbers, they furthermore absorbed remnants of other tribes.[59]

In addition to decimating the Native American populations in the Mississippi Valley, imported diseases quickly destroyed the very fabric of aboriginal culture. Sacred traditions and rituals were often lost, and the new social structures developed by Native Americans in the region proved much weaker: by the time of European colonization, only few Indians of the Mississippi Valley could recall their past. For example, the origin of temple mounds had become unclear for many Native Americans. At the same time, the collapse of the Mississippian agricultural system resulted in substantial reforestation of abandoned Indian fields throughout the South, including the Yazoo-Mississippi Delta. The diminished Native American populations in the Southeast turned to a less agricultural subsistence base, with hunting and fishing again becoming more important. Plant food gathering remained significant for most Native Americans until dependency on European foodstuffs was forced upon them. At the same time, cultural interaction with the French, Spanish, and English explorers and colonists resulted in Native Americans adapting various Old World crops; watermelons (*Citrillus vulgaris*) and peaches (*Prunus persica*) were grown by the Natchez as early as the late seventeenth century. Meanwhile, certain indigenous cultigens, such as the sumpweed, were abandoned.[60]

Along with the growth of European trade and settlement, southern Indians experienced enormous socioeconomic changes. "Dependency," a term often encountered in studies of modern world economy, is classically defined as "a situation in which the economy of certain countries is conditioned by the development and expansion of another economy to which the former is subjected."[61] Richard White has successfully applied the theoretical framework of dependency in his study of three Indian nations with very different subsistence economies, namely the Choctaw, Pawnee, and Navajo, and explains how the traditional subsistence systems of these Native American nations collapsed as they were integrated into world markets. The Indian nations' increasing reliance on the capitalist world system was accompanied by great sociopolitical change within the societies. The original ability of Native American nations to feed, clothe, and house themselves diminished as they gradually resorted to the whites for food, clothing, and various manufactured items. What began as a reciprocal exchange turned into an economic system wholly dictated by whites. Native American dependency did not, however, result from a single economic process but rather from a complex interplay between various environmental, economic, political, and cultural factors.[62]

The Native American participation in the transatlantic fur trade introduced various Old World commodities to the region.[63] Firearms, metalware, and manufactured textiles became esteemed items among Native Americans and transformed Indian material culture. Increasing reliance on European tools and consumer goods decreased traditional self-sufficiency and changed the Native American way of life for good. European traders were furthermore eager to exchange beaver pelts or deer skins for distilled liquor and found in Native Americans an insatiable market for such commerce. Initially a monopoly of the chiefs, guns became more common among the Choctaw and revolutionized not only the warfare but also the hunt. Competition over access to new trade partners often led to intensive warfare between Indian nations and furthermore disrupted the survival of traditional culture. Heavy European demand for pelts led to wide extermination of fur-bearing animals all over the Southeast by natives increasingly eager to obtain European goods. Writing in 1805,

Rushworth Nutt captured the essence of the change in Native American subsistence culture along the Mississippi: "The Indian game never grew scarce before the europians introduced a trade with them & encouraged a peltry & fur trade. They now made havock among the innocent animals of the earth merely for the skins, & the flesh wasted. From this time the game has been on the decline & now they are compell'd to resort to husbandry & raising live stock. The game up the Mississippi & Missouri will soon be abolished & destroyed as they have caught the spirit of trade. Thousands of buffaloe, deer, bear & beaver are shot down yearly solely for the skin."[64] The introduction of credit and liquor into the Indian trade by the English in the late eighteenth century greatly stimulated the amount of trade. Liquor, which may have comprised 80 percent of the Choctaw trade by 1770, typically produced debt and enabled the trader to secure more deerskins. Under the earlier system of reciprocal gifts and exchanges, Native Americans had merely satisfied some of their basic needs on a replacement level by receiving certain European products in return for deerskins. In addition to growing indebtedness, the soaring consumption of alcohol resulted in the breakdown of social order among many Native American nations.[65]

Vast trade in deerskins, initiated by Europeans, played a decisive role in the diminution of white-tailed deer populations in the Southeast. It has been suggested that, at the height of deerskin trade during the middle decades of the eighteenth century, up to one million deer were killed annually by Native Americans.[66] It seems probable that the southern Indians' widespread belief in the reincarnation of killed game animals made "conservationist" attitude toward deer foreign at a time of rapidly evolving consumer needs. By the 1790s, commercial hunting had depleted the deer populations over most of present-day Mississippi. Game still remained more abundant in the Delta, but it could not fill the needs of all Native American hunters who, by the turn of the century, had to cross the Mississippi in order to find deer. The crossing of the Mississippi inevitably resulted in warfare between the Choctaw and Native American nations west of the river.

The long history of tribal warfare among Native Americans offered a convenient excuse for land appropriation. James Kirke Paulding, a former Secretary of the Navy under Martin Van Buren, writing in 1842, eloquently captured the *Zeitgeist:*

> There is scarcely a tribe which has not preserved some tradition of having conquered the former possessors of the land they occupy, or having been itself driven from its ancient inheritance by some hostile invader. The subject is worth a more extensive examination than I can give it here, if only to relieve our forefathers, ourselves, and our posterity from some of those unfounded charges that have been urged against them, by the over zealous advocates of the race of red men, who in the fervor of their philanthropy seem to have forgotten the just denunciation against the foul bird that pollutes its nest. The contact of civilized with savage man has everywhere been productive of immediate results which can only be reconciled with the dispensations of a just and beneficent Providence by their ultimate consequenses. If mere existence be, as I have no doubt it is, in spite of all its drawbacks, a universal source of happiness, then the cultivation of the land, and all those arts of civilization which conduce to the multiplication and subsistence of the human race, is doubtless in accordance with the will of that great Being whose command it was to "go forth, and multiply and replenish the earth." [67]

By the late eighteenth century, mixed-blood families (having both Native American and European bloodlines) had become common among the Choctaw and Chickasaw. The growing mixed-blood community had far-reaching effects on tribal life as their very existence confused and disturbed the traditional system of family relationships. At the same time, matrilineal descent provided most mixed-bloods full participation in the political life of the tribes. Over time, traditional values were corrupted as tribal affairs came to be dominated by the mixed-bloods, who could more effectively accommodate the emerging new order. The new leaders remained tribal "nationalists," but believed that vast cultural change was necessary; the adoption of individualism, a market ethic, and Christianity surely coincided not only with their own interests but also with those of their nation.

Facing the wildlife depletion, a growing number of Native Americans turned to a new form of agriculture for the bulk of their subsistence. The livestock and poultry introduced by the European traders, as well as slaves of African descent, were enthusiastically received among the Choctaw and Chickasaw, especially among the mixed-bloods. As domestic animals and black slaves belonged to specific owners, the new economic system divided the Native American nations into identifiable rich and poor families for the first time. Black slaves and livestock were most commonly owned by the mixed-blood natives and their white neighbors resident in the Indian nations. The presence of black slaves changed the Native American culture in many ways. Slaves now performed the labor of clearing agricultural land, often for European cash crops such as cotton. Slave ownership could furthermore be used in emulation of the ways and values of white planters, generating stronger notions of superiority and social stratification within the tribes. As many of the slaves were fluent in English, they made communication with Europeans easier, accelerating the acculturation of Native Americans.[68]

By the beginning of the nineteenth century the bison had disappeared and bear and deer had become increasingly rare over much of Mississippi. Some Choctaw did not wish to desert the traditional way of life and settled west of the Mississippi where deer had remained more abundant. Soon the depletion of deer populations in Arkansas and Louisiana forced them to return to Mississippi or turn to begging or stealing for subsistence; in the absence of deer it had become impossible to practice the traditional farming-hunting economy. Choctaw moving successfully into the Yazoo-Mississippi Delta (and other borderlands) were nontraditionalist stock-raisers of mixed-blood heritage. They began to utilize the region's canebrakes and grasslands as pastureland.

By the 1820s, some of the Choctaw, especially those of mixed-blood heritage, had expropriated much of the most productive land in the former borderlands region of the nation and become truly "civilized," producing cotton, beef, pork, and horses and purchasing a wide variety of goods from the market. At the same time, female Native Americans of these families had, like their white neighbors, largely abandoned agricultural labor and were now engaged mainly in household affairs while former warriors and hunters had become planters, cattlemen, and frontier entrepreneurs. During the early nineteenth century, land also became a commodity that could be bought, sold, and leased, at least among the most acculturated Native Americans. Among the most successful Choctaw settlers in the Delta were members of the mixed-blood LeFlore family who had hundreds of acres in cultivation along the Yazoo by 1830. The preeminence of property and commercial exchange was furthermore reflected in the new laws of the Choctaw nation, which were much concerned with the protection of private property. Missionary work among Indians had also proven successful as biblical texts were being translated into native tongues using the English alphabet and creating a previously unknown form of communication: written language. The often successful acculturation of the Choctaw and Chickasaw, however, did not ease the mounting pressure on these members of the "Five Civilized Nations" for opening up their lands for white settlement.[69]

After the Spanish evacuated to the south of the thirty-first parallel in 1798, the United States began to put the Treaty of Hopewell into effect, regulating the trade and licensing the traders. The government instituted, stocked, and staffed a so-called factory system of trading houses in the Indian nations. The system was strengthened by the Trade and Intercourse Act of 1802, which made native dealings with traders unlicensed by the United States illegal. A rather blunt aim for increasing the trade with Native Americans was suggested by President Thomas Jefferson:

The Indians being once closed in between strong settled countries on the Mississippi and Atlantic, will, for want of game, be forced to agriculture, will find that small portions of land, well improved, will be worth more to them than extensive forests unemployed, and will continually be parting with portions of them for money to buy stock, utensils, and necessaries for their farms and families.

On the Mississippi, we hold at present from our southern boundary to the Yazoo. From the Yazoo to the Ohio is the property of the Chickasaw, a tribe the most friendly to us, and at the same time the most adverse to the diminution of their lands. . . . The method by which we may advance towards our object will be, . . . To establish among them a factory or factories for furnishing them with all the necessaries and comforts they may wish (spirituous liquors excepted), encouraging these and especially their leading men, to run in debt for these beyond their individual means of paying; and whenever in that situation, they will always cede lands to rid themselves of debt.[70]

During the early nineteenth century, Native American lands in the territory and later state of Mississippi came to be surrendered piece by piece through numerous treaties with the federal and state governments, until only small portions remained by the time the U.S. Congress passed the Indian Removal Act in 1830. Totally disillusioned with the United States government and its failure to comply with the existing obligations to protect their lands from intrusion and appropriation and traumatized by internal division into traditional and mixed-blood factions, leaders of the Choctaw and Chickasaw nations signed various cession treaties, culminating in the final removal treaty at Doaksville in 1837.

The southernmost part of the Yazoo-Mississippi Delta had been seized from the Choctaw nation by the Treaty of Doak's Stand in 1820 (the so-called second Choctaw cession). Central and northernmost portions of the Delta changed hands through the Treaty of Dancing Rabbit Creek in 1830 (the third Choctaw cession) and the Treaty of Pontotoc in 1832 (the Chickasaw cession). In return for the cessions, the Choctaw and Chickasaw were to receive new tribal domains in the newly established Indian Territory in the West. They were also to be given some compensation for removal expenses as well as aid during the early resettlement. The removal was nearly complete by midcentury, though some Choctaw elected to remain, received allotments, and became subject to state law. Descendants of these Choctaw have survived in the east central part of Mississippi to this day. In practice, however, the traditional Native American had ceased to play a major role in the transformation of the state's landscapes.[71]

Some mixed-blood families had been successful in integrating themselves into the new socioeconomic order, but as Richard White points out, for the vast majority of the Choctaw, "trade and market meant not wealth but impoverishment, not well-being but dependency, and not progress but exile and dispossession. They never fought the Americans; they were never conquered. Instead, through the market they were made dependent and dispossessed."[72] The Yazoo-Mississippi Delta, largely abandoned since the fall of the Mississippian cultures and supporting the vegetational growth of decades, if not centuries, on most of its rich soil, was now "an unbroken and inundated wilderness" ready to be opened for large-scale settlement by people of European descent.[73]

On January 7, 1819, the British naturalist Thomas Nuttall surveyed the depopulated forests along the Mississippi River near the mouth of the St. Francis and wondered "[h]ow many ages may yet elapse before these luxuriant wilds of the Mississippi can enumerate a population equal to the Tartarian deserts! At present all is irksome silence and gloomy solitude, such as to inspire the mind with horror."[74] The ages were not to be many.

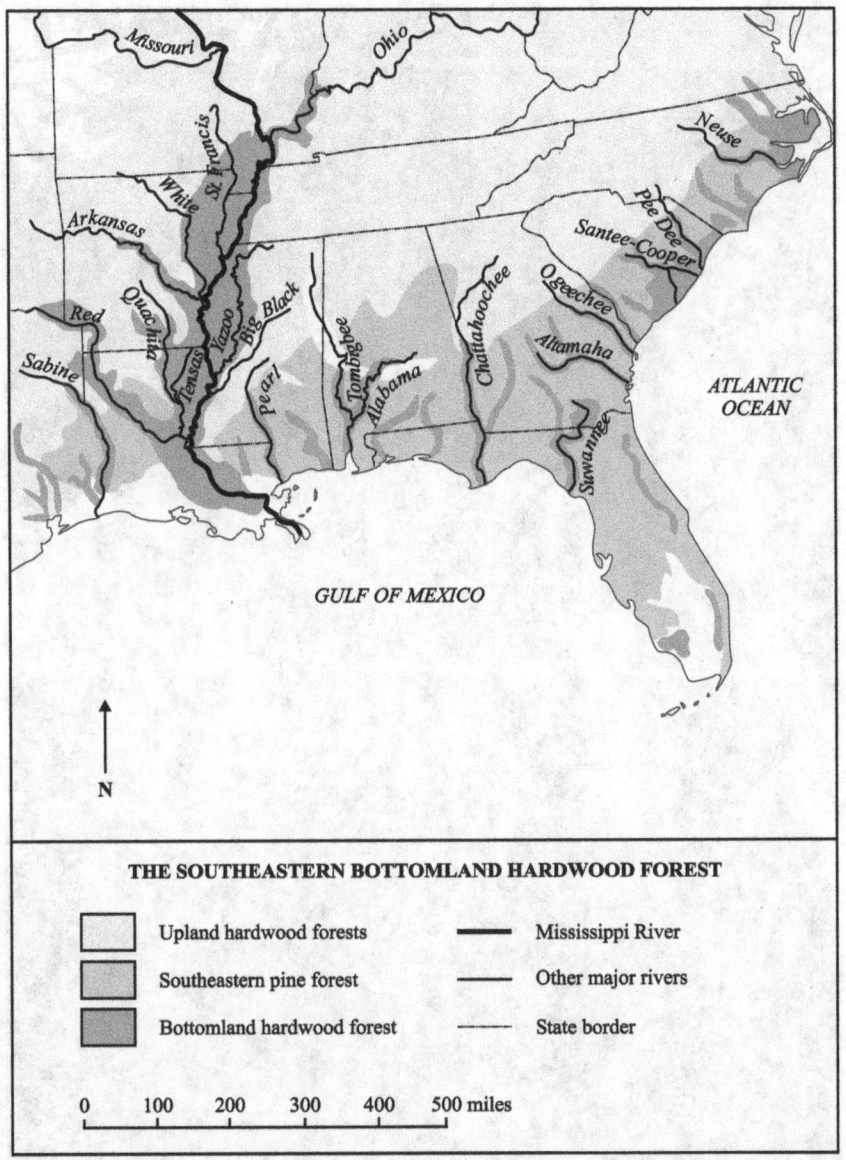

**THE SOUTHEASTERN BOTTOMLAND HARDWOOD FOREST**

| | | |
|---|---|---|
| Upland hardwood forests | —— | Mississippi River |
| Southeastern pine forest | —— | Other major rivers |
| Bottomland hardwood forest | ---- | State border |

0    100    200    300    400    500 miles

The extent of the bottomland hardwood forest in southeastern North America at the time of European conquest. The bottomland hardwood forest was confined to the floodplains of southeastern rivers. Only the broadest belts of the forest type are shown.

Old-growth sweetgum trees photographed in the 1950s at the future site of the Delta National Forest in Sharkey County. U.S. Forest Service photo courtesy of the Forest History Society.

The ivory-billed woodpecker, like the Carolina parakeet, cougar, and black bear, was among the original fauna of the Delta forests. This adult male at nest was photographed in the Louisiana bottomlands by Cornell University ornithologists in 1935. Reproduced with permission from *The Auk* 54 (April 1937).

During the late nineteenth century, the Delta evolved into a "New South Cotton Kingdom" and remained one well into the twentieth century. This Coahoma County sharecropper was photographed by Dorothea Lange in 1937. Library of Congress, Prints & Photographs Division, FSA-OWI Collection, LC-USF34-017305-C.

The predominantly black plantation work force remained under close control during the twentieth century, as illustrated by this picture of "Mr. Jones, one of the owners of Marcella Plantation, weighing in cotton by cotton house in field" in Holmes County. Photograph by Marion Post Wolcott in 1939. Library of Congress, Prints & Photographs Division, FSA-OWI Collection, LC-USF33-030582-MI.

Since the arrival of first humans to the Delta, fish found in the rivers and bayous of the floodplain have provided an important source of food and recreation. In October 1939, Marion Post Wolcott photographed these "Negroes fishing in creek near cotton plantations outside Belzoni," Humphreys County. Library of Congress, Prints & Photographs Division, FSA-OWI Collection, LC-USF34-052293-D.

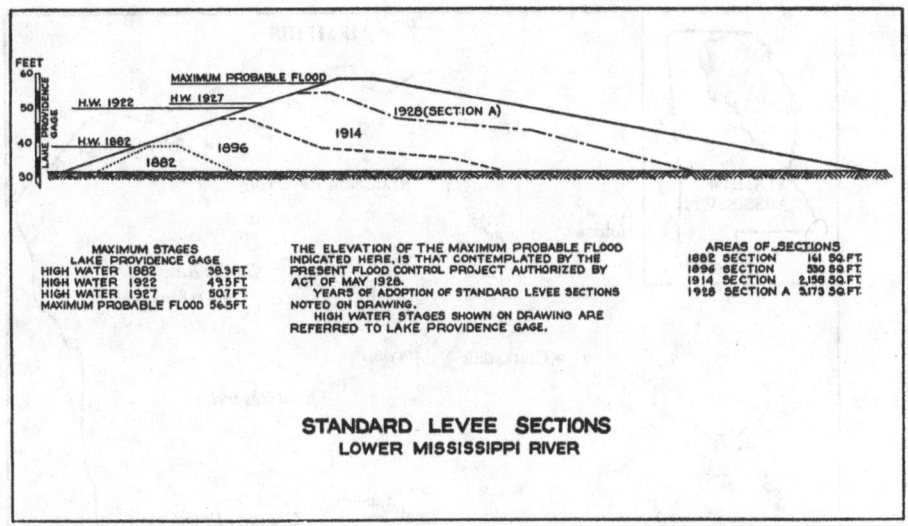

FEET
60
50
40
30

MAXIMUM PROBABLE FLOOD

H.W. 1922   H.W. 1927
1928(SECTION A)
H.W. 1882        1914
1896
1882

LAKE PROVIDENCE GAGE

MAXIMUM STAGES
LAKE PROVIDENCE GAGE
HIGH WATER  1882    38.3 FT.
HIGH WATER  1922    49.5 FT.
HIGH WATER  1927    50.7 FT.
MAXIMUM PROBABLE FLOOD  54.5 FT.

THE ELEVATION OF THE MAXIMUM PROBABLE FLOOD
INDICATED HERE, IS THAT CONTEMPLATED BY THE
PRESENT FLOOD CONTROL PROJECT AUTHORIZED BY
ACT OF MAY 1928.
YEARS OF ADOPTION OF STANDARD LEVEE SECTIONS
NOTED ON DRAWING.
HIGH WATER STAGES SHOWN ON DRAWING ARE
REFERRED TO LAKE PROVIDENCE GAGE.

AREAS OF SECTIONS
1882 SECTION     161 SQ.FT.
1896 SECTION     530 SQ.FT.
1914 SECTION   2,158 SQ.FT.
1928 SECTION A 3,173 SQ.FT.

STANDARD LEVEE SECTIONS
LOWER MISSISSIPPI RIVER

Standard Levee Sections along the Lower Mississippi River, 1882–1928. The "levees only" policy, adopted in the late nineteenth century, substantially increased the height of flood crests and eventually resulted in the Great Flood of 1927. Courtesy of the Mississippi River Commission.

Log train of the Lamb-Fish Lumber Company of Charleston, Tallahatchie County, loaded with old-growth hardwood logs. The company was among the Delta's largest hardwood operators at the turn of the twentieth century. Courtesy of the Mississippi Department of Archives and History, Archives and Library Division, Special Collections Section, Mosby Postcard Collection.

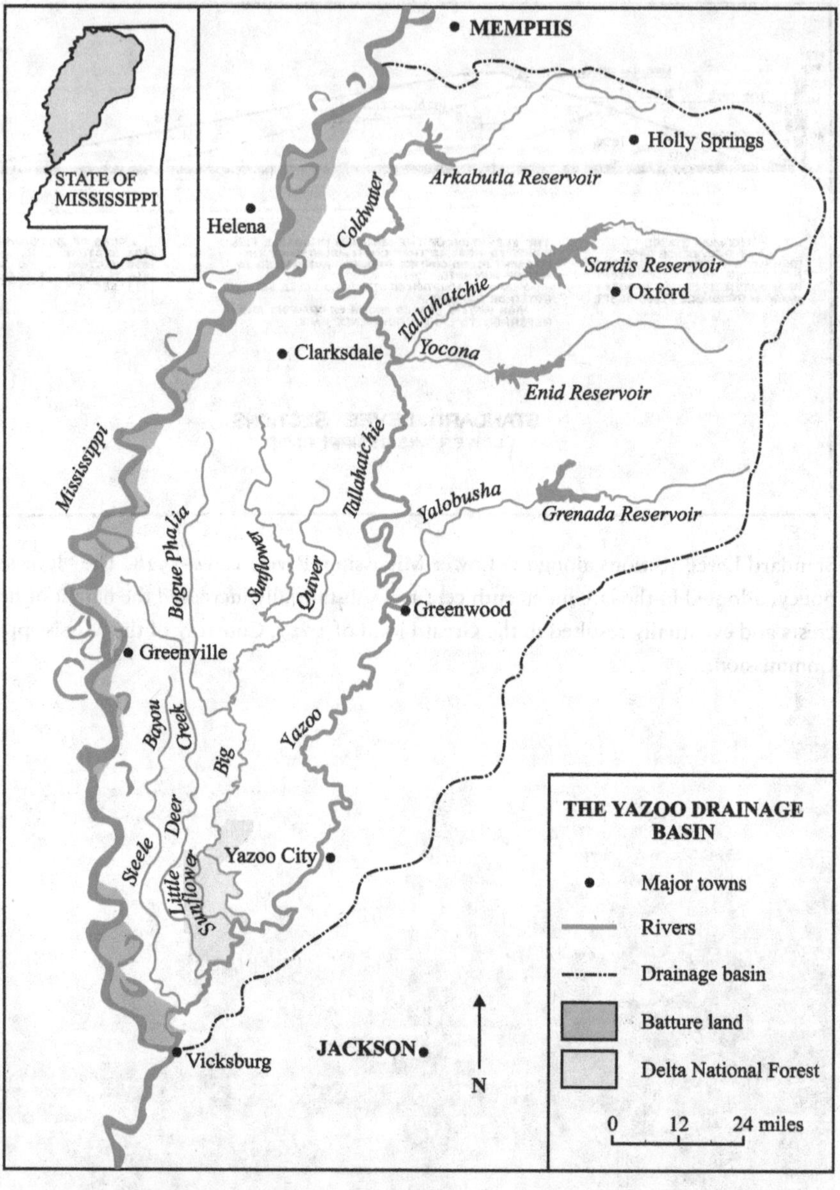
MEMPHIS
STATE OF MISSISSIPPI
Holly Springs
Arkabutla Reservoir
Coldwater
Helena
Sardis Reservoir
Oxford
Tallahatchie
Clarksdale
Yocona
Enid Reservoir
Mississippi
Tallahatchie
Yalobusha
Grenada Reservoir
Bogue Phalia
Sunflower
Quiver
Greenwood
Greenville
Bayou
Creek
Yazoo
Big
Steele
Deer
Little
Sunflower
Yazoo City
THE YAZOO DRAINAGE BASIN
Major towns
Rivers
Drainage basin
Batture land
Delta National Forest
0
12
24 miles
Vicksburg
JACKSON
N

The Yazoo drainage basin after the completion of the Arkabutla, Sardis, Enid, and Grenada reservoirs. During the twentieth century, a massive network of levees, dams, pumping plants, and auxiliary channels totally transformed the natural hydrological regime in the Delta. The Delta National Forest and batture land between the Mississippi River and the main levee line indicated by shading.

At first there had been only the old towns along the River and the old towns
along the hills, from each of which the planters with their gangs of slaves
and then of hired laborers had wrested from the impenetrable jungle of water-
standing cane and cypress, gum and holly and oak and ash, cotton patches
which, as the years passed, became fields and then plantations. The paths made
by deer and bear became roads and then highways, with towns in turn springing
up along them and along the rivers Tallahatchie and Sunflower which joined
and became the Yazoo, the River of the Dead of the Choctaws—the thick, slow,
black, unsunned streams almost without current, which once each year ceased to
flow at all and then reversed, spreading, drowning the rich land and subsiding
again, leaving it still richer.

—WILLIAM FAULKNER, "Delta Autumn," *Go Down, Moses*

CHAPTER FOUR

# *The Creation
of a Cotton
Kingdom*

WHATEVER THE SIZE of the pre-Columbian populations and
the amount of the acreage cultivated by them, people of European—
and African—origin were to practice agriculture in an unprecedented
scale everywhere on the continent, including the Yazoo-Mississippi Delta.
Agriculture has been the dominant factor affecting the historical development of southern life and institutions ever since the 1607 founding of the
first English settlement at Jamestown in Virginia. Southern agriculture,
as practiced by Euro-Americans, has often been seen as a combination of
monocultures—tobacco in Virginia, rice in South Carolina, and cotton in
Alabama and Mississippi. Although this simplistic view allows for little
detail or regional variation, the environmental implications of southern
agriculture become the clearest in connection to staple-crop economies.[1]

The majority of British immigrants were attempting to recreate a European agricultural economy and pastoral society in colonial North America, and the economic life of European colonists in the South was heavily based on agriculture since the seventeenth century. Agriculture differed somewhat in the various colonies, as it was influenced by soil, climate, and English trade regulations, but everywhere the use of the land was primitive. Crop rotation and fertilizers were hardly known, resulting in early soil exhaustion on less than optimal locations (such as loess versus alluvial lands). Virgin soil in apparently limitless reserves did not encourage scientific farming, and the colonial farmer seemed to have but one object: the plowing up of fresh land. Not surprisingly, southeastern bottomland hardwood forests with their rich soils were considerably affected by European agriculture from its very beginning in North America.

During the European conquest, land ownership became the foundation for economic development, whether in the form of subsistence farming or commercial planting. Consequently, a mixture of yeoman farms and plantations characterized the Euro-American economy in the South. Land companies soon appeared, speculating in immense claims of backcountry lands and encouraging ever-increasing immigration to the region. Commercial southern agriculture during the colonial and antebellum periods was based on plantations, typified by extensive holdings of land and a labor force made up of slaves. Plantations developed in areas with fertile soils suited for large-scale cultivation of subtropical staple crops for international markets. Plantation agriculture required extensive management, the organization of both supply and credit systems, and access to transportation facilities in order to bring the product to markets. In the beginning, plantations in the Tidewater region of Virginia and Maryland depended on white indentured labor. During the late seventeenth century, black slaves of African origin superseded whites as the labor force for southern plantations, and an agricultural system similar to the one established in the Caribbean took hold in southeastern North America.

To attain the status of "planter" in the antebellum South, one as a rule of thumb had to own more than twenty slaves; the number of acres possessed or amount of product grown and shipped were of secondary importance in attaining this venerable standing. At this level, an overseer was usually hired to supervise the labor unit. In 1850, some twenty-five hundred southern planters possessed thirty or more slaves, and only a handful owned more than a hundred. The development of slavery between 1790 and 1860 closely matched the westward expansion and growth of the cotton industry: the number of slaves in the South rose from about 750,000 to about 4 million, composing one-third of the southern population. Still, most of the southern agriculturists before the Civil War were nonslaveholding yeomen farmers; in 1860 only one-third of the one million white families in the fifteen slave states owned slaves. Of these, only some hundred thousand farmers owned more than ten slaves, with only ten thousand possessing more than fifty.[2]

Despite their relatively low numbers, planters largely directed the socioeconomic life of the antebellum South. Planters' dominance of the southern export economy and cultural life, combined with the low taxation of slave property, caused many small farmers and professionals to aspire toward such an esteemed status.[3] Although there were few legal impediments to the growth of the nonagricultural sector of the southern economy during the antebellum era, climate, topography, and the slavery-based labor system gave the region a comparative advantage in agriculture. Planting offered dependable profits with less risk, as compared to the higher profits of the still-developing manufacturing sector.

The plantation system has often been seen as a frontier institution, tied to and dependent upon "metropolitan" or "supralocal" capital, industry, and markets. Accordingly, antebellum plantation production in the South was geared to manufacturing and financial centers in Europe and New England. Sales and shipment of plantation products, as well as purchases of market goods for plantations, were arranged by agents acting as middlemen. Plantation agriculture undoubtedly created the greatest personal fortunes in colonial North America, and planters such as George Washington, James Mason, and Thomas Jefferson played a dominant role in the politics of the colonies and the fledgling United States.

Although the unquestionable center of capital accumulation in the antebellum South was the slavery-based plantation system, most southern agriculturists remained small, independent family farmers comparable to those of the northern United States. As cattle grazers and hog drivers, the poorest whites typically acted as forerunners of Euro-American civilization on the southern frontier. Some of them continued to move with the expanding frontier, while others settled down and typically became small farmers and—after the Civil War—members of the tenant class. In 1850, a representative southern farm of some 120 acres aimed to be a self-sufficient economic unit, existing at a subsistence level. The farm had perhaps twenty-five acres under cultivation, most of that in corn with several acres in other food crops and in hay crops. Typically three to five acres in cotton provided for the farm's small cash income with which the farm was tied to the markets.

Plantations, on the other hand, could have hundreds or even thousands of acres under cultivation. The typical labor organization of an antebellum plantation consisted of the overseer in the role of manager of various field gangs, each supervised by a driver, who himself was a slave. Domestic slaves and craftsmen usually worked under the direct supervision of the planter and his family. Aside from the immense emotional distress experienced by slaves, the physical harshness of slavery was at least partly mitigated by their capital value, as illness and injury could greatly lower the value of the owner's investment. Besides, many planters had learned through trial and error that rewards worked better than punishments in getting the most out of their labor force. During the nineteenth century, a growing number of slaves were assigned individual chores carrying compensation for a well-performed task.[4]

The foremost crop encouraging the spread of European agriculture and its effects across the landscapes of southern United States during the nineteenth century was cotton (*Gossypium* spp.). The recently vacated Native American lands in the South were largely settled by agriculturists interested in raising this staple cash crop. The so-called Sea Island variety of this plant had superseded indigo as a cash crop on the Atlantic coast during the eighteenth century, but it could not be successfully cultivated farther inland. After the invention of the cotton gin in 1793, short-staple upland cotton supplanted the Sea Island variety and became an important cash crop for southern farmers. The land and climate in the South seemed ideally suited for upland cotton. Cotton culture required relatively little skill of the farmer and provided almost year-round employment; the product was also relatively nonperishable and possessed high value per unit of weight.

Production first centered in South Carolina and Georgia. After 1815, however, the cotton belt expanded quickly through Alabama and Mississippi and—as the soils in the older areas were increasingly exhausted—into Tennessee, Louisiana, and Arkansas. In Texas, cotton reached the hundredth meridian just before the Civil War. The growth in acreage increased production so rapidly that during the 1840s prices plummeted, but recovered in the following decade. By the 1850s, the modern type of upland cotton had been developed by crossing exotic varieties and applying selective breeding to the hybrids, providing increased yields and more uniform quality.

Apart from some minor technological developments, the methods of cotton culture remained basically unchanged until the twentieth century. Cultivation began in the spring as the land was prepared for planting:

> Some eight or ten negroes with their mule-teams, commenced ploughing to-day, in the field near our school house. Two of them went ahead and struck the field out into furrows, three or four feet apart. Others followed, turning furrows against these on both sides, till the intermediate spaces between the original furrows were all ploughed up. This forms ridges some four feet apart. After the field is thus ridged, a negro with a single mule, before a small plough, strikes a furrow on the top of each ridge. Another negro follows him with a sack of cotton seed strung around his shoulders, and scatters the seed thickly along in this furrow; he is followed by one of his fellows, with a mule before a small harrow, who drags over the seed, thus covering it up. And finally, to make sure work, a negress follows the last one, with a hoe, to cover up what seed may not have been covered by the harrow. This is cotton planting, which is done on the first of April. . . . The overseer is seen walking or riding, here and there over the field, whip in hand, inspecting the work.[5]

After the plants had emerged, weeding and thinning out continued until the so-called lay-by time in midsummer. Picking commenced as cotton bolls opened in late summer, and could continue into December.[6]

The crop was processed for sale after harvest by ginning at the plantation facility. The term "cotton gin" originally referred only to the gin stand that removed seed from the lint in the gin house. Seed cotton was fed into a gin stand and then processed into bales in a press. The bulk of the American cotton crop was packed into a press box fifty-four inches long and twenty-seven inches wide with a depth of about forty-five inches, producing standard "square" bales that weighed roughly five hundred pounds. Both the gin stand and the press were first powered by horses or mules, and processed up to three or four bales of cotton per day. The gin stand and press were found on most plantations until the 1880s, when the ginning system came into general use. Under this system, cotton was processed in steam-powered plants that contained several gin stands and integrated the ginning with the baling. These modern cotton gins were capable of turning out three to four bales of cotton in an hour.[7]

During the nineteenth century, cotton came to dominate the local economies of the South and greatly affected the nation's economic development by its impact upon westward expansionism, slavery, and the growth of a credit system, shipping, and manufacturing. Between 1790 and 1860, cotton production expanded from some three thousand bales, valued at about $500,000, to almost 4.5 million bales, valued at $270 million. Cotton had become the chief export of the nation, inducing secessionist leaders of the South to assume that industrialized countries in Europe, especially England and France, had become wholly dependent upon a continuing supply of this southern product. In 1858, Senator James H. Hammond of South Carolina made the famous statement that "cotton is king" and "no power on earth dares to make war upon it." Contrary to southern expectations, France and England did not extend prompt recognition to the Confederacy and failed to oppose the northern sea blockade. Instead, Europe remained neutral during the conflict and accelerated its search for substitute supplies from Africa and Asia, proving the belief in "King Cotton" to be in error.

## An Agricultural Frontier

In October 1808, Henry Ker speculated about the economic future of alluvial bottomlands along the lower Mississippi:

> I this day passed a beautiful tract of wilderness, affording a plentiful subsistence to the animal creation, and inviting, by enchanting prospects and luxuriant growths, the free sons of Columbia to inhabit and share with them the beautiful gifts of nature. I imagined myself holding a discourse with these stately timbers,

and would answer that it only wanted time to make this a country of wealth and happiness: that though at present the attractions of the settled parts were more enticing, your hills and vallies, believe me, will one day be the resort of men of genius and enterprise, who will add to your future glory.[8]

Development of wilderness resources was closer than Ker could probably imagine, as significant alteration of the Delta natural environment by people of Old World origin commenced within a decade or two. By the mid-1830s, Native Americans had ceded all land in the Yazoo-Mississippi Delta to the United States. The federal government moved quickly to open the cessions for sale to incoming settlers, but in order to be sold, the former Indian land had first to be surveyed according to the laws of the new nation. Throughout much of the South, the old survey systems, such as the "metes and bounds" applied in the original thirteen colonies, came to be replaced by a cadastral system. The new approach to land division was based on a rectangular grid and presented a radical departure from the preceding irregular patterns of land apportionment.

The Land Survey Ordinance Act of 1785, masterminded by Thomas Jefferson, created the rules for the survey and distribution of public lands acquired from the individual states and Indian nations. A base line, also called a surveyor's line, ran east to west and was crossed by north-south meridian lines at regular intervals of six miles. The survey began with a zero base line and a zero north-south coordinate, creating units known as "Townships" when measurements were made north and south of the base line and east and west of the principal meridian. The townships created by the process could be identified by their location in relation to the zero baseline (Township *n* north or south) and the zero meridian (Range *n* east or west). The new federal mapping applied the statute mile of 5,280 feet and rigidly divided much of the continent into squares, largely ignoring the actual terrain. The basic unit of a township of 6 square statute miles was divided into 36 "Sections" of 640 acres each. A numbering system for the sections within a township was adopted in 1796; section number 1 was always located in the northeast corner and section 36 in the southeast corner of the township. Four sections (numbers 8, 11, 26, and 29) were reserved for the federal government and one (number 16) was intended for the maintenance of a public school in each township. Later legislation authorized the division of sections, creating rectangular areas ranging from a 320-acre "Half Section" to a 40-acre "Quarter-Quarter Section." Originally, townships sold whole were to alternate with ones divided into sections.[9]

Land surface measurement of the new nation was regulated by instructions issued by the surveyor-general of the United States. There were four base lines and seven principal meridians used in the South, and most townships of the Yazoo-Mississippi Delta came to be situated north of the base line in central Mississippi (T9–30N), and east and west of the Choctaw Meridian (R1–10W and 1–2E). Land surveys along the main stem of the Yazoo began as early as 1827 and continued through 1841, with the majority of township surveys completed during the 1832–34 period. Most of the area was therefore measured under the regulations of 1831 and 1834, which, if carefully followed, should have provided a reasonably accurate account of the floral cover in the region at the time of the survey. Under the regulations, two to four "bearing trees" were to be identified as markers at the intersections of all township and range lines, and at every section corner and quarter section post. Under these rules, each township survey should yield a minimum of 364 floral samples. A general description of each surveyed mile with special notation of prominent physiographic features was furthermore to be included.[10]

Robert M. Thorne and Hugh K. Curry examined the surveyor's notes and township plat maps for fourteen townships on four different ranges along the Choctaw Meridian in the easternmost Yazoo-Mississippi Delta. Unfortunately, the surveyors' possible biases, various misapplications, or even complete neglect of the regulations severely restrict the usefulness of the data in reconstructing the exact vegetation patterns of the early nineteenth-century Delta. Some of the reasons for this could be that topographic features such as lakes, streams, grasslands, and cane thickets would not provide tree specimens, whereas the distance from the nearest tree to the survey point could have occasionally been considered too great to be referenced. The data is, however, quite instructive and strongly supports the contention that mature bottomland hardwood forest was prevalent in the region. Well over 60 percent of the total of 3,563 specimens recorded for the fourteen townships consist of different oaks, gums, hickories, and elms. The most common species, with 568 specimens represented in the surveyors' notes, is sweetgum.[11]

My own examination of surveyor's notes for just one Delta township (TI4N, R6W), the area that today comprises Panther Burn, Nitta Yuma, and Delta City in Sharkey County, similarly reveals a mature deciduous forest. The 201 samples from this township are dominated by ash and sweetgum (56 percent of the total number of samples) with plenty of oak, elm, and mulberry (21 percent). Other species represented by five or more specimens are boxelder, [honey]locust, and persimmon. Baldcypress and gum (presumably black tupelo) yielded four specimens each. The three examples of dogwood from the township suggest that certain localities were excluded from frequent flooding.[12]

Surveyors' field notes from the eastern Delta also document the occurrence of tracts of open land throughout the survey area; these are referred to as "prairies" and were presumably more or less unforested and covered with grass. The prairies were located on higher elevations of the Yazoo Basin. The existence of such grasslands was also acknowledged in early French accounts of the region. Victor Collot had noted that "[t]he land above Cold River is no longer swampy, and the higher you advance the more fertile it is found. There are even some points which have been cleared by the Indians."[13] The portrayal of grasslands as remains of Native American clearings is not supported by Thorne and Curry because of the large size of the treeless areas (typically $1/4 \times 1/2$ mile, or some 80 acres). Instead, they propose the open areas to have evolved naturally and been maintained by Indian practices. This claim is supported by palynological evidence: enough pollen from the herbaceous members of the Chenopodiaceae family and the genus *Amaranthus* (the so-called Cheno-Am pollen) has been present during the last sixteen thousand years to indicate continuous existence of open areas within the bottomland hardwood forests of the Delta.[14]

According to the early surveys, dense stands of cane were also found on higher locations of the Delta, providing further openings in the forest; the extent of such canebrakes is, however, difficult to determine. According to Collot, "[f]rom the mouth of the Yazoo to Cold River, the country is covered with bamboo canes of a considerable height; from thence to its source is wood of different kinds, but neither the cedar, the pine, nor the green oak."[15] Similarly, Thomas Nuttall observed that "[t]he whole country, generally speaking, along the [Mississippi] river, appears uninhabited, though vast tracts of cane land occur in the bends."[16] In the 1830s, much of the northern half of T14N, R6W was described as "[o]pen woods subject to overflow" and supporting canebrakes.[17]

Hazel R. Delcourt's study of early-nineteenth-century forest conditions in the northeastern corner of Louisiana employs data from the General Land Office Survey of 1813–14. Her analysis of the surveyors' reports indicates the presence of a forest similar to the Yazoo-Mississippi Delta in Louisiana's bottomlands. Alluvial forests in the region were at the time dominated by various oaks and hickories, cypress, and sweetgum. Creating a picture resembling that provided by travel accounts and land surveys in the Delta, Delcourt suggests that the crests of the natural levees along the lower Red River supported extensive stands of cane, which could often outshade tree seedlings.[18] A similar situation evidently existed in parts of the Delta, as Thomas Hutchins had noted in 1784 that "further up this [Yazoo] river, the canes are less frequent and smaller in size, and at the distance of 20 miles there are scarcely any. Here the country is clear of underwood and well watered, and the soil very rich."[19] Accounts such as Major A. G. McNutt's 1836 observation that "[a] great portion of the [Deer Creek] tract is Cane Land, and but thinly timbered, and [ . . . b]etter adapted to the culture of Cotton than any other in the Southern country" would not go unnoticed among potential planters pouring into the state of Mississippi.[20]

The greatest obstacle to Euro-American settlement of the Delta was posed by the hydrological regime of the floodplain. Rushworth Nutt had in 1805 surmised that "all the country, or nearly so between the Yazoo & mississippi to the mouth of the Yazoo overflows annually & renders no value." Still, there were "some few places in the bottom land of the Mississippi on [w]hich there are vestages of ancient settlements, which do not overflow."[21] Portions of these lands on high natural levees had been occupied by the mixed-blood Choctaw settlers who had arrived in the Delta a few years earlier, and the later white settlers followed their example. The high ridges were, of course, the areas that had been most recently cleared by Mississippian Native Americans. Consequently, these were the easiest to make productive in the shortest amount of time with the least effort. During flood years, inundations took place relatively early in the spring and drained off rapidly enough from the ridges for hand planting, which could take place even in muddy fields.[22]

In 1837, a German observer emphasized the allure of cane-covered ridges for new settlers: "[A]lmost unpenetrable cane is appreciated as a sign of rich soil and is used for pasture or cultivated for cotton planting."[23] In addition to seeking a combination of rich lands and relative safety from flooding, early settlers in the Delta preferred tracts with easy access to riverboat traffic. As a result, the first European settlements in the Yazoo-Mississippi Delta concentrated along the banks of the Mississippi and the Lower Yazoo; the upper tributaries of the Yazoo and inland ridges were settled later. The waterways remained the sole means of access to the Delta settlements for decades, as the only things resembling roads were the few remaining Native American traces.

Euro-American settlement in Washington County, adjacent to the Mississippi, commenced around 1825, and the Greenville area was inhabited by 1829. The Delta towns of Yazoo City and Greenwood, adjacent to the Yazoo, were founded as speculative ventures in 1830 and 1834.[24] New settlements were similarly developing at other Delta locations. Charles Clark, who later became a Confederate general and the governor of Mississippi from 1863 to 1865, began his career as a lawyer and planter in the blufflands of Natchez and Fayette. In the 1830s, Clark "won his great law suit that stirred the whole State as well as Congress, for it was an Indian claim against large tracts of land, in Bolivar and Sunflower Counties, that had been pending for years." According to his daughter, "Pa saw the wonderful future" for the "rich, black soil" of the Delta's "almost unbroken" forests. Clark undoubtedly understood the agricultural promise of the bottomlands and was accordingly willing to take chances, as "[b]efore winning the law suit, Pa had purchased a large tract of this land and, with a Mr. Newman, had established negro settlements and planted cotton successfully. Large additional holdings were added, for his fee was paid in lands." Many of Clark's relatives also "moved to the Delta and bought wild land and cleared it for cotton." The Delta's Choctaw heritage, however, was not entirely forgotten, as Clark's plantation was named "Doro" after the legal pseudonyms John Doe and Richard Roe, used by two Indian plaintiffs in the lawsuit.[25]

During the 1830s, alluvial lands in Mississippi were widely bought and sold with the standard price of $1.25 per acre eagerly paid by the incoming settlers. Many of the migrants arrived from the older cotton states, especially Virginia. President Andrew Jackson's policy of accepting only gold and silver in payment for government land aimed to curb the nation's booming land speculation, which came to an abrupt end with the financial collapse of 1837. Alluvial land values plummeted to as little as twenty-five cents per acre, and development in the Delta was largely halted for almost a decade. Relatively few settlers were now willing to take the risk of farming the bottomlands, infested as they were with malaria and subject to nearly annual overflows. Instead, agricultural activity concentrated in the northeastern and central parts of the new state.[26]

Land sales picked up again around 1846 and reached a peak at the turn of the decade. The Mexican War had ended, and many war veterans decided to try their luck in the potentially productive bottomlands. Already by 1847, roughly half of the government land had been purchased in the riverfront counties of Coahoma, Washington, and Bolivar. Newcomers included planters from other parts of the state drawn to the Delta by the rich soil. Among them was the future governor of the state, Benjamin Grubb Humphreys, who later reminisced: "My plantation on Big Black Hills was fast failing and I determined to transfer my planting interests to the Yazoo Valley. I brought a tract of land on Roe Buck Lakes in Sunflower County then in the woods. I gave it the name of Itta Bena, i.e. 'Home in the woods,' and commenced preparing it for a permanent home in 1846." Most of the new settlers, however, were members of the slave-owning planter class from the older cotton states who were in constant search of fertile land.[27]

Among such established families to enter the Delta were the Hamptons of South Carolina. The founder of the planting dynasty, Wade Hampton I, had been involved already with the Yazoo Companies and was considered at the time of his death in 1835 the greatest slaveowner in the United States. In addition to the family's South Carolina holdings, his heir, Wade Hampton II, owned and operated a 2,529-acre plantation in Issaquena County, while Wade Hampton III owned the 835-acre Wild Woods plantation on Lake Washington in Washington County, where he in 1850 produced 5,000 bushels of corn and 453 bales of cotton. During the next ten years, the youngest Hampton extended his holdings to include more than 10,000 acres in five plantations.[28] Although making good money, an antebellum logging crew foreman in the Delta did not feel at home among the region's economic elite: "the people of this state are mostly rich cotton planters and do not turn their attention to much else[, with] a plenty of slaves to sup[p]ort them live a high life[. N]ot manny whites that do much labor in this country."[29] Construction of flood control devices and wresting agricultural land from the old-growth forest required capital and the ability to write off losses in case of a bad crop.

Both the conversion of bottomlands into cotton fields and the cultivation of the crop were extremely labor-intensive. In 1858, Paul C. Cameron had "64 Hands[,] 25 Mules[,] and 10 oxen to do the work on the place [in Tunica County]."[30] Wade Hampton II maintained 166 slaves on his Issaquena County plantation in 1850, many probably transferred from South Carolina, while another planter in Washington County had 115 hands working his fields in 1853. Wade Hampton III, who in 1850 had cultivated his fields with 177 slaves, expanded his work force to more than 900 within a decade.[31] Slaves continued to be in great demand in the Delta until the outbreak of the Civil War, as the following incident, witnessed by Frederick Law Olmstead in north-central Mississippi, indicates:

The next morning when I turned out I found [a young man from the] Yazoo [region] looking with the eye of a connoisseur at the seven prime field-hands, who at half-past seven were just starting off with hoes and axes for their day's work. As I approached him, he exclaimed with enthusiasm:—

"Aren't them a right keen looking lot of niggers?"

And our host soon after coming out, he immediately walked up to him, saying:—

"Why, friend, them yer niggers o'yourn would be good for seventy bales of cotton, if you'd move down into our country."[32]

Good overseers for the productive but primitive antebellum plantations in the Delta were hard to come by. The apparently competent Y. F. Griffin "had several offers but did not undertake." The plantation on the Sunflower River in Washington County where Griffin eventually began to work had experienced trouble with overseers in the past: "[T]he overseer, of this place Last year got into a difficulty and had to Leave he killed a negro, they got another he stayed 1 month and conc[l]uded that he could not [attend] to the business and quit with his own accord." The plantation was "a chois place" of "Splendid Lands and in tollerable good order, a heap of it fresh Cane Land." Griffin was paid an above-average seven hundred dollars and furnished with a good horse, some fifty hands, and a slave driver with long experience. In 1852, the place produced some five thousand bushels of corn and 640 bales of cotton. The overseer complained about the heavy rains that had halted his corn planting after the first one hundred acres, while the wet gum that slaves used as firewood was no match for the resinous pine used in the uplands. Still, he expected no further trouble with standing water on the fields, as ten miles of ditches had been completed around the plantation.[33]

Drainage ditches provided unexpected opportunities for fishing, and Griffin could boast to his nephew Micajah that "[t]he reigns has raised the waters and the fish has come up the Ditches in the Plantattions, till we have caugh[t] 500 lbs of a Day, the Buffalow Fish [*Ictiobus* spp.], they way from 15 to 30 lbs, this is no Fish story, we shut up the Ditches, and ketch them when we please."[34] Some two years later, while savoring a choice cut of "Water Mellon, not bad to take of [a] hot day," Griffin was happy to report that in July, due to "fine rains," the crops looked well: the corn was nearly ripe, while the cotton showed "fine av[e]rage from waist to he[a]d high."[35] In the absence of irrigation systems and without adequate flood protection, Delta fields remained highly vulnerable to changes in the weather, with cotton and corn crops varying greatly from year to year.[36]

The daily routine on an antebellum Delta plantation was carried out at a brisk pace, as explained by Griffin. The driver rang the bell forty-five minutes before dawn to wake up the plantation. The overseer's morning routine was probably a little more relaxed than that of his field hands: "[M]y fier is made[,] my nag is saddled, and hi[t]ched by my Door, my Boots is Blacked, and re[a]dy." Griffin was "generally in or about the plantation from morning till Knight." The overseer could have his meal at his "own Discretion," while the slaves' provisions were cooked and sent to them in the fields. Griffin was "furnished with a record, plantation Book prepared for the purpose[,] and ev[e]ry knight" he recorded "all the general transactions of the day[,] ev[e]ry thing . . . received on the place, any thing going off, anything such as Clothing, shoes, hats Etc. that is given out to the negroes, also the horses stack of all kinds[,] tool of all kinds, quantity and quality," in addition to the names and ages, as well as births and deaths among the slaves.[37]

The use of wild plants of the bottomlands for food, drink, cosmetics, and medicine had not passed with the Native American. The European and African newcomers to the Delta utilized the region's native flora in many ways, often adjusting their knowledge about Old World plants to the new surroundings. Herbs of the genus *Hypericum* figured prominently in southern folk beliefs. Known as High John the Conqueror, or John the Conqueror root, they were seen especially by blacks as capable of warding off evil spirits and ensuring good luck. Bark from various trees and sassafras, for example, came into use as home remedies, especially among blacks. Slave women "made a heap o' dewberry [*Rubus* sp.] an 'simmon wine," while children "would gather black walnuts in de woods an' store 'em under de cabins to dry."[38] Wild pecans, a Native American favorite, continued to be gathered in the turn-of-the-twentieth-century Delta; for white schoolboys they provided additional income.[39] Another bottomland delicacy for the Delta settlers was honey provided by feral bees: "[e]very day in falling timber they would find a bee tree, and Frank would bring in three or four gallons of fine white honey."[40]

In addition to believing in the powers of various plants, many slaves maintained that certain animals had special powers. Owls carried a special significance, as explained by "Uncle" Hamp Kennedy for the Federal Writers' Project: "If we heard a little old shiverin owl [eastern screech owl, *Otus asio*] we'd th'ow salt in de fire an' th'ow a broom 'cross de do' fer folks say 'twas a sign of bad luck, an' a charm had to be worked fas' to keep sumpin' terrible frum happenin', an' if a big owl hollered, we wasn't 'lowed to say one word." Kennedy's recorded anecdote seems to lend truth to this belief, as "one time Joe an' Green, two niggers on our place, et dey supper an' run 'way at night an' afte' dey was kotched, dey tol' us dat when dey was passin' through de woods dat night a great big gran'daddy owl flopped his wings an' Joe said 'we'd better turn back.' I allus heard hit was bad luck fer to hear a owl floppin' lack dat, but Green said 'twant nothin', jes a old owl floppin', but he jes naturally flopped diffrunt dat night, an' Green walked on 'bout 15 steps an' somebody shot him dead."[41]

Fishing had provided a supplemental source of high-quality protein for the Delta's Native American population, and the Euro-American settlers to the region soon learned to utilize this abundant resource, especially during hard times. In 1868, Thomas Gale of the Abydon Plantation in Yazoo County could comfort his concerned daughter with news that "Edward just now is engaged in fishing in the Lake below Ewings[;] he went down in the skiff with More & Alec. a few days ago and caught over 100 Perch and Bass [Centrarchidae], very fine[;] so you see we live splendid." In another letter, Gale maintained that "we have fine Fish for the catching" while "of course I do not Fish myself."[42] In the absence of traditional European game fish of the family Salmonidae, fishing in the Delta—unlike hunting—came to be conceived as a plain subsistence activity and therefore most suitable for young boys, poorer whites, and blacks. During the nineteenth century, the Mississippi supported a large group of professional fishermen on the western edge of the floodplain who stayed on the river year-round "catching huge '*catfish*' that often weigh 200 to 300 pounds, as well as '*buffalofish*' and other varieties, which are marketed directly or processed for oil."[43]

Cotton agriculture in the Delta soon diminished the canebrakes, as the destruction of cane by cutting and burning was considerably easier than clearing the forest for cultivation. Among the first tasks for the overseer A. J. McNeill of the Panther Burn Plantation in Washington County in 1859 was to supervise clearing of the canebrakes for cotton fields. In addition to clearing land and building houses and fences, slaves were kept busy cutting cordwood for the plantation's prized steam engine. By March 28, 1859, enough land had been improved that McNeill had "comenced to plant cotton on Panther Burn." McNeill noted that the planting was the first ever on the land and wished that Panther Burn would "always be a lucky place to make cotton." On the warm and clear September 12, picking began as the rich soil of the former canebrake yielded its first cotton crop.[44] Paul C. Cameron of Tunica County had similarly kept all his "hands engaged in cutting cane, bolting timber and building cabins" and could report "cotton stalks higher than my head on horse back."[45]

Cane was not only an indicator of good soil and a hindrance to cotton and corn cultivation, it furthermore provided nutritious fodder for the settlers' livestock that roamed freely around the plantations. The first fences in the Delta were erected to keep the animals from entering the fields, not to enclose them in pasture lands. "We have no manner of enclosure around our house, and the hogs from all Blantonia rendesvous here," complained an early settler.[46] By the late nineteenth century, "[t]here were lots of them all through those woods—domesticated hogs that had strayed off from home and gone wild and raised young. All of them were big, fat, and very, very dangerous. They would come . . . into the corn and do lots of damage." Feral swine joined deer in providing nineteenth-century Deltans with a constant supply of meat: "When they wanted to kill a hog, they just went to the woods and shot what they wanted to dress in one day, for the woods were full of hogs."[47]

A Michigan visitor to the antebellum Delta noted that "cane affords pasturage for the cattle in the winter. The planter raises no grasses, no clovers. What little fodder he needs is supplied by the blades of corn his negroes pick from the corn stalks, and the corn 'shucks' which he feeds his cows."[48] In addition, cane gave "milk and butter an agreeable taste."[49] Continuous clearing and grazing, however, was beginning to decimate canebrakes in the vicinity of new settlements: "[t]he cane [still] grows in luxuriance all through the woods; but the cattle and deer feed it down."[50]

Amanda Dougherty Worthington, living on a Washington County plantation, marveled at the productivity of the Delta soil for cotton and corn. Home gardening, however, was not easy on buckshot soils, but planted strawberries seemed to thrive in the bottomlands. Unfortunately, so did the indigenous mosquitoes: Worthington's mosquito-bitten baby was "as badly marked as one just recovering from the small-pox." Altogether, the Delta would have been "the most delightful country in the world if it were not for the months of June, July, August & Sep."[51] As a rule, the Delta was considered "verry unhealthy . . . in the summer season."[52]

Malaria and yellow fever were the most feared diseases associated with the bottomlands; miasma rising from the swamps was blamed for their occurrence. Only in 1898 did it become clear that malaria is caused by various parasites of the genus *Plasmodium* transmitted by *Anopheles* mosquitoes. The transmittance of yellow fever, an arbovirus of African origin, similarly involves an insect vector, the *Aedes aegypti* mosquito. This species of mosquito typically breeds in puddles and containers holding rainwater and shows disdain for swamps and lakes as breeding grounds. Yellow fever would rage in urban settings because the population concentration sustained the epidemic; the disease made only occasional forays into rural areas. Although Greenville lost around four hundred of its inhabitants to "Yellow Jack" in 1878, most of the Delta settlements did not experience epidemics comparable to this devastation or the ones that killed over 10 percent of the populations of Memphis, Tennessee, and Holly Springs, Mississippi, during the 1870s. Malarial chills and fevers, on the other hand, were all-too-common symptoms of a *Plasmodium* infection among the nineteenth-century Delta inhabitants; quinine, the only effective medicine known at the time, could be afforded only by the wealthiest settlers. "Swamp fever" was brought under control only in the mid–twentieth century with screened housing and new insecticides.[53]

The diseases affected everyone, so prospective planters had to include increased mortality among the valuable slave force in their cost analyses. Scouting for new cotton lands in 1833, James H. Ruffin reported that "[i]t is calculated that in the county of Yazoo lying between the Yazoo and Big Black river, the negroes die off every few years, though it is said that in that time each hand also makes enough to buy two more in his place."[54] Other epidemics, such as measles outbreaks, could furthermore impair the planters' valuable work force, as explained by Wade Hampton III: "There have been two deaths since my arrival: one a fine young man, and the other the youngest child of Lizzy, making in *all 37 deaths, this year*. I am greatly distressed at this mortality."[55] The bottomlands were unhealthy not only for humans. In 1857, the chickens, hens, turkeys, and ducks at the Belle Air plantation in Washington County were almost wiped out by an epidemic, which left the plantation mistress "disgusted with poultry" and eggs after seeing "too much of the sick chickens."[56]

Although the transmittance of malaria and yellow fever was not clearly understood at the time, the danger of infestation was very real in the bottomlands where mosquitoes abounded: "[t]he musquitoes in the valley annoy one very much, morning and evening, during the summer. At these hours there is no relief from them, unless you are enveloped, like Jupiter, in a cloud of your own creating."[57] Nevertheless, the impression of the Delta as "damp and unhealthy" was "fast losing ground, and the cotton planters, deserting the rolling uplands," were "fast pouring in upon the 'swamp.'" The reason was, of course, the soil, which possessed "all the strength of the prairie lands, without their sticky, adhesive and corrosive nature."[58] By the 1850s, it had become common knowledge that the Delta possessed "the richest soil of the South. . . . [T]hese valleys that have produced a rich crop of cotton, year after year, for more than half a century, are as fertile to-day, and yield as large a crop of cotton, without fertilizing, as they did when first cultivated."[59] In a family letter written in 1855, Wade Hampton II could report that "[o]ur crops are good, beyond all description, my cotton is now yielding upwards of two bales to the acre, notwithstanding the loss by rain. I have eight hundred and fifty bags picked out, and at least one thousand in the fields, now open."[60] The pine hill region of Mississippi posed no competition to the Delta in attracting settlers, as explained by F. Cuming: "On approaching the pine woods, the fertility of the soil ceases, but the climate becomes much more salubrious—that will however never draw inhabitants to it while a foot of cane brake land or river bottom remains to be settled. . . . The river bottom lands generally yield from eighteen hundred to two thousand pounds [of cotton] to the acre, the uplands about a thousand."[61] The unhealthy aspects of bottomland habitation were customarily downplayed, and it was no longer "considered quite an adventure to remove, and an exhibition of great heroism to reside [in the bottomlands along the Mississippi]. Now many families have made it their permanent residence; many who were to the *manor* born, as well as in the more humble walks of life; and they have beautified the wilderness."[62] By the mid-1850s, a traveler to the region could accordingly testify that "[t]he land along this beautiful river [the Yazoo] has improved incomparably in agriculture and wealth during the last few years."[63]

Before the Civil War, settlements in the Delta continued to cluster along the Mississippi riverfront and the Yazoo adjacent to the hill line in the east. Still, a small number of settlers had penetrated the forested interior via the smaller watercourses. By 1860, the total population of the five Delta counties bordering the Mississippi (Tunica, Coahoma, Bolivar, Washington, and Issaquena) amounted to 44,953, of which 39,357 were blacks. Sunflower County in the center of the floodplain was settled by 5,019 inhabitants, of whom only 1,102 were white. The population density of the core counties was thirteen inhabitants per square mile; the border counties, twenty-one.[64] (See table 1.)

In addition to claiming natural and Indian-made clearings, new land for the expanding agricultural economy in the antebellum South was obtained by clearing the forested areas. The colonists had promptly adapted from Native Americans the use of fire and girdling as means to transform forest to fields.[65] Combined with the settlers' sophisticated array of tools and craving for productive acreage, forest clearing resulted in a phenomenal alteration of the southern forest during the eighteenth and nineteenth centuries. By the mid–nineteenth century, results could be seen even in the more remote areas of the Old Southwest. In 1842, James Kirke Paulding noticed that along the lower Mississippi,

> [t]he more recent plantations are invariably indicated by masses of dead trees, presenting an abrupt and disagreeable transition from the rich verdure of the living forest, to the dreary aspect of decay and ruin. Here grow the fields of cotton and corn in all their primitive luxuriance, on a soil of unequaled fertility, unparalleled by any region of equal extent on the face of the globe. The people of the United States have been reproached with their indifference, or rather

*Table 1.* Population of the Core Counties of the Yazoo-Mississippi Delta, 1840–1930

| County | 1840 | 1850 | 1860 | 1870 | 1880 |
|---|---|---|---|---|---|
| Bolivar | 1,356 | 2,577 | 10,471 | 9,732 | 18,652 |
| Coahoma | 1,290 | 2,780 | 6,606 | 7,144 | 13,568 |
| Humphreys | — | — | — | — | — |
| Issaquena | — | 4,478 | 7,831 | 6,887 | 10,004 |
| Leflore | — | — | — | — | 10,246 |
| Quitman | — | — | — | — | 1,407 |
| Sharkey | — | — | — | — | 6,306 |
| Sunflower | — | 1,102 | 5,019 | 5,015 | 4,661 |
| Tunica | 821 | 1,314 | 4,366 | 5,358 | 8,461 |
| Washington | 7,287 | 8,389 | 15,679 | 14,569 | 25,367 |
| Total | 10,754 | 20,640 | 49,972 | 48,705 | 98,672 |

| County | 1890 | 1900 | 1910 | 1920 | 1930 |
|---|---|---|---|---|---|
| Bolivar | 29,980 | 35,427 | 48,905 | 57,669 | 71,051 |
| Coahoma | 18,342 | 26,293 | 34,217 | 41,511 | 46,327 |
| Humphreys | — | — | — | 19,192 | 24,729 |
| Issaquena | 12,318 | 10,400 | 10,560 | 7,618 | 5,734 |
| Leflore | 16,869 | 23,834 | 36,290 | 37,256 | 53,506 |
| Quitman | 3,286 | 5,435 | 11,593 | 19,861 | 25,304 |
| Sharkey | 8,382 | 12,178 | 15,694 | 14,190 | 13,877 |
| Sunflower | 9,384 | 16,084 | 28,787 | 46,374 | 66,364 |
| Tunica | 12,158 | 16,479 | 18,646 | 20,386 | 21,233 |
| Washington | 40,414 | 49,216 | 48,933 | 51,092 | 54,310 |
| Total | 151,133 | 195,346 | 253,625 | 315,149 | 382,435 |

Sources: *Statistics of the Population of the United States at the Tenth Census*, 67–68 (table 2); *Population, Vol. 1*, 583–84 (table 3).

antipathy to trees. The feeling is hereditary, and arises naturally from the peculiar circumstances in which their forefathers were placed first coming to the new world. Trees were the great obstacle to cultivation, and the first enemies to be conquered. It is the same with pioneers of the new settlements, whose first and indispensable object is to get rid of them in some way or another. The labor of cutting them down, and removing the growth of gigantic trees, such as are only found in primeval forests, would amount to perhaps ten, or sometimes twenty times the original cost of the land itself; and if prepared for the market, the distance is so great, and the quality of timber either for fuel or mechanical purposes so inferior that it would not pay the cost of transportation. The trees are, therefore, killed by girdling and by the application of fire, and thus remain standing till time and elements prostrate them to the earth; and nothing can be more dreary or unsightly, than a new plantation, bristling all over with scraggy dead trees, like a hedgehog.[66]

The basic techniques for felling the trees remained practically unchanged from the arrival of the first white settlers and their slaves to the 1940s, when chainsaws replaced axes and handsaws. When the timber was removed, the stumps were usually left in the ground to rot out. In the spring of 1857, Orville M. Blanton of Washington County was busy enlarging his cultivated acreage by having a part-time overseer from Warren County, "a man by the name of Richardson deadening 600 acres of land for him."[67] For the cultivation of cotton, in comparison to sugar cane, it was not necessary to clear the land completely before planting; the first cotton crops in the Delta were customarily planted on poorly prepared fields where deadened trees could still be standing.[68] As a rule, "the fire did its work" well, "for it burned saplings six inches through to ashes, and whole dead trees burned up till you could hardly see a piece of them."[69] Clearing of the bottomland forests by burning could, however, turn dangerous. Paul C. Cameron's fledgling plantation was almost destroyed by his neighbor's careless application of fire:

All of yesterday we were kept at hard work to save our Houses from being burnt up. My neighbor a mile to the south of me got his place on fire and the wind being pretty strong he lost the controul of it, swept his fencing[,] burnt a large new Cotton Gin[,] his well filled cribs of corn stables and crossed into my place and the fencing we saved by pulling it down and brought the fire to a halt by cutting down the trees as fast as they caught, but for the energy of the overseer and negroes, I should have lost my Houses as did my neighbor *Winston*.[70]

Annie Jacobs similarly remembered a great fire that burned the slave quarters of a Bolivar County plantation and made her think that everyone would "be burned alive like the old martyrs."[71]

Incomplete statistics make it difficult to estimate how much land was cleared in the coterminous United States before 1850, but it is probable that some 113,740,000 acres of land had been "improved" before that date.[72] The overwhelming part of this land was carved out of the eastern forests, with only a small amount coming from natural clearings. All the evidence points to a big upswing in the rate of clearing after 1840; and in any case between 1850 and 1860 the amount cleared rose by a total of 39,705,000 acres, equivalent to approximately one-third of all clearing carried out during the preceding two centuries.[73] Clearing for agricultural reasons had also begun to make significant inroads into the southeastern bottomland hardwood forests of Alabama, Mississippi, Arkansas, and Louisiana. In comparison with tobacco, cotton, and sugarcane, the other crops cultivated in the South did not deplete the soil as extensively. Their cultivation, therefore, did not increase the need for land clearing along the rivers so expansively as did the farming of these three crops.

In the Yazoo-Mississippi Delta, however, the staple crop was cotton, and the bottomlands were rapidly being transformed from forest into agricultural stronghold. The bottomlands were clearly turning into, in the words of William Faulkner, a "doomed wilderness whose edges were being constantly and punily gnawed at by men with plows and axes who feared it because it was wilderness."[74] The Delta consequently provided an outstanding example of the national boom in agricultural clearing. Between 1850 and 1860, the amount of improved land, planted overwhelmingly in cotton, surged on floodplain locations adjacent to waterways. In Faulknerian terms, there was truly "a commodity in the land now which until now had dealt first in Indians: then in acres and sections and boundaries:—an economy: Cotton: a king: omnipotent and omnipresent:

a destiny of which (obvious now) the plow and the axe had been merely the tools."[75] In most Delta counties, acreage in cultivation doubled or even tripled during the 1850s; in Tunica County, the increase was almost five-fold. Concurrent with the rapid increase in cotton production, land values swiftly rose—as did the number of slaves working in the newly cleared bottomlands. The 717 bales of cotton produced in Tunica County in 1849 multiplied to over 13,000 within a decade, while the labor force of less than a thousand slaves multiplied over three times. Similarly, Bolivar County's output of cotton increased from less than 5,000 bales to over 33,000 bales during the 1850s, while the number of slaves rose from some two to nine thousand. A ten-fold increase in land values over ten years was not uncommon in the antebellum Delta.[76] On the threshold of the Civil War, the floodplain was seriously beginning to challenge the standing of the Natchez region as the future center of cotton culture in Mississippi. The war, however, was to put an abrupt—if temporary—end to these and many other aspirations of the Delta planters.

The Civil War marked a watershed in southern and Delta agriculture in many ways.[77] The year 1865 witnessed the devastation of the old economy in the South as some four billion dollars worth of slave property ceased to exist, the value of farm property declined by almost three quarters, and stocks, bonds, and Confederate currency became more or less worthless.[78] Cotton cultivation had collapsed, as agricultural activities had by necessity concentrated on the production of foodstuffs. Much of the southern transportation network was in ruins, and private capital proved insufficient to repair the damages. Not surprisingly, deficit funding by the southern states continued and even expanded from 1865 to the late 1870s during the Reconstruction.

## Agriculture in a New Era

After the Civil War, farm prices in the South evaporated and nonfarm prices, such as the cost of credit, rose dramatically. This trend, evident in the scarcity of cash and credit, continued through the rest of the nineteenth century and contributed greatly to the collapse of yeoman-style subsistence farming and to the rise of sharecropping and the crop lien system in the South. The crop lien was a mortgage on a future crop, the only asset a southern farmer had after the Civil War. One of the foremost historians of the New South, C. Vann Woodward, has called the crop lien arrangement "one of the strangest contractual relationships in the history of finance: the seeker of credit usually pledged an unplanted crop to pay for a loan of unstipulated amount at a rate of interest to be determined by the creditor."[79]

Many historians have blamed the crop lien system for the spread of monocultures in the postwar South.[80] The monopoly held by the lien merchants enabled them to control their customers in many ways. By forcing their tenants to plant just one crop (usually cotton) instead of different food crops, the merchants simultaneously acquired an easily marketable crop and increased their business, as the tenant farmers had to purchase their food from the merchant instead of raising their own. In most regions of the postwar South, such practices resulted in a tremendous increase in cotton production and a sharp decline in the production of foodstuffs.

In addition to illustrating the dire economic situation of the post–Civil War South, the crop-sharing and lien-financing systems reflected the limitations of the agricultural technology of the day. Before the large-scale mechanization of agriculture in the mid–twentieth century, a large labor force of unskilled workers was a prerequisite for the successful cultivation of cotton and many other crops. The war had ended slavery, but ownership of large entailments persisted in the southern states. In fact, in 1880, roughly half of all southern plantations were owned by the same families as in 1865.[81]

In December 1868, Wade Hampton III had to petition for voluntary bankruptcy for the first time. In a resigned letter written in 1875, the fallen magnate fostered some hope for saving the remains of his Delta property, but had to confess that "I do not greatly fancy the idea of becoming an overseer. Nothing can be made, however, unless I remain on the place, and after all, it will make little difference in the long run where the rest of my life is spent."[82] The following year and the end of the Reconstruction government in South Carolina, however, reversed Hampton's fortunes once again; in 1876 the former Confederate general was elected governor of his home state for the first time. Despite his later duties as a railroad commissioner and U.S. Senator from South Carolina, Hampton regularly returned to his plantation in Issaquena County for bear hunts.[83]

Lacking cash and unable to obtain enough credit for agricultural operations, many large landowners in the South and the Delta leased their holdings to investors arriving from the victorious North. For example, the Tokeba plantation, situated two and one-half miles above Yazoo City and containing nine hundred acres of "open land," was rented out for "seven dollars per acre per annum for a term of three years, one-half of the annual rental to be paid in advance."[84] There were also other, more unorthodox approaches to plantation management in Reconstruction Mississippi. Prior to the Civil War, Joseph E. Davis owned some four thousand acres of Mississippi bottomlands on his Hurricane and Brierfield plantations at Davis Bend (today Davis Island) in Warren County, just south of the Delta proper. Developments on these two plantations in 1867 offer interesting insights into the general problems faced by Delta landowners and freedmen after 1865, even if some elements of plantation life on Hurricane and Brierfield were far from typical.

Not surprisingly, as the brother of Jefferson Davis, the president of the ill-fated Confederate States of America, Joseph Davis seems to have experienced considerable harassment from the occupying Union forces during and after the Civil War. Consequently, Davis filed numerous complaints against the Freedmen's Bureau claiming that his plantations had been wrongfully confiscated and his former slaves evicted for no reason. As the pursuit of agricultural operations by the original owner seemed impossible in the new situation, Davis made elaborate schemes with his ex-slave, Benjamin Montgomery, to keep the freedmen on the land as tenants. In 1866, Davis even arranged for the sale of the two plantations to his articulate companion, who headed a colony of freedmen.[85]

Assuming leadership at the plantations, Montgomery faced enormous difficulties "[w]ith the necessity of a considerable outlay for the repairs of levees, . . . & a constant downward tendency in price of cotton." The war had largely destroyed the existing levee system, and in the spring of 1867 a flood delayed planting and destroyed many buildings. Maintaining the labor force necessary for operations proved difficult because of "the persuasion of various parties anxious to secure labor elsewhere." All this "rendered the impossibility of meeting the payments as they respectively fall due so plain to any one capable of thought, that it seemed to me nothing more than my duty to ask what must be done. A duty that I considered to you[r]self and to your children as well as to those of my own, who, with myself manifest much concern for our condition, present & future[.] Hope of success being so necessary a stimulant to man in his every engage must, that it is not reasonable to suppose that any person or class of person can or will use the required perseverence without it." [86]

Despite the hardships, Montgomery was able to report in mid-November that the weather had been favorable for cotton harvesting and "so far the time has been very well used in picking the little we have." The freedmen had shipped over three hundred bales but had "made but little corn comparatively, though considerab[ly] was planted. The expense being heavy we felt bound to do all we could in the way of making cotton." Still, Montgomery and his group were able to hang on while some other operators had to quit because of the "[h]eavy expenses and short crops." In a letter written to Davis on December 11, Montgomery noted dire financial conditions and labor shortages faced by all plantation operators in the region. As for the next year's prospects and tenants' terms, Montgomery confessed: "I am undecided as to what rent I shall be able to obtain for acre. It will have to be arranged according to locality[, and] the price cannot be high." [87]

On December 19, Montgomery had to admit that "[a]ll that I have made the present and past year goes to meet the liabilities which is insufficient even to accomplish that object." He was not alone, as "[c]omplaint[s] of failures" were "general alike o[n] hill and river plantations," and laborers in the region had "not received a dollar as yet and [all] but very few are likely to being also in debt for a part of their necessary supplies." Asserting his "willingness to do as near right as the teachings of reason[,] justice and experience dictates[,]" Montgomery accused a neighboring planter of asking too high a rent from his tenants and feared that if such charges were adopted, "less land will be cultivated than other wise might be." Planters had to realize that though "the present depreciation in the price of land is temporary it nevertheless does exist and the most wise cannot see the end of it. Some little time must elapse before recovery can reasonably be expected." As for the Hurricane and Brierfield plantations, he remained "unable to say to the people what the land can be rented at." By the end of the month, Montgomery was able to inform the freedmen that their lands would "be rented the ensuing year at $4 to $6 per acre[, the] rates varying according to locality, that is to say[,] for the lowlands, sloughs etc. that are likely to be lost by water[,] the charge will be less." In renting such land Montgomery tried "to induce the lessees to ditch in order to secure as much of the slough as possible."

As on other nearby plantations, some tenants were dissatisfied with the rental terms and proposed "looking out for more favorable arrangements." Advising freedmen of a neighboring plantation against leaving, Montgomery noted that "[i]f they will remain together and work[,] they can make something as well as save the place from going to distruction." "Feeling satisfied of having made just offers and dealt fairly with the people[,]" Montgomery "entertain[ed] no fear of the result." Besides, "Gen[eral] Gillum's order concerning vagrants" had been read to the freedmen on the plantations. The only difficulty Montgomery feared for the year 1868 was "that of obtaining necessary supplies until we can raise something."[88]

In addition to comprehending the financial aspects of running a plantation, the energetic Montgomery showed a surprising degree of ecological understanding by the standards of 1867, suggesting the use of birds as a form of pest control in farming: "As I have before said the abundance of distructive insects may be attributable to the scarcity of birds, and if such a supposition is at all consistant with the views of others of more judgement & experience than myself in such matters, I do hope that some steps may be taken to extend some protection to them (the birds) which when more abundant the insects were less troublesome. Even now, the small portions of the fields that are subject to the range of chickens, are not troubled by the insects."[89] Despite the attempts at biological pest control, the idealistic experiment on the Hurricane and Brierfield plantations collapsed for lack of funds during the 1870s, and most of the inhabitants—led by Benjamin Montgomery's son, Isaiah—moved deep into the Delta bottomlands in pursuit of land of their own.[90] Montgomery's followers constituted only a trickle of the stream of new settlers to the Delta, black and white, who hoped to transform the bottomland hardwood forests into a setting for the Jeffersonian society of small, independent yeoman farmers. In 1868, a Yazoo City planter noted that "[N]othing now interests Black or white Folks but Cotton buying and renting land &c."[91]

At the termination of the Civil War, the bottomland hardwood forest had reclaimed much of the cleared area in the Delta, and cane once again covered the fertile ridges. The planters' emancipated work force had scattered, and the prospects for economic development were meager, as explained later by the Mississippi governor elected in 1865, Benjamin Grubb Humphreys: "The soil, rich and productive belonged to the debt incumbered white race, but was utterly worthless without regulated labor."[92] Land forfeitures due to unpaid taxes became common, while conflicting titles effectively hampered land sales for the next fifteen years.[93]

In the summer of 1866, W. I. Hindman, a logging crew foreman working in the Delta, notified his Natchez employer of his backup plan in the tumultuous postwar times: "I are going to make a little cotton an[d] corn whare I live next year if I get a little assistens so if thar did any thing Happin[,] I wood have som thing to fal[l] back own but my farming will not interfer[e] with my Rafting buisness. . . . I suppose there is money in the buisness but it wood be a new buisness to me an[d] wo[ul]d take me som[e] time to get acquainted with[.] It is hard to tell what to dew for I think that war will be own this continent a gane." Still, it seemed to the logger that "this time ther[e] wil[l be] som[e] thing for the South to fight for."[94] In December, he further informed his employer that he was "a bout to get som[e] [darkeys?] to gou own the tract of land to clear up an[d] build." The foreman wanted "to get it in a condition to rent for that will be the best Buysness a [going?] now." Hindman accurately anticipated that, within a short time, the Delta would "be som[e]what like Ireland[:] ther[e] will be lords an[d] tenants."[95]

With the demise of slavery, the Delta was ready for the entrance of the small farmer, and a great influx from the other parts of the state to the region took place during the 1870s. Not surprisingly, the rich bottomlands of the Yazoo-Mississippi Delta drew not only ambitious whites, but also a considerable number of freedmen in hope of better life. The black population in the Delta increased over 60 percent between 1860 and 1880; the white population almost doubled but remained a clear minority. Accordingly, the population density in the core counties rose to over twenty inhabitants per square mile.[96]

During the Civil War, close to eight hundred thousand acres of abandoned and confiscated land in the South had been amassed under the administration of the Freedman's Bureau. Much of this acreage had been cultivated by blacks under the impression that after the war they would be able to own and cultivate their own farmsteads. "Forty acres and a mule" was an aspiration for most freed slaves in the South, and the Delta was no exception. The dream, however, seldom materialized, as explained by Berry Smith: "I hear'd a heap o' talk 'bout ever' Nigger gittin forty acres an' a mule. Dey had us fooled up 'bout it, but I never seen nobody git nothin'."[97] Even those lucky enough to obtain fertile land in the South soon learned that carving a farmstead out of pristine bottomland forest called for economic means that most freedmen lacked. As explained by Isaac Stier, "[d]ey promised us a mule an' forty acres o' land. Us ain't seen no mule yet. Us got de lan' all right, but twant no service."[98] James Lucas offered a more elaborate explanation for the failure of many black farmers:

> Dey gimme 160 acres of lan', but twant no 'count. It . . . was low an' swampy. Twant yo' lan' to keep lessen you lived on it. You had to clear it, dreen it, an' put a house on it.
>
> How I gwine a-dreen an' clear a lot o' lan' wid nothin' to do it wid? Reckon somebody livin on my lan' now.[99]

Battling the overwhelming odds posed by poverty, lack of education, and institutionalized racism, a number of blacks were still able to acquire land in the post–Civil War South, including in the Delta. Despite the overall failure of the federal government to enact viable resettlement and purchase plans for the former slaves, the dream of an independent black farmer class endured for a few decades. Black land ownership in the South peaked in 1910 with more than fifteen million acres in the possession of freedmen and their descendants. There were some 175,000 farms fully owned by southern blacks at the time, their size averaging between ten and twenty acres. Even these modest holdings usually had been obtained in parcels. The vast majority of black farm operators, however, had become tenants participating in the debt peonage system created by sharecropping.[100]

The end of legal bondage forced the rural labor system in the South to undergo drastic changes, as former slaves now had to be financially compensated for their contribution of agricultural labor. In a situation in which former slaves and landless whites wanted access to land, landowners needed laborers but lacked money for wages, and financial institutions gave inadequate farm-production credit, tenancy emerged.[101] Landowners now allowed their workers to farm small plots of ten to forty acres on a crop-sharing basis. Tenants bought food and other necessities during the crop season with credit secured by a lien on their portion of the future crop. During the years 1865 and 1866, laws were passed in all southern states authorizing merchants to provide farmers with supplies necessary for growing the year's crop in return for a portion of the future crop. Larger landowners, however, often came to furnish their tenants directly through plantation commissaries.

The investor, usually a cotton factor or broker but sometimes a bank, advanced money to the planter in return for a mortgage on an expected crop and its marketing rights. The planter, who typically could not afford wage labor, contracted with his farm workers or tenants to farm in shares. According to an 1876 U.S. Department of Agriculture report, the Delta proprietors in Washington County typically put "the place in order" and furnished everything, while the laborer had "the garden and house rent free" and got one-half of all he produced.[102] From his crop receipts, the sharecropper then had to deduct the living expenses charged to his account at the plantation store, where the prices were always higher than in town.

Some landless farmers were able to operate as independent cash renters, but the vast majority became share tenants or sharecroppers. A bad crop, however, could easily turn a cash renter into a share tenant or sharecropper. The difference between share tenants and sharecroppers lay in the farmers' reliance on the landlord and/or the creditor. Share tenants typically owned mules and other equipment and could even supply some of the needed seed. Depending on their needs for furnishings, they could receive as much as two-thirds or three-fourths of the crop. From this amount, the cost of furnishings and interest were then deducted. Sharecroppers typically contributed only labor and depended on lien credit for their living necessities. Whereas some share tenants could make independent decisions concerning their farming practices, sharecroppers worked under closer supervision of the landlord. During the Christmas season, sharecroppers made their year-to-year verbal contracts, most often limiting their share to 50 percent of the crop grown, advances and interest to be deducted.[103]

Crop lien merchants have been widely blamed for using the system to exploit farmers: merchants often held a local monopoly, which they capitalized on ruthlessly since the farmers had no option to buy supplies or market their produce elsewhere. In addition to high interest rates, profits were increased by the widespread use of a two-price system, in which the credit price could be over 50 percent higher than the cash price. Dishonest weighing and measuring practices were common, and customers were often incapable of examining the account book in which their debts were recorded. Exploitation of tenants was built into the system, as the landlord, troubled by mortgages, taxes, production costs, and falling crop prices, could hardly profit without getting as large a portion as possible from the tenants' shares. Furthermore, the superior position of the landowner, obtained by extending credit and evidenced by close supervision, effectively robbed the tenants of incentives to develop competency in their farming practices. Mounting indebtedness, combined with easily terminable agreements on sharecropping, hardly encouraged stewardship of the land on the part of the tenants.

Falling commodity prices forced the landlord to respond by increasing the cultivated acreage in order to secure a crop lien comparable with years past, thus creating a seemingly inescapable dilemma of increased production depressing prices. The result was a culture of rural poverty, as tenants received some of the lowest incomes in the United States. In some years, they received no net income, as their share of the crop had been consumed at settlement time by the furnishings debts, which had carried exorbitant interest rates. In many cases, sharecroppers sank into a state of peonage, bound to the land through mounting debts. Even the landowner could not escape the problems of the sharecropping and crop lien systems, often ending the crop year himself more deeply at debt than in the beginning. Not surprisingly, the loudest opponents of the crop lien system during the late nineteenth century were the large landowners or planters who felt that the merchants were infringing on their power over land and tenants, as the investor's power to refuse credit came to determine which tenants remained on the land. The system, of course, provided the investor considerable profits from the crop grown on the planter's land, and sometimes even allowed him to drive out an inefficient planter and acquire the plantation.

On many plantations, tenancy came to resemble slavery: tenants lived in shacks provided by the landowner, received their supplies on credit from the plantation commissary, and worked the fields from dawn till dusk under the supervision of a riding foreman, often an armed sheriff's deputy. The antebellum slave plantation had transformed into a so-called fragmented plantation, where the new status of agricultural workers resulted in a changed plantation occupancy pattern. Instead of living in the centrally grouped slave quarters of the antebellum era, the labor force was now housed in cabins on the subdivided tenant farms. The extremely modest habitations, situated in the middle of the fields, often lacked basic necessities such as glass windows and wells. Often possessing no livestock or gardens, many sharecroppers seldom consumed milk, eggs, or fresh meat and vegetables, and instead relied on a diet of salt pork and flour obtained on credit from the plantation store. Cash renters' diet was often better, as they could support a few pigs and chickens and tend a vegetable garden. Malnutrition and primitive living conditions made chronic illnesses such as pellagra, hookworm infection, and malaria common among tenants. Sharecroppers' children attended school only sporadically, as public schools were customarily closed for weeks during the cotton harvest.

The tenants' lack of education, manifested by widespread illiteracy and ignorance of even rudimentary mathematics, made them extremely vulnerable to cheating by both the storekeeper at the plantation store and by the landlord on the settlement day for the year's crop. Faced with a loss during a given year, merchants and landlords could always raise prices at the plantation store or manipulate the tenants' accounts. Disputes between the bookkeeper and the tenant were common, and southern folklore abounds with stories of the manipulation of tenants' accounts. Because of an unwritten law dictating that a black could never question the word of a white man without subjecting himself to a great danger, cheated black tenants could not take legal recourse. White tenants could challenge the landlord but seldom did, as the odds were overwhelmingly stacked against them. A sharecropper in debt had only two options: try harder the next year or leave. Many chose to leave, hoping to locate a kinder landlord who would offer better terms and a bigger piece of land to farm. The search was often in vain, and many tenants were forced to move again after a few years in a new place. Naturally, from the landlords' point of view, the situation was different. According to a postbellum Mississippi planter, "the great problem in the South to-day was the labor question. The negroes could not be depended upon."[104]

An unorthodox solution to the labor shortage problem was provided by the Delta planter Edmund Richardson, who in 1868 struck a deal with the federal authorities to employ black felons for the grueling work of land clearing, levee building, plowing, and hoeing on his vast holdings. In return for providing accommodations for approximately 150 prisoners, Richardson received eighteen thousand dollars per year from the state and all the profits derived from their labor. During the late nineteenth century, the abusive system of convict leasing became firmly established in Mississippi while the number of black prisoners proliferated. The exclusive right to lease state convicts, including children, guaranteed large profits for its holder and became Mississippi's most coveted political prize. In the 1870s Colonel Jones S. Hamilton, a wealthy Jackson businessman, began subleasing his prisoners to plantation and railroad interests around the state and made a fortune in the process.[105]

Convict leasing obviously could not match the growing demand for labor in the Delta, and many large landowners harbored hopes about the replacement of black labor altogether by ethnic groups viewed as more industrious and reliable. Chinese laborers were brought to the South from

California and the Caribbean. On the Yazoo-Mississippi floodplain, as elsewhere in the region, the Chinese newcomers did not supersede the black sharecroppers and convicts as the labor force for cotton cultivators. Instead, they found an economic niche in the Delta as small independent merchants, operating small grocery and supply stores and providing goods and services, mainly to blacks. At the same time, they were careful not be identified with their clientele and formed a small but distinct social class between blacks and whites. In a similar attempt to replace black farm workers with more desirable people, Italians were recruited in the 1890s to work the cotton fields in Arkansas bottomlands. The experiment on the Sunnyside Plantation, owned partly by the prominent Delta planter LeRoy Percy, soon failed. Italian authorities were informed about the miserable working conditions of the imported workers and a diplomatic conflict developed between the United States and Italy.[106]

The large tenant force on cotton plantations was supplemented by seasonal workers brought in for the fall harvest. Some of these would stay and become permanent residents, taking the place of disillusioned tenants leaving the plantation in search of a better sharecropping contract. While many sharecroppers moved from one plantation to another within the Delta, sometimes only a few miles apart, some would travel considerable distances in search of a better arrangement. The best prospect for a sharecropper was to find an honest landlord and work hard. Provided the year was good for cotton, he would make some money in the fall and survive until the next season by doing odd jobs around the plantation. This modest dream drew thousands of blacks from other parts of Mississippi to the Delta to try their luck as sharecroppers. The pattern of movement into the Delta was not, however, a random activity: a network of friends and relatives distributed information on the character of Delta landlords and their holdings for potential tenants. For example, a large number of settlers, both black and white, arrived in the vicinity of Drew in Sunflower County from the Crystal Springs area in Copiah County of southwestern Mississippi.[107] Migrants from other parts of Mississippi could, however, find the conditions among Delta cotton cultivators too harsh to endure. In the 1920s, Berry Smith "picked cotton in de Delta awhile, but de folks, white an' black, is too hard. Dey don't care 'bout nothin!"[108]

In 1890, approximately 84 percent of the farms in the ten core counties of the Delta were operated by tenant farmers. By 1900, the percentage had risen close to 90 percent, and by 1910, over 92 percent of the Delta farms were operated by tenants, in plots averaging only twenty-three acres. Over three-fourths of the crop land in the core counties of the Delta was devoted to cotton cultivation, and the fertile soil was worked almost exclusively by black tenant farmers. Blacks made up almost 80 percent of the rural population, and over 95 percent of the tenants were black. The level of tenancy attained in the Delta stood in great contrast to the national situation: 37 percent of all American farmers were tenants in 1910, and even in the state of Mississippi they constituted "only" two-thirds. Cash and share renters in the Delta typically had close to thirty acres in cultivation, while sharecroppers averaged only 19 acres. Almost all the land in the small tenant farms was under cultivation, in comparison to farms operated by owners. The average size of independent farms was considerably bigger: 123 acres, of which less than half was improved. The dream of the independent black farmer had rapidly faded in the Delta; by 1910 only 5.5 percent of the black farmers in the core counties owned the land they worked.[109]

The extremely high rate of sharecropping arrangements in the Delta was easily explainable. As summed up in a 1916 U.S. Department of Agriculture report, the Delta landlord gained "a great deal more by higher yields in the case of share croppers" than he did in the case of other tenants. The sharecroppers, on the other hand, were "not so greatly benefited by a higher yield as [were] the other types of tenants."[110] In 1929—before the initiation of cotton acreage adjustment programs by the federal government—close to 80 percent of the cropland in the core counties of the Delta was still devoted to cotton. While the amount of farmland operated by tenants had fallen to 62 percent by the beginning of the 1930s, the proportion of tenant farmers of all farmers remained well over 90 percent between 1910 and 1935. Furthermore, almost 70 percent of the farmland in the core counties was in plantations of at least four hundred acres.[111]

The decline in black land ownership all over the South accelerated after 1910, and a growing number of independent farmers were forced to join the ranks of tenants. Many chose to leave the land altogether and became a part of the exodus to the North between 1910 and 1960, when close to five million blacks left the region.[112] The Delta similarly experienced a massive black out-migration in the years following World War I. The shattering of the black dream of a good life in the New South after the turn of the century had found a cultural expression in the development of a new musical form, the blues. Reflecting the dissatisfaction of black southerners, the blues came almost entirely from the existing black musical tradition, with only minimal elements from formal Western music included. A secular music, early blues was largely based on field hollers, improvised work songs sung by manual laborers working in the cotton fields and in logging and levee construction. Remarkably, much of the early blues emerged among the agricultural laborers of the Yazoo-Mississippi Delta, as evidenced by the places of birth and residence of recorded blues singers of the early twentieth century.[113]

Although the most common theme in the blues from the beginning was the relationship between a man and a woman, most aspects of the black experience were commented on by its practitioners. Many themes reflected recurring events in the lives of the blues performers. In addition to themes such as illness, crime, and migration, a number of early recorded blues commented on issues germane to environmental history. Considering the overpowering presence of cotton cultivation in the lives of so many early blues performers, surprisingly few blues on the subject were recorded. Still, natural phenomena from floods and droughts to agricultural pests were among the topics addressed by blues recordings of the 1920s and 1930s.[114]

In 1880, the German traveler Ernst von Hesse-Wartegg pointed out that, despite the influx of settlers, much of the Delta remained "primeval forest, bottomland, and in the lowest parts, swamps. Americans call it '*virgin soil*,' inhospitable and uninhabited today." The German aristocrat was not alone in comprehending the vast potential of the entire Delta to become "a productive and populated agricultural district."[115] Already in 1873, the *Greenwood Times* had published a marketing plan with the aim to attract as many "hard working, good, and honest people" from the North to the Delta as possible, "the small farmers especially."[116] Calls for railroads, capital, and immigration intensified during the 1870s and raised hopes of higher property values.[117] According to an overly optimistic Greenwood newspaper editor, land values in Leflore County, estimated at $15 to $25 per acre in 1873, were expected to rise to around $100 within five years with the best lands reaching $300 within a decade![118]

By the late 1870s, some semblance of order had been brought to the land title question in the Delta, and large land sales resulted, though not targeting the small farmers nor securing the high prices anticipated only a few years earlier. In 1881, approximately two million acres of the region's bottomlands were transferred to the first real Delta railroad, the Louisville, New Orleans & Texas Railroad. In another transaction some five years later, the Delta and Pine Land Company obtained some six hundred thousand acres originally sold to the Georgia Pacific Railroad, and soon began an expansive program for land sales and settlement.[119]

The first north-south railroad line through the Delta, the Louisville, New Orleans & Texas Railroad, was constructed from Memphis to Vicksburg between 1881 and 1884. It was soon consolidated with another company, the Yazoo & Mississippi Valley Railroad, under the name of the latter line. After the Yazoo & Mississippi Valley Railroad became a part of the mighty Illinois Central Railroad Company in 1892, the rail system in the Delta began to expand enormously: mileage rose from 235 miles to 816 miles by 1903. By this time, branch lines of the Illinois Central extended into nearly every county of the Delta. Furthermore, the Columbus and Greenville Railroad had crossed the floodplain from east to west near its middle. By 1910, the thousand-mile mark in railroad mileage had been surpassed.[120] As the railroad system developed, river traffic for personal transportation and for shipment of goods declined in the interior of the Delta. Already by the mid-1880s, traffic on the Yalobusha was more or less restricted to the floating of timber. At the same time, the new railroad towns, such as Clarksdale in Coahoma County, began to supplant riverfront settlements as centers for commerce.[121]

At the turn of the century, lumber companies released large areas of cutover timberlands in the lower Mississippi Valley and developed schemes to attract buyers and transform former bottomland hardwood forests into cotton fields. Myriad pamphlets were produced, advertising "Choice Pickings in the Yazoo Valley" and describing the ease of reclamation and the possibility of harvesting two crops a year in the "Garden Spot of America."[122] The reclamation promoters found a receptive audience in many recent immigrants and second-generation Americans who had settled on marginal lands in the Midwestern states. To many of these home seekers, the undeveloped lowlands along the lower Mississippi seemed the last opportunity to obtain a fertile farmstead; the western frontier was fast coming to a close.[123] Past yellow fever epidemics and malarial infections were largely forgotten by southerners, even by pioneer settlers of the Delta: in 1909, the octogenarian Orville M. Blanton reminisced how the floodplain had changed from "a wilderness of Immense forests & cane breaks" into endless "expanses of Cotton plantations" during his lifetime, attributing his great "age & Physique" to the region's healthiness.[124]

The boosters aimed some of their promotion efforts at the small farmers of the southern pine hill region, but were first and foremost enticed by the good reputation of Midwestern farmers, widely viewed as hard workers and successful cultivators of small plots. Besides, they were of northern and western European stock. Blacks or other "non-Aryans," whether northern or southern, were not included in the targeted pool of potential bottomland pioneers. Working together with railroads, lumber companies wishing to dispose of their logged-over forestlands came up with elaborate schemes to provide transportation for the new settlers. To induce prospective settlers to visit the advertised lands, railroads commonly offered special tickets at low cost. A one-way ticket was known as the "colonist's fare," while the less adventurous had the option of purchasing a round-trip "homeseeker's excursion" ticket.[125]

The promotion of agriculture by railroad companies was by no means restricted to the alluvial lands along the lower Mississippi. Throughout the United States, railroads had immense tracts of land for sale. From the 1850 land grant to the Illinois Central Railroad to 1911, railroads obtained from the public domain some 115.5 million acres—more than the total land area of the seven Atlantic states from New York to Virginia. Originally, the U.S. Congress donated land to individual states, which in turn ceded the land to the companies; later grants were made directly. Land sales for agricultural purposes not only brought in cash but supported the railroads' main business: some 10 percent of the total freight revenue in the United States in 1910 was derived from carrying farm produce.[126]

Land offices advertising the fertility of the Yazoo-Mississippi Delta were established in Chicago and other northern cities. In many cases, the companies cooperated with various city and county commercial interests in the Delta in their promotional efforts. Probably the most influential of the bottomland boosters was the Southern Alluvial Land Association, which devoted much of its efforts to advertise the fertility of the Delta. The association's efforts culminated in the handsome 1919 publication, *The Call of the Alluvial Empire,* which—coinciding with the rapid spread of the cotton boll weevil—advocated the region's suitability for diversified agriculture. Promotion of the Delta bottomlands, however, remained somewhat less sensational than that of the Arkansas and Louisiana swamps. Established settlers in the Delta strove to develop their region as plantations rather than in forty-acre lots.[127]

Nevertheless, there were plenty of prospective bottomland cultivators, both independent farmers and sharecroppers. Population growth in the Delta was consequently phenomenal during the last decades of the nineteenth century. The population in Washington County almost doubled from 25,367 to 49,216 between 1880 and 1900, and similar growth figures were attained in the other Delta counties. Cotton production in the Delta rose accordingly: Washington County tripled its output from 30,362 bales in 1880 to 90,423 bales in 1897. Sharkey County almost matched this growth rate while cotton production in Bolivar County nearly doubled.[128]

As the bottomlands became increasingly cleared, more labor was needed for cotton cultivation; a typical family could hardly handle more than forty acres of the crop. After a successful harvest, it was only natural for an ambitious and enterprising settler to acquire new land with further gains in mind. Nevertheless, the question of labor had to be resolved.

While the large plantations had begun to employ freedmen as sharecroppers, the smaller landowners hired blacks as tenant farmers. A common arrangement was to contract tenants to prepare the cutover timberland for cultivation in return for two or more rent-free crops on the newly cleared land. The great productivity of the soil meant more bales of cotton for the tenant's share of the crop than in the hill country, so the migration of blacks to the Delta intensified during the last decades of the nineteenth century. In Washington and Bolivar Counties, the black population more than doubled between 1880 and 1900.[129]

By the turn of the century, land values in the Delta had begun to reflect the growth and speculation. Not surprisingly, land values in the region were claimed to have at least doubled during the last five years of the nineteenth century. High-quality cultivated land commanded prices from $35 to $50 per acre, while unimproved lands could be obtained "very much cheaper."[130] Between 1880 and 1910, the estimated value of farmland and buildings in the core counties of the Delta more than tripled, averaging from a little over $14 per acre to more than $47 per acre. The real boom, however, was still to come: prices for bottomland acreage along the lower Mississippi were to increase tenfold between the 1880s and the early 1920s. An outstanding example of corporate interest in the rich bottomlands is the evolution of the Delta and Pine Land Company during the 1910s. The company had originally been founded as a speculative venture in 1886, when restrictions on corporate landholdings in Mississippi were nonexistent. Its charter was purchased in 1911 by a group of American and British investors who wanted to retain their newly acquired thirty-eight thousand acres in the Delta without breaking the new state law forbidding landholdings larger than ten thousand acres. One and a half million dollars were subsequently invested in improvements on the company's two Delta plantations in Bolivar and Washington counties.[131]

The planters were not the only investors in the Delta at the turn of the century. Mississippi now abandoned the widely criticized system of convict leasing and founded instead an enormous penal farm in Sunflower County for state prisoners. The twenty-thousand-acre Parchman farm was to provide the convicts with better living conditions while channeling the profits from their labor to the state coffers. In the latter respect, the venture was an undisputable success: the prison–cotton plantation remained highly profitable until the Great Depression and the collapse of farm prices. The prisoners did not fare as well.[132]

According to the 1910 census, the fertile farmland in the core counties of the Delta was valued between $25 and $50 per acre, as compared to the average of only $14 per acre for the entire state of Mississippi. By 1920, the rise in cotton prices, land speculation, and the influx of small farmers had driven the land values close to $200 per acre. The rising land prices were not limited to improved acreage; in the turn-of-the-century Delta, cutover land was obtainable at $10 per acre or less, but by 1920 prices had gone up to $75 to $90 per acre. Reflecting the falling price of cotton, Delta land values began to decrease during the 1920s and were estimated at less than $90 per acre by 1930. The trend continued during the Depression: by 1935, the value of farmland had sunk to 1910 levels.[133]

Postbellum plantations, whether operated by individuals or corporations, played a constantly changing role in the history of land clearing in the Delta. During times of economic prosperity, plantations sought to expand their holdings by purchasing as much of the available bottomland acreage as possible. Small farmers with more limited means were unable to compete with the plantations, which secured vast amounts of both developed and undeveloped land. Downturns in the general economy could, however, reverse these roles. Planters started to downsize their operations, beginning with disposal of their undeveloped acreage. These in turn could be purchased by family farmers, often on the basis of contracts for deferred payments. In many cases, the newly developed farmsteads failed within a few years and returned to plantation ownership.[134]

The autobiography of Mary Hamilton, written and edited in the early 1930s but published only in 1992, offers an intimate glimpse into the everyday life of the turn-of-the-century Delta settlers. Although heavily edited and somewhat romanticized, the autobiography contains detailed and apparently accurate information on the natural environment, lumbering practices, and agricultural development in the region from the rare viewpoint of a white working-class woman.[135]

Hamilton arrived in the Delta in 1896 to accompany her husband, Frank, who worked in the bottomlands as a logger. At the turn of the century, the Hamiltons were able to buy "seventy acres more or less" on the east bank of the Big Sunflower River in Sunflower County, where a "settlement sprang up overnight like a mushroom." Inexperienced in cotton cultivation, the family initially "decided it was best not to farm at all, just have a good garden truck patch and about two acres of corn to feed hogs on," while Frank Hamilton "was making staves out of his own timber and timber he bought here and there." After clearing more land, the family successfully grew cotton and corn as cash crops for a while. Resurvey of the Hamiltons' property lines was, however, demanded by their neighbor. The new survey drew the lines "a good hundred yards below the old government survey at the river, cut through our cultivated land, and took in part of our orchard. We . . . were left with fifty acres." Their place now "practically worthless," the family "knew it was a game of freeze-out, but we had no money to take it to court, and the man with money and power behind him, then as now, rode roughshod over the little fellow. So we sold out to" the neighbor. [136]

Deprived of their livelihood, the family eventually moved to northeastern Arkansas, where "[i]t didn't cost half as much to live as in the Delta." However, Frank Hamilton died as a result of a work-related accident, and Mary and her children returned to the Delta to make their living as tenant farmers in Minot, Sunflower County. For the surviving Hamiltons, like so many others, social mobility in the early-twentieth-century Delta proved an elusive dream: "Our work was steady as clockwork, and by all doing our share we made a good crop; fifteen bales of cotton, four hundred bushels of corn, plenty of syrup and potatoes, and our own meat and lard. This Delta land is so rich, and there was no boll weevil then; but that was 1914 and cotton dropped to five and six cents a pound. So at the end of the year, after we had paid our rent, we had less than one dollar left for our year's work." [137] A more successful settler to the Delta was William "Will" Dockery, who arrived in Cleveland, Bolivar County, in 1888. He began his career as a storekeeper, but soon entered the lumbering business, and, eventually, turned his attention to cotton agriculture. After 1895, he acquired nearly ten thousand acres along the Big Sunflower River and carved an immense plantation out of the heavily forested land. [138]

Tenancy originated as an answer to the problem of getting the former slaves back to work on the land, but during the late nineteenth century it also became the last refuge of a growing number of poor southern whites. Many of these whites had been independent farmers who had lost their farms because of their inability to compete with the cheap labor of the plantations. Between 1865 and 1930, tenancy increased steadily as many independent southern farmers, both black and white, tried in vain to produce enough staple crops to pay for both the supplies and the interest on crop liens. They became increasingly indebted and were forced into the ranks of share tenants and sharecroppers, who in 1930 comprised nearly half of the southern farm population. After the turn of the century, so many of the politically disfranchised blacks had fled the repressive southern countryside for urban areas in the North that by the 1930s most southern sharecroppers were white. The Delta, however, remained a world apart. Despite their massive out-migration from the area, blacks still made up approximately 87 percent of the tenant force in 1934.[139]

By the beginning of the twentieth century, the crop lien system and the primitive farming operations of individual tenants and sharecroppers were proving increasingly disastrous, as cotton production remained economically unviable for its impoverished producers. In many regions, careless methods of cotton cultivation furthermore destroyed much of the topsoil and created a gullied landscape. Because of the immense depth of the topsoil, erosion did not constitute an immediate problem in the Yazoo-Mississippi Delta, but a new danger for southern cotton agriculture emerged after the turn of the century. The cotton boll weevil (*Anthonomus grandis*) had entered the United States from Mexico in 1892 and reached Mississippi in 1907. Cotton fields had been subject to attacks by various pests and disease before; the cotton bollworm (*Helicoverpa zea*) and the budworm (*Heliothis virescens*) had especially damaged crops in varying degrees for decades. Ernst von Hesse-Wartegg, the German visitor to the Delta, offered this competent description of the ravages brought about by these two pests in 1879:

The bolls ripen, turn yellow, and burst into a snowy-white tuft. At that moment the plantation's mortal enemy attacks.

The cotton worm [budworm] wants nothing but the encapsuled cotton and, multiplying fast, can devour the crop in a week. Fortunately, the first hatch appears at a time when, with torches and other fire at night, the planter can hunt down the moths of the adult stage and thereby preclude the second and third hatches. The ally of the cotton worm, the bollworm, can be seen [in the adult stage] flitting about the fields in the summer. Swarms of small and delicate moths hasten from bloom to bloom, laying an egg in each. Larvae appear in three or four days and gobble their way through the boll and into the stem, destroying it.[140]

The boll weevil, however, proved even more destructive than these pests, and could destroy well over half of the crop. The weevil deposited its eggs in the cotton square, and the grubs effectively prevented the development of fiber locks, turning the bolls into empty shells and demolishing the crop. Early planting with fast-maturing varieties could reduce the damage caused by the boll weevil, but many cotton-growing communities panicked on the arrival of the pest.[141]

The boll weevil came to play a great role in black folklore, as evidenced by the eminence of the insect in the blues repertoire. Paradoxically, black sharecroppers could even express a certain sympathy for the pest destroying their crop, presumably because they identified with the boll weevil's endurance under relentless persecution. The following lines from Kokomo Arnold's recorded version of "Bo-Weavil Blues" leave no doubt about the sharecropper's position on the arrival of hard times:

Says I went to my captain, and I asked him for a peck of meal, [twice]
He said, "Leave here Kokomo, you got boll-weevils in yo' field."
. . . . . . . . . . . . . . . . . . . . . . . . . . .
Now Mister Boll-Weevil, if you can talk why don't you tell? [twice]
Say, you got poor Kokomo down in Georgia catching a lot of hell.[142]

The appearance of the boll weevil intensified the plight of the tenant farmer by throwing him out of work, but not out of his debt. When cotton was not planted for fear of the boll weevil, many tenants and sharecroppers had to move to noninfested areas in search of new farmland. Intensifying the hardships encountered by the tenant farmer, the boll weevil greatly contributed to the exodus of blacks to Northern industrial cities, already seen as the "Promised Land" because of the greater economic opportunities and personal freedom there. Furthermore, the arrival of the pest coincided with the evolution of the gasoline tractor, which decreased the need for manual laborers in cotton cultivation: machines now assumed the hard work of breaking the land for planting.[143]

Coping with the boll weevil problem accelerated regional shifts in cotton production. Faith in cotton as a certain source of income was shattered, and the dangers associated with monocultures became more evident. Many southern landowners came to realize the long-term danger presented by the persistent pest to cotton agriculture and became more receptive to new crops. Fertile lands and colder winters could reduce the damage caused by the boll weevil, which made the Yazoo-Mississippi Delta more attractive for cotton growing than many other areas in the South. To hasten the growth of cotton and thereby curb losses caused by the boll weevil as much as possible, Delta planters now began to apply nitrogenous fertilizers to their lands.[144]

Application of calcium arsenate as insecticide curbed some of the losses in the South after 1919, but the increased cost of the prevention of boll weevil infestation accelerated the shift in cotton production to better soils. Large-scale cotton cultivation began to move west, especially to the states of California and Arizona, where the boll weevil presented fewer problems and the farm labor consisted of wage workers. Sea Island cotton had proved extremely vulnerable to the boll weevil, and U.S. production of the variety plummeted from 116,000 bales in 1916 to less than 2,000 bales in 1920. At the same time, cultivation of the Egyptian variety expanded from 7,000 to 100,000 bales on the irrigated lands of Arizona and California.[145] The Yazoo-Mississippi Delta, however, remained a successful enclave for cotton cultivation, and the abandonment of cotton as a cash crop in many other parts of the South emphasized the Delta's position as a refuge for traditional cotton economy.[146]

At the beginning of the twentieth century, the United States had for years been the greatest cotton producer in the world; in 1891 the country had produced some nine million bales of cotton while the second-largest grower, India, amounted to less than two million bales. Although the cotton crop of 1866 had amounted to less than two million bales, which was less than half the 1859 crop and only a little greater than the production in 1839, the subsequent recovery of the cotton economy was rapid, and production exceeded five million bales for the first time in 1875. With new acreage under cultivation in the western part of the Cotton Belt and extensive application of fertilizers in the older regions, production doubled between 1879 and 1898.[147]

Not surprisingly, the southern fertilizer industry greatly suffered from the Civil War, but already by the 1870s new factories were springing up throughout the region. The new plants were often backed by northern capital attracted by cheap labor and newly discovered phosphate fields in the South. The heavy use of fertilizers was concentrated on the sandy soils of the Atlantic Coast and the Piedmont. By 1927, there were over six hundred plants manufacturing or mixing fertilizers in the South. Fertilizers became the most important factor after labor in the costs of cotton cultivation in the Atlantic states. West of Alabama, the use of fertilizers remained limited in the 1920s, and the alluvial lands of the Delta were still thought to possess the richest soil on earth. However, experiments carried out at the Delta Branch of the Mississippi Agricultural Experiment Station after 1921 indicated that the land was beginning to lose some of its fertility; the boll weevil was not the sole reason the Delta and Pine Land Company commenced applying nitrogenous fertilizer to its cotton fields.[148]

Westward expansion of cotton continued into the twentieth century with subsequent increases in acreage and production: in 1911 and 1914, over fifteen million bales were produced in the United States.[149] Between 1919 and 1921, the country produced some 52 percent of the world's cotton crop. During the same time period, the combined yield of China and India amounted to 37 percent of the total, followed by smaller producers, such as Egypt and Brazil, in single digits. The importance of cotton for the American agricultural system was summarized in 1921 by the U.S. Department of Agriculture as follows:

> Cotton is the great crop of the South. It is the chief and often almost the only source of income to a large proportion of the farmers in the Southern states.

It is so important that low prices or any other factor which greatly reduces the profitableness of the crop greatly disturbs the economic life of the Southern States. When the cotton crop is good and brings good prices the South is prosperous.

There is a division of labor between the States of the North and those of the South by which the North depends upon the South for cotton clothing or the raw materials out of which to manufacture the clothing and for products of the cotton seed, and the South in turn buys many of the products of farms of the North. It follows, therefore, that when the South is prosperous it furnishes a good market for corn, flour, meat, and dairy products, and that a prosperous North makes a good demand for cotton and cotton products.[150]

Hundreds of different varieties of *Gossypium* were cultivated around the world by the late nineteenth century. In the early 1920s, close to forty principal types were recognized at Liverpool, England, then the chief cotton market of the world. Commercial cotton varieties at the time could be grouped into five general classes, largely according to the length, strength, and fineness of fiber. The Sea Island and Egyptian varieties of *Gossypium barbadense* possessed the longest and finest fiber and commanded the highest prices. The long-staple and short-staple varieties of upland cotton (*Gossypium hirsutum*) were, however, the types most commonly cultivated in the United States since the early 1800s. By the early 1920s, the bulk of the American long-staple cotton crop of 1.5 million bales was grown in the Delta and on the irrigated fields of the Southwest. In the beginning of the 1930s, the Delta was recognized as the largest producer of long-staple cotton in the nation, and the city of Greenwood in Leflore County was known as the world's primary center for the marketing of the variety. The type most widely grown in the country, however, was still the short-staple, which, in hundreds of different strains, constituted over 90 percent of all American cotton. The ten million bales of short-staple cotton produced yearly in the United States made up roughly half of the world's total cotton crop.[151]

Much of the acreage for the production of this enormous crop had been carved from the southern forest during the nineteenth century, reflecting national trends in land clearing. During the 1860s, the amount of land newly cleared and settled in the United States had understandably plummeted by almost 50 percent, but during the next decade the settlement of the Great Plains raised it again to its highest total in any intercensal period of almost fifty million acres.[152] The shift in the center of pioneer activity from the east and the south to the central and unforested parts of the continent after the 1860s did not mean a diminution of the human impact on the eastern forests. Uncleared areas in the South were largely brought under cultivation. For example, in Florida the acreage of land in farms almost doubled between 1870 and 1900.[153]

The Yazoo-Mississippi Delta, of course, offered an outstanding example of the tremendous rate of land development attained in the United States after the Civil War. As late as 1880, Ernst von Hesse-Wartegg had claimed that

> [p]lanters are always calling this *"bottomland"* the richest, the most fertile agricultural land in the [Lower Mississippi] valley; yet it remains primitive and uninhabitable. . . . In the mostly oak, sassafras, cypress, and silver-poplar forests you will find many deer. Panthers and small black bears dwell in low-lying reed thickets. On numerous crescent-shaped lakes . . . , pelicans, swans, and countless other birds, and even alligators, disport themselves. In a word, this is a wilderness . . . , with the unique features and the advantages and disadvantages of such a wild place. Yet its days are numbered. In time, with the growing incursion of northerners, the land will be cleared and expanses brought under the plow.[154]

The wilderness described by Hesse-Wartegg was, however, already undergoing significant human-induced change, as over 10 percent of total land area in the core counties of the Delta was classified as improved. Still, the amount of land cleared and cotton produced in the Delta was to undergo its greatest expansion in the following decades.[155]

After 1880, a marked increase in agricultural development and population took place in the Delta. Within a decade, the proportion of improved acreage of the total land area almost doubled, while the population—and the value of farmland acreage—rose by 50 percent. The immense growth of Delta agriculture continued during the following decades. By the end of the nineteenth century, Delta production of cotton had far surpassed

*Table 2.* Amount of Cultivated Acreage (Improved Land) in the Core Counties of the Yazoo-Mississippi Delta, 1850–1930

| County | 1850 | 1860 | 1870 | 1880 | 1890 |
|--------|------|------|------|------|------|
| Bolivar | 16,973 | 85,188 | 39,629 | 74,072 | 161,337 |
| Coahoma | 11,478 | 39,139 | 28,959 | 52,490 | 95,019 |
| Humphreys | — | — | — | — | — |
| Issaquena | 27,631 | 56,596 | 35,286 | 32,928 | 68,837 |
| Leflore | — | — | — | 40,981 | 80,182 |
| Quitman | — | — | — | 5,714 | 15,827 |
| Sharkey | — | — | — | 24,824 | 44,994 |
| Sunflower | 5,966 | — | 30,264 | 14,170 | 35,587 |
| Tunica | 6,015 | 29,341 | 14,141 | 39,558 | 58,796 |
| Washington | 59,126 | — | 70,119 | 99,887 | 199,001 |
| Total | 127,189 | 210,264 | 218,398 | 384,624 | 759,580 |

| County | 1900 | 1910 | 1920 | 1925* | 1930* |
|--------|------|------|------|-------|-------|
| Bolivar | 185,746 | 251,595 | 291,324 | 289,117 | 342,464 |
| Coahoma | 121,905 | 172,389 | 185,614 | 200,329 | 214,596 |
| Humphreys | — | — | 97,452 | 81,031 | 111,707 |
| Issaquena | 55,052 | 54,154 | 54,697 | 43,845 | 49,538 |
| Leflore | 117,013 | 173,595 | 166,733 | 168,198 | 221,207 |
| Quitman | 23,363 | 58,982 | 102,128 | 95,523 | 122,535 |
| Sharkey | 61,115 | 82,573 | 68,724 | 67,795 | 76,552 |
| Sunflower | 73,696 | 156,906 | 220,497 | 237,598 | 334,722 |
| Tunica | 93,438 | 111,963 | 117,239 | 112,595 | 125,037 |
| Washington | 197,896 | 192,882 | 230,317 | 188,035 | 218,367 |
| Total | 929,224 | 1,255,039 | 1,534,725 | 1,484,066 | 1,816,725 |

Sources: *Seventh Census*, 456 (table 11); *Agriculture of the United States in 1860*, 84; *Statistics of the Wealth and Industry of the United States*, 184 (table 4); *Report on Productions of Agriculture as Returned at the Tenth Census*, 156–57 (table 7); *Report on the Statistics of Agriculture in the United States at the Eleventh Census*, 217–18 (table 6); *Agriculture, Vol. 5, Pt. 1*, 284–85 (table 19); *Agriculture, 1909 and 1910, Vol. 5*, 799–800; *Agriculture, Vol. 6, Pt. 2*, 522–29 (county table 1); *Agriculture, Vol. 2, Pt. 2*, 1050–55 (county table 1).

*Does not include plowable pasture.

those of the Natchez region and the black prairie belts in eastern Mississippi and south-central Alabama. The proportion of farmland of the total land area in the core counties, less than 32 percent in 1880, rose to 44 percent in 1890, and had passed the 50 percent mark in 1910. By 1930, over 65 percent of the total land area in the Delta was in farms, rising in the following five years to more than 70 percent. In the thirty years between 1880 and 1910, the proportion of improved land of the total Delta acreage increased from 11.5 percent to almost 40 percent. In 1920, almost 45 percent of the Delta had been improved, and by 1930, well over half of the total land area was classified as such. The intensification of land usage was naturally reflected by a rise in the proportion of farmland classed as improved: between 1880 and 1910 the amount almost doubled, from 36.2 to 66.6 percent. By 1920, over 80 percent of the farmland in the Delta heartland was classed as improved. Accordingly, the proportion of farmland in cotton rose steadily between 1880 and 1930, from less than 22 percent to 62 percent. At the same time, production of cotton in the core counties increased from 17.5 to 35.6 bales per one hundred acres of farmland.[156] (See table 2.)

The rise of a New South cotton kingdom on the Yazoo-Mississippi floodplain can largely be explained in two words: flood control. Land clearing for cotton production and other purposes was closely connected to the growing human control of the hydrological processes of the floodplain, a development gradually attained during the nineteenth and early twentieth centuries. There was a close relationship between the agricultural expansion, the amount of land cleared, and the growth of flood control and drainage systems in the Yazoo-Mississippi Delta.

He knew that the wild water on which the skiff tossed and fled flowed above
no soil tamely trod by man, behind the straining and surging buttocks of a mule.
That was when it occurred to him that its present condition was no phenomenon
of a decade, but the intervening years during which it consented to bear upon its
placid and sleepy bosom the frail mechanicals of man's clumsy contriving was the
phenomenon and this the norm and the River was now doing what it liked to do,
had waited patiently the ten years in order to do, as a mule will work for you ten
years for the privilege of kicking you once.

—WILLIAM FAULKNER, "Old Man," *The Wild Palms*

CHAPTER FIVE

# *Taming the Rivers*

THE IMPORTANCE of hydrological conditions for human habitation and subsistence is hard to overestimate for a floodplain bordering
a river that draws water from 42 percent of the continental United States,
accommodates the runoff from the entire Coldwater-Tallahatchie-Yazoo
watershed, and receives more than fifty inches of annual rainfall. Flooding
posed a physical threat to human inhabitants and many of their subsistence activities in the Lower Mississippi Valley and the Delta from the
very beginning. In many ways, the annual flooding of the Delta bottomlands was an unpredictable process for the early settlers of European origin, resulting in heavy economic losses but simultaneously boosting the
productivity of the land. The settlers soon realized that transformation
of the floodplain's hydrological system was the prerequisite for economic
growth and prosperity. It was the relief from flooding—a natural phe-

nomenon of the floodplain—that made the development of agriculture, infrastructure, and industry possible in the lowlands of the Delta. Similarly, improved drainage increased crop yields and allowed for the development of previously unimproved lands while freeing the settlers from many of the autochthonous diseases of the floodplain.

## A Hydraulic Civilization?

Karl A. Wittfogel's classic theory of "hydraulic civilizations" emphasizes the continuous socioecological dialectic of people with their natural surroundings, using the management of water in the subsistence economy of certain agrarian societies as an example. In its agriculturally most precious occurrence—such as rivers in arid regions—the availability of water defined the extent of agricultural activities. In order to cultivate water-deficient areas, humans had to create large-scale enterprises for water management, which resulted in a new type of agrarian economy. Wittfogel calls it "hydraulic agriculture" in contrast to traditional rainfall farming and the small-scale irrigation he calls "hydroagriculture." His main thesis is that rainfall farming and hydroagriculture encouraged the evolution of a multicentered society, whereas hydraulic agriculture required substantial and centralized works of water control, operated by a central government, and led to the monopolization of political power and societal leadership by government representatives. Wittfogel's hydraulic civilizations are consequently characterized by a combination of hydraulic agriculture, hydraulic government, and a single-centered society.[1]

Wittfogel's theory, referring most directly to ancient civilizations in China and India, has been successfully applied to modern environmental history. Donald Worster's *Rivers of Empire* convincingly shows that the history of reclamation (referring to large-scale irrigation of arid and semiarid areas for agricultural purposes) and hydropower in the American West, controlled by the federal government, shares common features with Wittfogel's hydraulic civilizations. In the more humid South, reclamation as a term has had a different usage. Over much of the region, settlement and agricultural expansion similarly required major transformation of the natural hydraulic regime. The problem, however, was primarily how to keep excess water off agricultural land rather than bring water to it.

This was accomplished by building flood-control works and by providing drainage. It should be noted that Wittfogel's "hydraulic agriculture" is not limited to the emergence of big productive (referring to irrigation) water works. According to Wittfogel, the advent of big protective (referring to flood control) works frequently accompanies the former, and may even surpass it in magnitude and urgency.[2] The concept of hydraulic society is worth remembering when examining the history of water control along the lower Mississippi and in the Delta.

Increasing governmental involvement, both state and federal, in flood control and water resource development in the Delta evolved during the nineteenth and twentieth centuries with far-reaching effects on the floodplain's natural hydrological regime. In the beginning, the burden of flood control was placed directly on the riparian landowners by the state government and later extended to the riverfront counties. As the inadequacy of this approach for successful flood prevention became evident, state and federal governments began to assume more responsibility in the region's evolving water management, resulting in a complicated relationship between local interests and government agencies.[3]

From the standpoint of flood control, the most remarkable characteristic of the lower Mississippi proved to be the great variation in the volume and elevation between high and low water stages: the mean low water discharge along the river could increase tenfold during a major flood. During historical times, gage readings along the riverfront in the Yazoo-Mississippi Delta have varied more than fifty feet between low and high water. In the mid–nineteenth century, the natural channel of the Mississippi River was estimated capable of carrying one million cubic feet of water per second. During seasons of heavy rainfall, however, the runoff from its drainage area could amount to more than two million cubic feet per second, resulting in flooding as the flow from the natural channel spilled onto the river's floodplains. The amount of water entering the Delta during major floods was enormous, turning the region into a virtual inland sea as witnessed already by the de Soto expedition. Vast areas were inundated, but usually the water levels remained relatively low; however, a major flood in the early summer of 1858 covered almost all of the Delta to a mean depth of some three feet. Seasons of heavy rainfall and flooding happened, on average, every 2.8 years.[4] Major flooding of prolonged duration was, however, sporadic. Early overflows of extreme proportion were recorded in the Lower Mississippi Valley in 1782, 1828, and 1858. The latter was not of the same magnitude as the first two, but exceeded the floods of 1844, 1850, and 1851 in extent and depth. Considerable flooding took place also in 1862, 1865, 1867, 1882, 1883, 1884, 1890, 1897, 1903, 1912, 1913, and 1922, culminating in the Great Flood of 1927. The great eighteenth- and early-nineteenth-century floods, however, coincided with the end of the Little Ice Age, a period that probably experienced increased rainfall in the headwater drainage area of the Mississippi. It is therefore possible that these floods are not truly representative of Mississippi flooding in the context of *la longue durée;* such excessive overflow may not have been present during periods of warmer climatic conditions.[5]

Localized headwater flooding along the Yazoo and its tributaries frequently combined with flooding stemming from overflows of the Mississippi River. To some extent, the natural levees bordering the Mississippi, Yazoo, and Sunflower Rivers prevented overflows. Prior to levee construction, numerous channels (the so-called passes) in the natural levees, however, connected and interconnected the Mississippi with the drainages of the Yazoo and its tributaries; overflow from the Mississippi could enter the floodplain through these and easily transform the tributary system into a distributary one, even before full flood stages. Probably the best known of these points was the Yazoo Pass, which could at times connect the Mississippi with the Coldwater, normally some eighteen miles apart. Through such channels, water from the Mississippi more or less regularly entered the Delta drainage systems of the Coldwater-Tallahatchie-Yazoo watershed, Sunflower River, and Deer Creek, passed into the main stem Yazoo, and returned to its "right" channel at the mouth of the Yazoo, on the southern edge of the floodplain.[6]

From the beginning of European settlement, there were attempts to control floods along the Mississippi river and its tributaries and to improve riverine navigation. The early French settlers from present-day Missouri down to the Gulf of Mexico quickly realized that southeastern rivers could top their natural levees and flood the settlers' fields, and responded by raising the ridges with artificial embankments. Dams and levees played an inestimable role in the socioeconomic and environmental history of the Lower Mississippi River ever since the founding of New Orleans in 1718 by Bienville. The French engineer Vitrac de La Tour had understood that the new settlement was prone to periodic flooding and opposed locating the city at its present site. Bienville, however, overruled La Tour's objection, and the engineer had to design a 5,400-foot-long and 18-foot-wide earthen embankment along the Mississippi, completed in 1727, to protect the city. The French word *levée* for such structures, as well as *crevasse* for breaks in them, were later adopted by all European settlers along the lower Mississippi, and became a part of the regional vernacular.[7]

Under the French administration, attempts at flood control were tied to land grants: landowners with riverfront property were responsible for building and maintaining levees, and the government intervened only to enforce the requirements with occasional inspections. This policy generally continued after the Lower Mississippi Valley became a part of the United States, emphasizing local responsibility and the ideals of Jacksonian Democracy. The early landowners in the Delta by necessity restricted their flood-control efforts to building small levees, which provided at least some relief from the nearly annual overflows. The early levees were only two to three feet high and constructed by field hands with no specialized equipment for the task. Despite the continuous extension of the levee system, problems with this approach soon became evident. Individual efforts resulted in the construction of a scattered and unconnected levee system.[8]

The flood of 1828 destroyed many of the early levees along the lower Mississippi. Crevasses opened in the frail embankments "generally not more than two or three feet wide at top, and ten or twelve at the base," as described in Louisiana by Basil Hall:

The river was tumbling through the opening with a head or fall of four or five feet, in a tumultuous manner, resembling one of the St. Lawrence rapids. This boiling, or rather surf-like appearance—for it rose and fell in snow-white ridges or short waves—did not spread itself far to the right or left, . . . but gushed nearly at right angles to the parent river straight forward, across the cultivated fields, into the forest growing in the boundless morass lying beyond the cleared strip of land. There was something peculiarly striking in this casual stream—a mere drop from the great Mississippi—which in many other countries would have claimed the name of a river, leaping, and writhing, and foaming along, with a sound exactly like that of breakers on a reef, through the middle of a village, amongst trees, over the tops of sugar plantations, and at last losing itself in a great cypress swamp.[9]

The flood of 1828 caused considerable damage to riverfront property but could also provide some aesthetic pleasure, at least for a passing traveler: "There was something finely contrasted with all this wretchedness [caused by the flood] in the magnificent foliage, and enormous stems of the trees bordering the Mississippi."[10] For settlers with limited agricultural interests, flooding along the Lower Mississippi provided an "opportunity of traversing the woods in canoes for the purposes of procuring game, and particularly the skins of animals, such as the deer and bear, which may be converted into money." According to Audubon, the squatters combed "the low ridges surrounded by the waters, and destroy[ed] thousands of deer, merely for their skins, leaving the flesh to putrefy."[11]

In the 1830s and 1840s, American settlers began clearing the rich alluvial lands in Arkansas and Mississippi in earnest. Already by the mid-1830s, the reliance upon riparian landowners to provide levees for community needs had proved insufficient in Mississippi, and flood-control projects were placed under public officials in each county. Known as the Board of Police, the county officials largely determined levee lines and were responsible for securing the cooperation of the landowners in levee construction and cost sharing. If flooding threatened, levee inspectors could order planters to send out their slaves for levee work. Funding the levee system remained a constant problem and construction standards varied widely.[12] The Yazoo-Mississippi Delta had already demonstrated its potential for becoming "the storehouse, the granary, the Egypt of the surrounding country," with only one drawback—the "liability to overflow from freshets in the Mississippi river." The danger, however, was believed to be "diminishing every year, and as population increases, levees, good and substantial, will be built."[13]

Contradicting the boosters' claims, the levees still could not resist the power of the floodwaters, and severe floods continued to occur. Combined with the growing economic importance of the Lower Mississippi region, this gradually led to federal involvement in flood control. In 1835, Henry Clay campaigned in the U.S. Congress for a survey of the possible leveeing of public lands on the west bank of the Mississippi, but Congress objected to such an expenditure of federal funds. A great flood in 1844 intensified demands for national help. The following year, John C. Calhoun declared flood control a federal problem on the basis of the "inland sea" philosophy. If the Mississippi were considered a sea, levee building and other improvements on the river could be interpreted as issues of national defense, and not of local or state responsibility. By 1847, the issue had gained national attention because of the attendance of public figures such as Calhoun, Thomas Hart Benton, Abraham Lincoln, and Horace Greeley at conventions advocating the federal government's involvement in levee building. As most of the floodwaters originated from outside the Lower Mississippi Valley, proponents argued that flood control was not a local problem.[14]

There were congressional attempts at flood protection, such as the Swamp Lands Acts of 1849 and 1850, but lack of coordination among different states and levee districts resulted in an ineffective program. The Swamp Lands Acts donated to Arkansas, Louisiana, and Mississippi all unsold swamp and overflowed federal lands within their boundaries to be used to raise funds for flood control. In the following three decades, states undertook various flood control and drainage programs that became plagued with frauds and scandals. The state of Mississippi had been granted close to four million acres of federal land, most of it in the Yazoo-Mississippi Delta. Unlike Arkansas and Louisiana, Mississippi did not develop a central state organization for the development of a levee system; land scrip was divided among the river counties, which were free to decide on the details of the reclamation program. There was some standardization of levee construction and state control in the area, but the system remained largely inadequate and most levees were still modest earthen embankments only a couple of feet high. Still, as numerous openings along the lower Mississippi levee system enabled the diffusion of floodwaters into the backswamp areas, the river did not attain the heights it later did with the more comprehensive levee program.[15]

Prior to the initiation of steamboat traffic on the Mississippi in 1811, Delta produce could move only downstream on rafts, flatboats, and keelboats. Steamboats gradually replaced the more primitive vessel types in southern river traffic, and between 1830 and 1845 steamboat traffic expanded enormously as it became clear that rivers could effectively serve as arteries for trade. Natchez was soon rivaled as a cotton market by Vicksburg, which had become an important river port for exporting Delta produce. By 1857, there were 150 steamboat landings on the Yazoo between Vicksburg and Greenwood alone. However, trees deposited in the river by erosion inconvenienced and endangered the growing traffic.[16] For an expanding nation relying heavily on waterways for transport, easy navigation was of utmost economic importance, and substantial federal appropriations were approved for clearing the channels of southern rivers already during the presidential administration of Andrew Jackson in the 1830s. The series of federal actions aimed at facilitating navigation on the Mississippi and its tributaries resulted in the accumulation of a vast amount of scientific data that later proved helpful also for flood-control interests.[17]

Cotton lands increased in value during the 1850s, and it became evident that with adequate flood control, the Delta could turn into the envisioned garden. Hundreds of thousands of land scrip, issued to the Delta counties under the provisions of the Swamp Lands Act of 1850, were sold. Incoming planters were willing to invest in flood prevention, and between 1852 and 1858 an estimated two million dollars was spent on levee work in the region. Levee contractors were also willing to accept land scrip as payment for levee construction. By the late 1850s, the levee line bordering the Mississippi in the western Delta had become more or less continuous with an average height of four feet. Still, the degree of flood control in the Delta was widely viewed as inferior to that elsewhere in the Lower Mississippi Valley.[18]

The disastrous flood of 1858 justified the criticism of the less-than-uniform levee management procedures under the system of County

Boards of Police. Wade Hampton III was among the many who lost their cotton crop and complained that "I could have saved nearly all the land I have, had I not depended on the river levees. It will be a lesson to me however and I will now go on with my own levees."[19] The Mississippi legislature now decided to change its policy, abolished all the county systems for levee building, and established a common levee district to manage all flood-control work in the Delta in relation to the Mississippi River. The new program was to be financed with a uniform tax of ten cents per acre for five years on all lands that flooded in the district. Known as the 1858 Levee Board, the new organization conducted extensive surveys and created new standards for levee construction.[20] The first president of the Board of Levee Commissioners—the highest paid state official with an annual salary of $6,000—was James Lusk Alcorn, a Coahoma County planter and state legislator who later became the governor of Mississippi and a U.S. Senator. Alcorn was closely identified with the planting interests of Mississippi. On July 17, 1850, he valued his estate, including seventeen slaves, at $16,625. By 1855, the number of slaves had increased to thirty-nine and the value of the estate to $65,000. In 1860, he owned ninety-three slaves and valued his estate at $250,000. Alcorn had long "insisted that the reclamation of [Delta] lands and their protection was in the nature of public internal improvements," and that the Delta counties should cooperate with the state and federal governments "in the appropriation of public funds for this great industrial and agricultural enterprise."[21]

Under Alcorn's direction, over sixteen hundred miles of levee survey was conducted. It was noted that many of the early levees had been located on natural ridges too close to the river bank and were often destroyed by bank caving. New embankments were therefore to be built a safe distance from the river. The reversal of the levee line on many occasions left considerable tracts of improved crop land of the best quality on the river side of the levee, which led to numerous—but unsuccessful—complaints by the landowners affected; such conflicts continued well into the twentieth century. Close to 150 miles of new levee had been built in the Delta according to the improved standards before taxes for levee work were suspended by the legislature at the commencement of the Civil War in 1861.[22]

Slave labor had initially been used for levee construction, but with the price for slaves continuously on the rise, various immigrant groups, especially the Irish, were recruited to replace the precious slaves in the grueling and dangerous work carried out in malaria-infested swamps.[23] Walter Sillers Sr. described the rough lifestyle of the immigrant levee workers in the antebellum Bolivar County: "The levee . . . was built by Irish laborers under a Scotch contractor named Bain. . . . He worked Irish exclusively and his white tents were along the front of my father's plantation. When paid off every Saturday night, the workers invariably got drunk and caroused and fought and howled like Bedlam turned loose. Sunday mornings found them badly bruised, with many eyes closed from fighting. The levee contractors kept barrels of whiskey on hand and, at stated periods of the day, the laborers were given 'jiggers' of whiskey."[24]

Severe flooding in 1862 and 1865, combined with the destructive effects of the Civil War, caused extensive damage to the new levee system. Many of the new structures had been swept away by the floods, and military action and neglect had devastated the remaining levees. The capture of heavily fortified Vicksburg, protected on the west by high bluffs and the Mississippi River and on the north by swamps, presented the Union General U. S. Grant with a great tactical challenge during the Civil War. He planned to get his troops into position to attack the town by destroying the levee at the Yazoo Pass in 1863, hoping to gain access to and conquer Vicksburg via the interior streams. In addition to Grant's soldiers, Confederate guerrillas damaged levees along the Mississippi, aiming to hinder the movements of Union troops. At the end of the war, the bottomlands of the Delta were flooded once again, and the prosperous cotton economy of the last decade seemed but a dream.[25]

Taxation was the only source of funds for levee rebuilding, but few could afford to pay the taxes. During the early Reconstruction period, total annual taxes for Delta lands were frequently above their sale value. Consequently, formal forfeitures of land to the state and various levee boards became common. The heavily indebted 1858 Levee Board had been disbanded but the debts remained; the debts had accumulated considerable interest during the war and amounted to over 1.5 million dollars. In 1867, the state organized the Liquidating Levee Board to meet these debts. As almost every acre in the Delta had at some point been forfeited to one or more of the levee boards, a series of extremely complicated land title cases followed, and the rebuilding of the levee system was delayed. The Liquidating Levee Board was created solely to undertake the management and eventual payment of the debts, not to build or repair levees. The tax titles of the new board became entangled with the state tax titles of various county levee boards and created a turmoil of controversy and litigation. By 1872, almost all Delta lands had been forfeited to the Liquidating Levee Board.[26]

After the war, the states of the Lower Mississippi Valley responded differently to the problem of flood control: Arkansas issued state bonds in order to finance the construction work, Louisiana contracted a private company to rebuild the levee system, while Mississippi returned to local initiative. The rebuilding of the levee system was entrusted to levee districts with power to sell bonds based on the future value of the previously overflowed land. In the dire economic conditions following the Civil War, the 1858 Levee District was not revived, and prioritizing various rebuilding schemes important to the Delta economy proved problematic. Besides, the concern for an efficient flood-control system varied markedly between different parts of the Delta; natural conditions for the upper and lower counties had become substantially different as a result of the destruction of the levee system. Planters in the upper Delta, for the time being, could choose to ignore repairing the levees because most of the Mississippi floodwaters entering the region through crevasses flowed away through the interior streams and left the better agricultural lands in a reasonable condition for farming. In the lower part of the Delta, the interior streams, already full of water, could not hold the additional input from the Mississippi, and much of the cropland was continually subject to serious flood damage.[27]

During the late 1860s and early 1870s, there were largely unsuccessful efforts in Bolivar, Washington, and Issaquena counties to restore portions of the 1858 levee. Funds were to be secured from an annual tax of ten cents per acre on lands west of the Sunflower not held in trust by the state. In 1877, the Board of Levee Commissioners for Bolivar, Washington, and Issaquena Counties was reorganized into the Board of Mississippi Levee Commissioners by the state legislature. The new organization concentrated its efforts toward rebuilding the levee line through the lower Delta, while the riverfront in the northern counties of DeSoto, Tunica, and Coahoma remained relatively unprotected. Between 1865 and 1882, the lower Delta had invested some 3.5 million dollars in the rebuilding of the levee system, money raised by heavy taxation on land and the amount of cotton produced. The levees built by the Mississippi Levee District prior to 1882 averaged about seven feet in height. Irish laborers continued to be hired for levee construction, but now also freedmen were employed by levee contractors, especially as mule drivers for scrapers packing down each layer of built-up earth. Convict labor was also experimented with.[28] The levee work was grueling but its objective was probably clear for everyone involved in construction, as later expressed in song by "Kansas" Joe McCoy:

If it keeps on rainin', levee's goin' to break, [twice]
And the water gonna come and have no place to stay.

. . . . . . . . . . . . . . . . . . . . . . . . . . . . . . . .

Oh, crying won't help you, praying won't do no good [twice]
When the levee breaks, mama, you got to move.
I works on the levee, mama, both night and day, [twice]
I work so hard to keep the water away.[29]

Overall, little lasting flood protection resulted from the different approaches adopted by Arkansas, Louisiana, and Mississippi, and during the 1870s, the federal government came to recognize the need for a coordinated control effort to boost the southern economy. As a result, Congress in 1879 established the Mississippi River Commission to supervise all federal public works on the river. Initially these projects aimed at aiding navigation on the river, but the U.S. Army Corps of Engineers soon included flood control in their hydrological design: they could approve levee construction having a direct impact on maintaining the navigation channel. The River and Harbor Act of 1881 for the first time provided direct federal assistance to levee construction, but lack of funds limited its effectiveness primarily to maintenance of levees and navigational channels already in existence. Local levee districts continued to bear most of the financial burden, while the Mississippi River Commission served a coordinating function.[30]

A major flood in 1882 "put all the unprotected lowlands under water, from Cairo to [the Mississippi's] mouth," causing Mark Twain to claim that the flood would "doubtless be celebrated in the river's history for several generations before a deluge of like magnitude" would be experienced.[31] The 1882 flood was indeed among the most severe ever recorded along the lower Mississippi and in the Delta. On March 23, 1882, Captain R. H. Bowman, on board the steamboat *Anita,* noted that "[f]rom Faisonia [in Sunflower County] to the mouth of Sunflower, the water is from two to three feet higher than ever before known." In addition to saving human lives, rescuing livestock became an important mission for the *Anita:* "We have on Board about 75 mules, principal[l]y belonging to those who think it better to get them to a pasture, so that when the water falls they will have them in condition to do good and effective work. We have also about 100 head of cattle, some of them taken from the water and from rafts, and to keep them from drowning or dyeing, and others sent to pasture."[32] The upper Delta suffered widespread damage, demonstrating once again that continuous economic development in the region depended on a high degree of control over the hydrological system. Such control called for immense investments and proved to be achievable only with the help of the federal government. For the time being, however, the state of Mississippi continued to emphasize the role of local initiative.

In 1884, the Yazoo-Mississippi Delta Levee District was created for the upper and interior Delta counties not included in the older Mississippi Levee District. The Board of Levee Commissioners for the new organization was appointed by the state governor. In addition to county representatives, the Board included one commissioner to represent the stockholders of the new Memphis and Vicksburg Railroad Company who had been vocal promoters of the new levee district. From the beginning, the railroad interests recognized the close correlation of their construction and maintenance costs with adequate flood control in the Delta. Disagreements soon arose between planters and railroad promoters, as railroad companies held vast acreage in the region and hoped to avoid property taxes for levee purposes. They consequently proposed a tax on cotton production as the basis for financing levee construction. After a brief experiment with a cotton tax, the planter faction was able to convince the state legislature to adopt a system of *ad valorem* property and various privilege taxes, with no mention of a tax on cotton production.[33]

Because of the good soils and relative safety from overflows, Native American and early Euro-American settlers had concentrated their farming activities on higher lands of the floodplain. Drainage of existing farmland consequently remained a relatively insignificant problem through most of the nineteenth century. The situation, however, began to change toward the end of the century as new settlers to the Delta, encouraged by the continuously expanding levee system, began to clear and cultivate the lower lands in the interior of the floodplain. Despite all the efforts to contain floodwaters both from the Mississippi and the hills, seasonal heavy rains still produced flooded fields because of the easily saturated soils and flat topography of the Delta.

As with building levees, providing drainage remained largely a private endeavor until the late nineteenth century. The early drainage systems were relatively simple, consisting of handmade ditches that carried rainwater and river overflow to the adjacent swamplands. Drainage programs carried out before the twentieth century proved generally ineffective, but the experience gained during the nineteenth century contributed greatly to later successful reclamation efforts in the region. Besides, systematic development of drainage in the interstream areas of the Delta would have been practically useless until protection from major flooding was reasonably assured.[34]

During the last years of the nineteenth century, the Mississippi and the Yazoo-Mississippi Levee Districts pursued their goal of a flood-free Delta by constructing higher and stronger levees along the riverfront. The estimated adequate height of the levees was two feet higher than the previous recorded high stage of flood waters. This "levees only" policy, however, had an inherent problem: closing riverfronts everywhere along the lower Mississippi increased the stages of subsequent floods. When the levee was raised by five feet, the actual high water line was estimated to rise by almost three.[35] The worst dreams of the Chief Engineer of the Yazoo-

Mississippi Delta Levee District were realized during the flood of 1897, when new protective works on the Arkansas side of the river raised the flood waters along the upper Delta by over two feet and caused a major crevasse. Flood control in the Lower Mississippi Valley had become "a desperate contest between opposite sides of the river . . . the side that could hold out the longer would win."[36]

The flood of 1897 demonstrated once again the vulnerability of the regional economy to a major overflow. The flood received wide national attention, and calls for a federal flood-control program intensified.[37] Already in 1890, a convention in Vicksburg had led to the formation of the Inter-State Mississippi Improvement and Levee Association for the purpose of attracting federal funds.[38] A revetment program to protect caving levees in harbors and other special localities was seen as so important in the Delta that the Yazoo-Mississippi Delta Levee Board contributed money to this federal activity.[39] Although the federal government still declined major funding, the levee system along the lower Mississippi continued to expand. The national economy had picked up after the financial depression of 1893, and land values in the Delta were again on the rise. New drainage districts were being organized, while lumber and railroad companies had begun aggressive marketing of their cutover acreage to agricultural interests. New debts and higher taxes could now be tolerated, as bigger and stronger levees were seen as the surest way to assure a prosperous future. Between 1903 and 1912, the average size of levees in the Yazoo-Mississippi Delta Levee District increased by one-third, from about two hundred thousand cubic yards per mile to about three hundred thousand cubic yards per mile.[40] Developments in levee construction were by no means restricted to the Delta; by 1912 the levee system along the Mississippi and its tributaries amounted to some fifteen hundred miles.[41]

Not all planters, however, benefited from the upgrading of the Mississippi levee system. The seven-thousand-acre Woodstock plantation, occupying a Mississippi bend in Washington County, had been purchased in the late 1840s by Alfred Grayson Carter: "The land was selected by Mr Carter on account of its great fertility, because of the transportation facilities afforded by the river, and because its situation [on a natural levee] made it, (at that time) secure from the annual floods." The plantation had survived the ravages of the Civil War but faced problems after 1890 as a result of the readjustment of the levee line:

> [T]he great "bends" were thrown out, and an effort made to shorten and straighten the line of levees. Consequently, for the first time, Woodstock was subjected to inundation. Captain W. F. Randolph, then in charge of the estate, tried to protect it by creating a private levee around the property. For years this effort was persisted in, great sums of money being annually spent for the purpose. Year after year the private levees were broken and the growing crops destroyed. To keep up the unequal struggle large sums were borrowed on the land, and finally, about 1910, Captain Randolph abandoned the place, and, with his wife and children, went to Virginia, where he had bought the place in Fauquier County known as Inness Hill. Woodstock became the property of a firm of Jews in New Orleans, who soon sold it, and it is now [circa 1930] virtually a deserted waste.[42]

Despite the constant development of the Delta levee program, serious flooding took place in 1903 and 1912.[43] Ex-slave Berry Smith had worked in the Delta cotton fields but left the Greenville area after experiencing a Mississippi flood. The former resident of the hills was rightfully awed by the power of the river: "I hear'd a noise like de wind an' asked dem Niggers, 'Is dat a storm?' Dey said, 'No, dat's de river comin' th'ough an' you better come back 'fore de water ketch you.' I say, 'If it ketch me it gwine a-ketch me on my way home. I ain't been back since.' "[44]

In 1916, yet another major flood occurred along the lower Mississippi. Levees in the Delta held, but devastation elsewhere along the river created enough momentum for the passage of the Ransdell-Humphreys Flood Control Act of 1917. The act authorized greater federal aid for the construction of levees and affirmed the policy of cooperation between local levee districts along the lower Mississippi and its tributaries. Under the new legislation, the federal government would pay up to two-thirds of the construction cost of new levees. Local interests remained responsible for acquiring the right-of-way and for paying no less than one-third of the construction cost and all maintenance costs. Construction costs plummeted around this time with the adoption of steam shovels for excavation and cableways for moving dirt.[45]

By concentrating on the construction of higher and stronger levees of a new standard, the legislation perpetuated the old "levees only" policy and ignored other solutions for flood control along the lower Mississippi. By the time the act was passed, levees in the upper Delta were already up to the new standards, so the Yazoo-Mississippi Levee Board received no federal assistance.[46] The Mississippi Levee Board, however, was able to upgrade the levees in the lower part of the floodplain with the help of the federal Mississippi River Commission. The strengthened levees successfully constrained a record flood in 1922. Consequently, the entire Delta was, for the first time, seen as adequately protected from flooding. Already at the turn of the century, it had been claimed by the chief engineer of the Mississippi Levee District that "three feet more in height [of the levees], with a relative strengthening of the base, will give protection from any flood that may be expected in the future." Levee work in the future was commonly expected to entail only maintenance of existing structures.[47]

The combined efforts of state and federal government to control the Mississippi river sometimes included manipulation of hydrological processes other than flooding. In 1876, the Mississippi River had changed its course near the mouth of the Yazoo, and the cutoff left the port of Vicksburg dry. With work commencing in 1892, the mouth of the Yazoo was relocated and a diversion canal was built to reopen the town for river traffic with help from the federal government. The project, valued at over one million dollars, was completed in 1905, and Vicksburg became reestablished as a port city.[48]

New developments were also taking place along the interior streams of the Delta. In 1886, the state legislature drafted a system for dealing with drainage problems in Mississippi: each county was to have a Board of Drainage Commissioners to conduct the affairs of all the districts in the county, but it proved impossible to raise enough money for the costly reclamation of the swamplands. Legislation in 1906 and 1912 created a similar board for each district and allowed drainage districts to levy improvement taxes based on the benefits the lands were expected to derive from drainage. Between 1906 and 1917, individual and local efforts to provide adequate drainage for agriculture evolved into hundreds of different drainage districts in the Delta alone, as every rise in cotton prices created new districts by the score.[49]

Providing drainage for the region, however, remained an *ad hoc* activity as land owners and drainage officials strongly preferred to maintain local control: no comprehensive planning or coordination of drainage activities was developed for the floodplain. The extreme complexity of providing drainage in the Delta resulted from natural conditions: drainage basins on the flat floodplain were not easily definable and frequently became interconnected during periods of overflow. Consequently, drainage activities in other districts could dramatically affect conditions in a given district. In 1908, an attempt to create the Tallahatchie Drainage District with over one million acres in the northern Delta was abandoned because of various conflicts of interest.[50]

Drainage districts came to play an increasingly important role in the agricultural development of remaining bottomland hardwood forest acreage. The drainage districts continued to view reclamation as a local problem, and, accordingly, officials administered the projects under county or other local jurisdiction. State legislation relating to reclamation was largely confined to standardizing cooperation among landowners, apportioning costs, and authorizing payments. States furthermore joined the lumber companies, railroads, and local interests in promoting settlement and established the legal foundation for special improvement districts. The drainage movement in the South was assisted also by federal agencies, especially the Department of Agriculture, which conducted various investigations and provided technical assistance during the planning process. Envious of the plentiful federal funds channeled to the reclamation of the arid lands in the West, southerners lobbied hard—but unsuccessfully—in the 1920s to include their region within the sphere of the Bureau of Reclamation.

The new drainage programs launched in the late nineteenth and early twentieth centuries were, of course, part of a national movement for attracting settlers and putting undeveloped areas into agricultural use. The U.S. Department of Agriculture was actively involved in the planning and promotion of many drainage enterprises in the Delta. However, the choice of the forty-acre family farm as the unit for the basis for drainage calculations soon proved to be a mistake. On the local level, much of the drainage was furthermore designed by people with no adequate training for the endeavor. By this time, lumber and railroad companies in the Delta had begun to release their cutover timberlands for agricultural settlement. Unlike many other southern floodplains, most of this improvable land in the Delta was grabbed by established landowners: plantation interests, encouraged by the relatively good protection from flooding, could afford to undertake the development of "virgin" soil. Many of the incoming small farmers, on the other hand, failed miserably and left the region after a few years, as establishing adequate drainage for the new fields proved hopeless with their limited means. The tracts sold piecemeal by lumber companies had been drained enough to facilitate logging, but could not be farmed without additional drainage. Plantation companies, on the other hand, were able to design independent drainage systems for their vast holdings.[51]

## When the Levee Breaks . . .

The shortcomings of the "levees only" approach to flood control and the inadequacy of the existing levee system to protect settlements along the lower Mississippi were tragically demonstrated in the spring of 1927. Prolonged heavy rainfall in the headwater areas swelled the Mississippi tributaries and added to the already high water levels in the lower Mississippi. In April 1927, waters began to rise phenomenally. Coming in two separate waves, flood crests in the contrived river system topped all the previous records. In late April and early May, high waters in the Mississippi along the Delta approached sixty feet above mean sea level. Not designed to hold such waters, levees along the lower Mississippi began to break. The Army Corps of Engineers later estimated that, had the flood been confined, water stages in some places would have come close to the seventy-foot mark, a level previously inconceivable. Even with numerous crevasses up the river, the confined discharge at Vicksburg was estimated at almost 2.3 million cubic feet per second. By the time the second flood crest reached the Gulf of Mexico, it had caused thirteen major crevasses and inundated some twenty-three thousand square miles of alluvial lands along the Mississippi. The mighty river had reclaimed its alluvial plain, and muddy waters once again extended across the bottomlands for dozens of miles.[52]

Flood waters entered the Delta through a single crevasse in Bolivar County, augmented by backwater flooding from the mouth of the Yazoo. The largest break on the east bank of the river, the Mound Landing crevasse, opened on the morning of April 21 and quickly attained a width over three thousand feet. Discharging close to five hundred thousand cubic feet per second, this one opening approached Niagara Falls in its volume and inundated over 2.3 million acres—the entire lower Delta. People frantically sought safety from the rising waters on rooftops and Indian mounds and in trees, while much of the levee line turned into a makeshift refuge camp.[53] A letter written by a Greenville resident to his wife and son who had escaped to the northeastern Delta described the devastation as it was happening:

> [T]his is hell down here. We haven't got a speck of land out except the big levee and there is about fore or five thousand he[a]d of stock on it[,] besides at le[a]st as many people. The boats have taken about 2000 of them away to the hills. . . . You know Herbert didn't beleive the levee would brake or if it did the water wouldn't git in town so his mules are standing on the protection levee in water to their knees and the current is so strong they can't get them[;] he has lost every thing but one cow.
>
> They have been getting people off of house tops & out of trees every since the brake. [T]he moter boats are working as hard as they can bringing them in. The people got on the top of houses and thought they were safe and the houses were washed away. The only houses that stood were the ones with heavy brick chimneys. There is at le[a]st between 600 & 1000 people drowned.
>
> We were trying to stop the water at the protection levee and you could here the people comeing up the road ahead of the water.[54]

Not surprisingly, the people hardest hit by the 1927 flood—and its predecessors—were black sharecroppers; probably more blues were recorded on the subject of heavy rainfall and inundation than any other natural phenomenon.[55] Even before the waters had receded along the lower Mississippi, a number of blues recordings about the flood were released. It is remarkable to find such a large number of performances describing the same historical event, since the emphasis of the blues is on individual feelings and perceptions. Texas bluesman "Blind Lemon" Jefferson, in his "Rising High Water Blues," could express the spectator's genuine sympathy for the flood victims, but it was the Delta native Charlie Patton whose desperate account, "High Water Everywhere," truly evoked the essence of the bottomland apocalypse:

Lord the whole round country, Lord, creek water has overflowed,
Lord the whole round country, man, is overflowed,
(*Spoken:* You know I can't stay here, I'm goin' where it's high, boy)
I would go to the hill country but they got me barred.

. . . . . . . . . . . . . . . . . . . . . .

Looka here the water now, Lordy, done broke out, rolled most everywhere,
The water at Greenville and Leland it done rose everywhere,
(*Spoken:* Boy, you can't never stay here)
I would go down to Rosedale but they tell me there's water there.
Now the water now Mama, done struck Charley's town, [twice]
We'll I'm going' to Vicksburg for that higher mound.

. . . . . . . . . . . . . . . . . . . . . .

Backwater at Blytheville, backed up all round,
Backwater at Blytheville done took Joiner town,
It was fifty families (*Spoken:* and children) some of them sink and drown.

. . . . . . . . . . . . . . . . . . . . . .

Ooh water was risin', families sinkin' down,
Say the water was risin', at places all around,
It was fifty men and children, come to sink an' drown.
Ooh Lord, women and grown men drown,
Ooh-uh, women and children sinkin' down.
(*Spoken:* Lord have mercy)
I couldn't see nobody's home, and wasn't no one to be found.[56]

Referred to as the greatest peacetime disaster in the history of the United States by Secretary of Commerce Herbert Hoover, the flood caused staggering economic losses and human suffering. Altogether over sixteen million acres in seven states had been inundated, with estimates of direct property loss varying from 236 to 363.5 million dollars. In some places, prospects for growing cash crops had been destroyed for years to come, as the productive acreage had been covered by a foot-deep layer of sand carried in by the flood waters. Situated close to the Mound Landing crevasse, the Delta and Pine Land Company plantations lost almost five thousand acres of agricultural lands to erosion and sand deposits, with total flood damage estimated at five hundred thousand dollars. Hundreds of lives were lost along the Lower Mississippi, and some 637,000 persons became homeless, many of them in the Delta. The American Red Cross, responsible for most of the relief work, provided food and shelter for more than three hundred thousand people in its refuge camps. As suggested by Charlie Patton's "High Water Everywhere," black refugees were not free to flee the Delta, as the planters were afraid of losing their work force. Once in the refuge camps, blacks were often coerced to perform manual labor while northern labor recruiters were run off by the Mississippi National Guard. In the aftermath of the devastation, Congress requested that the Army Corps of Engineers examine the flood problem in a national context. As a result, the chief of the Army Corps of Engineers, Lieutenant General Edgar Jadwin, came up with a three-hundred-million-dollar program for the development of the Mississippi and its tributaries.[57]

The "levees only" policy was literally blown out of existence on April 29, 1927, when the authorities were compelled to dynamite a levee at Caernarvon, Louisiana, downstream from New Orleans. The artificial crevasse eased the pressure on the levees protecting the Crescent City, but flooded Plaquemines and St. Bernard parishes, turning their inhabitants into refugees by federal action. In addition to a stronger and more comprehensive levee system, the Corps now proposed new approaches to flood control under the so-called Jadwin plan: deepening river channels with jetties and constructing floodways and storage reservoirs. Congress hastily passed the Jones-Reid Flood Control Act of 1928 authorizing the Corps to proceed with the plan and committing the federal government to a definite and comprehensive program of flood control along the lower Mississippi. The federal government now undertook the full cost of levee building and left to local levee boards only the tasks of obtaining rights-of-way and maintaining the completed levees. The objective of the new program—known as "Project Flood"—was to safely channel a hypothetical flood of unprecedented magnitude through the Lower Mississippi Valley to the Gulf of Mexico.[58]

Already during the late nineteenth century, people such as John Wesley Powell of the U.S. Geological Survey had begun to promote the principle of regional planning under local control, including the treatment of river systems as units. These ideas found enthusiastic response among leading conservationists of the time, such as Theodore Roosevelt and Gifford Pinchot, who advocated more rational exploitation of natural resources. During the first decades of the twentieth century, legislation approved by Congress reinforced the idea of multipurpose water developments under federal authority, as evidenced by the Reclamation Act of 1902. Another landmark law, the Weeks Act of 1911, recognized the connection between forest conservation and watershed protection. As president, Roosevelt strongly supported federal watershed planning and fiercely opposed private development of water power without federal supervision. During the 1910s and 1920s, Senator George Norris of Nebraska championed more federal involvement in water development. Norris's interests were not limited to reclamation of arid lands in the West; he maintained that water reservoirs on the Missouri would contribute to flood control along the Mississippi. The idea of comprehensive watershed planning gained momentum in Congress and was included in the Water Power Act of 1920 and the Rivers and Harbors Act of 1925. Such aspirations for regional development projects culminated with the launching of the Tennessee Valley Authority (TVA) in 1933.[59]

In the 1930s, significant legislation was passed authorizing structural control of flooding along the Mississippi tributaries. The Flood Control Act of 1936 affirmed flood control as a federal activity. Hundreds of projects commenced in the United States, including one of the world's most comprehensive flood control systems, generally known as the Mississippi River and Tributaries Project, in the Lower Mississippi Valley. More than twenty additional flood-control acts authorized "corrective" works in the area. These included new levees for containing flood flows, floodways for the passage of excess flows, and channel improvement in order to increase flow capacity. By 1948, the federal government had spent over fifty million dollars for levee work in the two levee districts of the Delta alone.[60]

The practice of placing revetment to stabilize caving riverbanks was expanded enormously as new—and expensive—techniques became widely available. Mile after mile of Mississippi riverfront was buried under an asphalt layer, while the underwater portion of the bank was blanketed with mattresses made of concrete.[61] To augment the passage of flood waters, the lower Mississippi river channel was deepened considerably. Already in 1808, Henry Ker had surmised that the extensive bends of the Mississippi could be straightened out:

> By cutting small courses across these necks, there would be no difficulty in uniting or straightening this mighty stream in many places; some of which are not more than half a mile, and are made up of loose earth. In many places it would save ten, in others twenty, and in one or two instances fifty miles. All these added, would be in course of time an object, particularly in ascending the river. It has been contemplated by some characters of enterprise, but fell through for want of funds. But as the country grows older, and its advantages are discovered, improvements will assist the workmanship of nature in instances of the above-mentioned kind. This country has far exceeded the most mature nations of Europe, in improvements and inventions which have surpassed in their practical results all the theories of the old world.[62]

Ker's dreams of conquest ultimately came true, as sixteen artificial cutoffs across the bends of the Mississippi shortened the river's course from Memphis to Baton Rouge by some 150 miles between 1929 and 1942. The purpose of the cutoffs, however, was not so much to shorten the traveler's journey as to lower the flood crests by speeding them to the Gulf of Mexico.[63]

The continuous development of the levee system along the Mississippi, however, did not end flooding problems for Delta agriculturists. The levee system stopped at the mouth of the Yazoo, and serious backwater flooding continued to occur, as well as overflows along the Coldwater-Tallahatchie-Yazoo river system from the runoff from the eastern hills. Actually, the expansion of the Mississippi levee system aggravated flooding in the southern Delta's traditional backwater area, a fan-shaped area of over one million acres extending northward from the mouth of the Yazoo. Prior to the large-scale levee construction, plantations in the backwater area had been subjected to roughly the same amount of flooding as other parts of the Delta. With the increase of levee mileage along the Mississippi, however, high river stages of longer duration became more common in the main stem, and the unprotected backwater lands flooded more often. Furthermore, the acreage prone to backwater flooding increased, and highly developed front lands at the edge of the backwater area were subjected to flooding. Because the backwater area continued to be flooded on average of once in five years, only 17 percent of the area was cultivated in 1930.[64]

The Yazoo Backwater Project, including levees, floodgates, and other works to provide protection from backwater flooding, was authorized by the Flood Control Act of 1941 and initially concentrated on the construction of a levee system. Studies carried out during the 1940s claimed that almost half a million acres of the backwater area could be developed for agricultural purposes.[65] Close to 300,000 acres of the potential agricultural area had already been cleared, while a little less than 200,000 acres were still timbered. Almost 226,000 acres of the backwater area was deemed marginal for agriculture, and it was this part of the Delta, in addition to the batture area between the Mississippi and the levee line, that retained some of its primeval character into the third millennium.[66]

Southern drainage districts, like most American institutions, were hard hit by the Great Depression. Tax delinquency became increasingly common and debts mounted, but New Deal programs, such as the Reconstruction Finance Corporation, provided some relief. By 1941, the 108 drainage districts in operation had built over three thousand miles of drains, covering approximately 2.4 million acres of the Delta. Reclamation in the South gained new momentum as important legislation relating to drainage was passed during the 1940s and 1950s. The 1944 Flood Control Act for the first time enabled the U.S. Army Corps of Engineers to engage in drainage projects not directly related to levee building or other flood-control works. The 1954 Watershed Protection and Flood Prevention Act authorized the U.S. Department of Agriculture to assist in the planning and implementation of small watershed projects, largely obliterating the gap between the duties of the Soil Conservation Service and the Corps of Engineers.[67]

With the continuous development of the Mainline Mississippi River Levee, awareness of the flood problems on tributary basins became more pronounced. During the twentieth century, tributary basins of the Mississippi became subjected to increased improvement activity in the form of new dams, reservoirs, pumping plants, and auxiliary channels. Large-scale federal flood-control projects commenced as early as 1929 along the lower Mississippi on the Atchafalaya, lower Arkansas, and lower Red River Basins, while similar work along the St. Francis, White, and Yazoo River Basins was initiated between 1936 and 1938.[68] Of these, probably the most controversial has been the modification of the Atchafalaya Basin. Originally the Atchafalaya acted as a safety valve for flooding on the Mississippi by conveying excess water from the main stem into the bottom-lands of southern Louisiana. Development on the Atchafalaya Basin by the U.S. Army Corps of Engineers in the form of deepening and straightening the river greatly restricted the basin's absorbency. Today, drainage into over twenty tributary bayous has been blocked, and an extensive system of high levees hurries floodwaters through the Atchafalaya to the Gulf of Mexico. In addition to the wide criticism of the immense ecological changes along the Atchafalaya caused by the flood-control measures, the safety of the project has become an issue. By changing the natural hydrological regime of the floodplain, the modifications may have created an unprecedented potential for disaster, should the levees ever be breached.[69]

Floods from the hill tributaries of the Tallahatchie, Coldwater, Yocona, and Yalobusha Rivers had greatly hampered agricultural activities in the eastern Delta from the arrival of the first white settlers. Local levee and drainage districts had over the years made significant efforts to contain floodwaters, but the disastrous floods of 1932 and 1933 proved these efforts in vain as more than one hundred thousand Delta residents fled their homes fronting the tributaries. Alarmed by the vast human suffering in the poorest state of the Union, Congress passed the Flood Control Act of 1936, which provided for a comprehensive flood-control plan in the region with full federal funding. Known as the Yazoo Headwater

Project, the plan included levees, detention reservoirs, and channel maintenance and commenced immediately. Flows from the hill country into the Delta became completely controlled by the operation of four storage reservoirs constructed between 1936 and 1955. The Grenada Reservoir on the Yalobusha, the Enid on the Yocona, the Sardis on the Little Tallahatchie, and the Arkabutla on the Coldwater came to regulate the flows for about 60 percent of all the drainage area above Greenwood. Combined with new levees and auxiliary channels in the Delta, the reservoirs dramatically reduced the occurrence of floods in the interior Delta. At the same time, the projects permanently inundated large areas of bottomland hardwood forest along the tributaries and reconstructed the natural hydrological regime of the floodplain beyond recognition.[70] The hunting alter ego of William Faulkner, Ike McCaslin, noted that with the completion of the Sardis reservoir in 1940, "[t]he Big Woods; the Big Bottom, the wilderness" had "vanished now from where he had first known it; the very spot where him and Sam were standing when he heard his first running hounds and cocked the gun and saw the first buck, was now thirty feet below the surface of a government-built flood-control reservoir whose bottom was rising gradually and inexorably each year on another layer of beer cans and bottle tops and lost bass plugs."[71]

There were railroads in the wilderness now; people who used to go overland by carriage or horseback to the River landings for the Memphis and New Orleans steamboats could take the train from almost anywhere now. And presently Pullmans too, all the way from Chicago and the Northern cities and the Northern money, the Yankee dollars arriving between sheets and even in drawing rooms to open the wilderness, nudge it further and further toward obsolescence with the whine of saws; what had been one vast unbroken virgin span was now booming with cotton and timber both. Or rather, booming with simple money: increment's troglodyte which had fathered twin ones: solvency and bankruptcy, the three of them booming money into the land so fast now that the problem was to get rid of it before it whelmed you into strangulation.

—WILLIAM FAULKNER, *Big Woods*

CHAPTER SIX

# *Bounties of the Bottomland*

THE ANTHROPOGENIC ALTERATION of the southern flood-plain was not restricted to the rebuilding of its hydrological regime and the extensive conversion of bottomland forests to cotton fields. European settlement in North America brought sweeping changes in land use all over the continent; in addition to the clearing for crops and pastures, there were the harvesting of trees for housing, fencing, lumber, and fuel and the manufacturing of naval stores and charcoal iron, all of which accelerated the depletion of the eastern forests. Technological developments, such as the introduction of the steam engine, which consumed wood for fuel—and came to wide use in sawing machines—assisted this attack on the forests from the early nineteenth century on. Lumbering had been widespread on southern plantations and along coastal waterways since colonial times, and vast areas of forest land in the South had been cleared for agriculture before the Great Lakes region was even settled. With European colonization, lumber manufacturing became the first southern industry. Commercial lumbering had already begun in Jamestown, where the colonists sawed clapboards for transportation to England. Until tobacco emerged as Virginia's main export, the English supply ships returned from the colony with timber as their main freight.[1] Similarly, lumbering was among the first industries in Mississippi; by the 1830s, it had become a business of considerable importance in the Delta.

The lumber industry in the United States expanded enormously during the nineteenth century; the lumber cut increased from less than 0.5 billion board feet in 1800 to more than 35 billion board feet by 1899.[2] This expansion was partly made possible by various technical developments. Improved circular saws and steam engines replaced the earlier up-and-down saws and waterpower. The improvement of local transport with the development of the log drive, and the evolution of a continental transport system that linked areas of timber surplus and deficiency were other critical factors.[3] In 1909, the all-time peak of lumber production in the United States was reached: over 44.5 billion board feet of lumber was sawed that year, with almost 20 billion board feet of it coming from the South and South Atlantic regions.[4]

The growing utilization of the bottomland hardwood forest resources of the Delta during the nineteenth and early twentieth centuries closely reflected these national developments. However, the history of hardwood and cypress exploitation in the South—and the Delta—differed in many aspects from the more familiar story of pine lumbering in the Northeast, the Great Lakes region, and even the southern "Piney Woods" region. Furthermore, the commercial harvest of cypress, pine, and hardwood resources in the state of Mississippi showed considerable variation, as different species were exploited at different times and for a variety of reasons.[5]

## The Birth of an Industry

Rivers and streams were the primary means of transportation of logs and timbers to the lumber mills in the antebellum era. Forests bordering the streams were the first to be exploited, and so, for obvious geographical

reasons, lumbering activities in early-nineteenth-century Mississippi concentrated in the western and southern sections of the state. High-quality
cypress located close to waterways was felled, collected during high water,
and rafted down the Mississippi River to mills at Vicksburg, Natchez, and
New Orleans. The Yazoo, Homochitto, and Big Black Rivers, their tributaries, and streams enabled the floating of cypress timber to market, while
the Pearl and Pascagoula river systems were utilized for transporting logs
in the pine forest region.[6] In addition to the harvesting of cypress, barrel
staves were cut from the white oak timber and sold to overseas markets
during the first decades of the nineteenth century. In 1837, a German
observer stated that along the borders of navigable streams in the state of
Mississippi, "inexhaustible quantities of building lumber and hardwoods
of all kinds are to be found, readily absorbed by foreign markets. . . .
Many [trees] are of considerable size, and practically all can, when cut
down, easily be floated and rafted on the rivers. Already much of the
better timber is shipped down the Mississippi River to New Orleans."[7]

The rapid growth of New Orleans in the 1830s and 1840s provided the
stimulus for the development of the lumber industry in the pine belt of
southern Mississippi; by 1850 there were some twenty-five small mills in
operation at the mouths of watercourses emptying in the Gulf of Mexico.
The expansion of sawmilling was furthermore facilitated by the adaptation of the steam engine as a source of power. The Mississippi lumber industry seemed to be expanding rapidly before the outbreak of the
Civil War.[8]

Steam engines consumed vast amounts of wood, which remained the
basic fuel in the United States until the late nineteenth century; domestic
coal consumption increased during the nineteenth century and surpassed
wood in total energy consumption around 1880. Of all timber cut in gross
volume, the amount cut for fuel still exceeded the amount cut for lumber
in the late nineteenth century.[9] Ever since the colonial period, southeastern woodlots and sawmills had concentrated in the proximity of streams
and rivers, where the transportation of lumber was easy and the dominant
trees were hardwoods that constituted cordwood of superior quality. In
order to save labor and expense, cutting was customarily restricted to
riverbanks and adjacent slopes so the logs could easily be rolled into the

streams.[10] Captain Basil Hall commented on the volume of wood consumption by the Mississippi steamboats in 1828:

> As the steam-boats on this river, and indeed all over America, burn nothing but wood, and as their engines are mostly high pressure, the consumption of this bulky description of fuel is so considerable, that they are obliged to call at least twice a-day at the wooding stations on the banks of the stream. The *Philadelphia* used about one cord of wood an hour, or 128 cubic feet. A cord consists of a pile eight feet long by four high, and four in thickness, each billet being four feet in length. Sometimes, when we were pushing hard, we burnt 30 cords in a day.[11]

With the rapid increase in river traffic, the gathering and selling of wood for steamboats became a lucrative business for many settlers of European origin. The "woodhawks" presumably paid little attention to the question of legal ownership of riverbank forests and must have inflicted a major impact on the Delta forests bordering the Mississippi. Little has been preserved in the official records about this primitive form of "squatter capitalism," but the 1840 census on forest products reveals that the upper Delta produced a disproportionately large amount of cordwood in relation to the number of inhabitants in the area.[12]

The clearing of land for agricultural purposes decimated much of Mississippi's hardwood forests outside the Delta already during the antebellum era, while the barren pineland soils of the state's southern part remained outside of agricultural attention. Consequently, the majority of Mississippi's pineland acreage remained the property of the federal government in 1850, as did most of the land not adjacent to watercourses within the Delta. The federal Swamp Lands Acts of 1849 and 1850 aimed to increase the amount of arable land and donated all federal swamplands to the individual states, under the condition that the money obtained from their sale be used to levee and drain them. The state of Mississippi acquired over three million acres of federal land through this legislation. Interpretation of the law proved quite liberal; thousands of acres of dry pine forest changed ownership in Mississippi under the Swamp Lands Act. During the 1850s, the state then sold its newly acquired land for prices ranging from five to fifty cents per acre.[13]

From the beginning of the European settlement in the South, there had been a market for the durable baldcypress, a tree characteristic of the bottomland swamps. Unlike other typical bottomland trees, cypress is a conifer and falls in the softwood category.[14] Cypress was highly valued for shipbuilding and roofing purposes, as the wood efficiently resisted damp. Cisterns, cooper's staves, rails, and fences were among other applications for cypress.[15] There had been sawmills specializing in cypress products in the vicinity of the Delta since the 1820s. The sawmill founded in 1826 at Helena, on the Arkansas side of the Mississippi, provides the earliest example of such activity. In 1828, the first sawmill in the state of Mississippi was founded in Natchez by the Scotsman Andrew Brown, while the first sawmill on the Yazoo was erected by one Stephen Howard in the early 1830s.[16]

A German traveler on the Mississippi described the traditional method of bottomland lumbering as practiced in the lowlands of Missouri and Arkansas by local woodsmen:

> Rifles across their backs, game bags at their sides, revolvers in their belts, and axes on their shoulders, they cruise the primeval forest, find the tallest, strongest and most attractive trees, fell them with axe and fire, cut them into logs, and note their positions. With the coming of spring and high water, when the river overflows its banks for hundreds of miles around, they gather their logs, now afloat, make *"rafts"* of them, and head them to the river's main course. In a few days the strong current carries them to Memphis and even New Orleans. There the rafts are sold for logs. The lumbermen return to the forest in summer and fall when the floods have receded.[17]

While cypress dominated the early professional lumbering activities, ash and oak were also commercially harvested by the late 1860s. The oaks growing on the ridges of Delta bottomlands, however, would not float sufficiently without deadening, making a logger muse that it "seams strang[e] that oke timber will not float [in the] South," while it did in the North.[18] The lifestyle of a nineteenth-century bottomland logger could be summed up: "Well[,] Do You Know the Life of a Raftsman[,] Thay Don[']t Follow It for health."[19]

During the antebellum period, much of the best cypress land along the lower Mississippi ended up in the hands of commercial operators through fraud involving surveyors and land agents. On many occasions, lumber companies proceeded to illegally remove the valuable cypress from nearby lands still in public ownership. A forester, writing in the 1880s, gave the following short account of the history of cypress lumbering in the Delta: "The traffic in cypress lumber in the Yazoo region dates from 1830. In 1838 it was commenced upon the Sunflower river and Deer creek, ten years after the first settlements were established upon the banks of these streams; since that time rafts have been sent regularly to New Orleans, and camps of lumbermen have been established in every direction, the forests, particularly those upon the public domains, being regarded as the undisputed property and lawful prey of the log-getter."[20] Without doubt, large portions of the finest cypress stands along the lower Mississippi were customarily cut in trespass on public lands. Custody of the public domain in the remote bottomlands proved extremely difficult; besides, standing timber was still considered practically worthless by state authorities. The major assault on the cypress stands, however, was delayed until after the Civil War because of the inaccessibility of many swamp areas.

The original method for cypress lumbering was that of float logging. Cypress trees were girdled several months in advance of logging to make them buoyant. Girdling, also called "belting," usually took place in the fall, the driest season in the bottomlands. The trees were then felled during periods of high water and floated out of the forest through "float roads" previously opened in the forest.[21] In an 1847 letter, a transplanted Minnesotan described the early method of cypress lumbering in the Delta. Although Charles Helme admitted that his spelling was bad, with "severral words misspelt," and urged the recipient to "corect misstakes," the letter is worthy of an expanded quote:

[W]ell i will say some thing a bout my self and How fast i am a get[t]ing rich[.] i ame at labor by the mounth[,] twenty dollars a month if my patrons pay me[.] i do not think that thare is enny dou[b]t of that[.] i ame cut[t]ing cypress timber on gover[n]ment land[.] it is a verry profitable buisness to my employers[:] a log of cypress 50 feet long is worth 12 dollars at the mill[.] it is cut in the dryes season and the timber is cut on awwrage from 12 to 15 and 20 feet frome the [—ground?,] and a hand can cut frome 6 to 8 trees a day and when the mississippi river overflows its banks[,] the swamps are lull of watter frome [?] to 15 feet deep and then wee float this timber to the river and forme lange rafts of it of 200 hundred trees in a raft[.] thay com[m]only are frome 40 to 50 and 60 feet long[,] verry seldome under 40 feet [is] the len[g]th of trees . . . all a man has to do to cut timber is to find grab and axes and he can cut as much timber as he pleas[es]. this is the third year that i have done more or less in the buisness and do not think that i ever will make a hand for anny man again in the swamp[,] for it requir[e]s but a little capital for a man to furnish him self in the swamp.[22]

Given the high value of cypress lumber, it is no wonder that, in addition to harvesting corn and cotton, the slaves at the Panther Burn Plantation in the Delta were occupied for weeks in the fall of 1859 with "hauling Cypress logs into the creek."[23] On this occasion, the timber was probably cut from the domain of the rightful owner.

The Civil War abruptly ended the first boom in southern lumbering, but soon the region was again on its way to becoming the nation's leading lumber producer, a status won by the turn of the century. The large-scale lumbering industry had begun in the Northeast, extended to the Central and Great Lakes states at midcentury, and reached the southern states during the last decades of the 1800s. As northern timber resources were depleted, the South emerged as the center for forest exploitation in the United States, and after a quick and massive transfer of northern capital, technology, and know-how, the production of southern lumber increased from 1.6 to 15.4 billion board feet between 1880 and 1920. In 1919, some 37 percent of American lumber was produced in the region, and new uses for the southern forest, such as paper and pulp production, were being actively promoted. It has been estimated that this assault on the southern forest as a whole reduced the original wooded area in the region from nearly 300 million acres to 178 million acres, or by nearly 40 percent. Agricultural clearing had played a significant part in this process, but there is no doubt that the greatest inroads were made by lumbering.[24]

The Civil War had destroyed much of Mississippi's existing lumber industry. Operations of some coastal mills had, however, remained almost unhampered by the war, and in 1865 these mills were shipping lumber to Mexico and the West Indies in addition to New Orleans. Small mills along the Illinois Central Railroad and in the northern parts of the state were also busy, sawing lumber for the rebuilding of towns destroyed by the war. As a rule, however, markets for lumber production in Mississippi remained limited in the 1860s.[25] By the 1870s, Mississippians had begun to recognize the commercial potential of their forest resources, and the state's pavilion at the Centennial Exhibition at Philadelphia in 1876 included "a rustic log cabin in the Swiss style of architecture," containing sixty-eight varieties of Mississippi woods, "its eaves being festooned with long, waving festoons of Spanish moss, and the whole presenting the most picturesque structure upon the premises."[26]

The rise of activity in southern lumbering during Reconstruction resulted largely from certain ongoing processes in post–Civil War America. Many in the defeated South saw emulating the ways of the victorious North as an answer to the evident problems of their old agricultural society and looked to industrialization as a savior of the region. The northern methods of industrial enterprise were, however, not so easily imitated in the capital-poor South: before the war, most assets had been tied up in land and slaves. Timber was one of the few additional resources the South had to offer boosters of economic development.

The concept of "New South" evolved in the aftermath of the Civil War from southerners who saw economic regeneration as the region's most pressing need and industrial development as the way to attain prosperity, social stability, and even racial accommodation. While the originator of the term, Atlanta newspaperman Henry W. Grady, was the most outspoken proponent of industrialism and economic independence in the post–Civil War South, he took great care to reassure the large landowners of the region that industrial development would not endanger the plantation system by depleting its pool of cheap, controllable labor. Despite lofty late-nineteenth-century rhetoric to the contrary, many scholars have pointed out that the Reconstruction South remained a classic example of an underdeveloped region with an unbalanced economy that plunders abundant resources in search of quick gain. Postwar southern enterprises, as a rule, offered little employment in more developed forms of industry or manufacturing, and turned out be exploitative of both resources and workers. At the beginning of the twentieth century, the modest industrialization that had taken place was largely controlled by northern and even international capitalists, while local ownership remained limited.[27]

Following the Civil War, many southern states were economically strapped because no taxes could be collected on federal lands, so the sale of these lands—in order to put them back on the tax rolls—soon became inevitable. The sale enabled northern lumber companies and speculators to take over vast areas of southern forest: millions of acres of forested public land were sold to ruthless northern buyers, often at minimal prices.[28]

Initially, the investors were not interested in the land itself but only wanted access to the timber growing on it. A Virginia planter explained some postbellum realities to a Delta landowner who was trying to sell some of her landed property: "[t]he people have not the money. There are but few who could pay for it, and those who have money, generally have abundance of land & do not buy *except* in cases to save money lent *on it*." A timbered piece of land could bring the same amount of money for cutting rights as for sale: "Abandon the idea of a sale. Most parties who *want it,* really want it for the timber & desire to make the *land clear. . . .* There is a saw mill immediately by it & my opinion is that they will buy the timber & that they would give just as much for it, as for *both land & timber*."[29]

Between 1866 and 1876, the only way to obtain land from the federal government was through the Homestead Act of 1862. During that period, the rapid expansion of lumbering in the United States created a strong demand for the sale of federal lands to individual purchasers in unlimited amounts. The four million acres of federal land in Mississippi were opened for sale without restriction in 1876 for the price of $1.25 per acre. Northern lumbermen and land speculators consequently obtained vast tracts of prime timberland in the state: between 1881 and 1883, over one million acres of federal land changed ownership.

In 1877, the amount of public land that could be purchased by an individual within one year was restricted to 240 acres. Lumbering interests, however, still obtained enormous acreage by using middlemen. Timberlands were also often purchased in the names of deceased or invented persons. Employing such methods, Phillips and Company, predecessor of the Delta Pine and Land company, was able to acquire some 180,000 acres of Mississippi pinelands over a three-year period. Such manipulations were not restricted to the state of Mississippi or pinelands; vast areas of mature cypress forests on public lands along the Mississippi, Red, and Atchafalaya Rivers passed via corrupt officials to lumbermen. In 1881, one operator purchased over half a million acres of Delta hardwood forest that eventually became property of the Delta Pine and Land Company. Huge purchases of federal and state lands by companies and individuals under the new land laws resulted in high concentrations of land ownership and widespread fear of monopolies. In 1888, the private entry law was repealed, and the remaining one and a half million acres of federal land were made available only through homesteading and the Soldiers and Sailors Act. Still, by 1905, practically all federal land in the state of Mississippi had become private property.[30]

In the aftermath of the Civil War, northern lumber companies were primarily interested in the pristine pine forests of the southeastern uplands. The variety of southern pines, the shortleaf, longleaf, loblolly, and slash (*Pinus echinata, P. palustris, P. taeda, P. elliottii*), made good substitutes for the recently depleted white pine (*P. strobus*) forests of the Great Lakes region. In the early 1870s, the manufacturers of railroad cars in the North faced problems as the price of white pine, commonly used for sills and decking in the industry, increased considerably as a result of the overcutting of the stock in the Great Lakes region. These developments gradually led to its substitution by the cheaper pines from the South and to the expansion of softwood lumbering in the southern states. Initial problems with the quality of southern softwood timber were solved by the adoption of dry kilns in the southern lumber industry. The kiln, used in Mississippi since the late 1870s, prevented bluing (fungal discoloration) of lumber and considerably increased its overall quality. It furthermore reduced the average weight of the product by roughly one quarter, which resulted in lower shipping costs. Consequently, the Mississippi pine industry expanded rapidly during the 1880s in areas where transportation facilities were available.

Mississippi sawmills had originally been erected near the timber supply, and the sawn product was transported to railroad stations by teams and wagons. Sawmills in the 1870s, on the other hand, were located in the vicinity of railroads and teams brought logs from the forest for processing. By the mid-1880s, however, the timber supply close to the railroads had been exhausted, and more elaborate ways of logging the pinelands had to be developed. Log transportation was made considerably easier by the development of tram-roads. The first tram-roads consisted of a track made of poles that were placed end to end. Wagons with grooved treads to fit the poles were pulled along the track by oxen or mules, which were later replaced by small steam locomotives. Technical innovations already in use by the northern lumber industry, such as endless chains, friction and wire feed, the so-called steam nigger, edging machines, and double circular saws, were commonly adopted in Mississippi during the 1880s and led to great increases both in the volume and quality of lumber processed in the state.[31]

In the late-nineteenth-century Gulf States region, methods characteristic of the old lumber industry were used: enormous mills, huge log drives, and aggressive pursuit of timberlands combined with "cut out and get out" policies resulted in vast deforestation. The new owners usually looked only for a quick profit and missed the fact that trees formed an essential protection to the soil, which could yield a valuable crop also in the future; no effort was made to replant the clear-cut forests. Railroad construction significantly spurred the timber economy and made it possible to harvest formerly inaccessible stands of virgin timber. "Logtowns" sprang up across the region, and logging drove the local economy until the supplies were depleted. The settlements and the cut-over lands were then abandoned—and dropped from the tax rolls once again. A modern sawmill, capable of processing some one hundred thousand board feet per day, required a vast local supply of timber. The local source of timber, however, had to last at least ten years in order for the sawmill to be economically viable, as transportation costs prevented the hauling of logs over long distances. The largest mill operations therefore clustered in the pine-producing region of the southern part of the state, where the largest and most homogenous timber stands were located.[32]

As a result of the almost complete depletion of white pine stock in the North by 1900, the Mississippi lumber industry's most rapid expansion occurred during the first decade of the twentieth century. Numerous wealthy lumbermen from the North began operations in Mississippi and elsewhere in the South. Their onslaught on the remaining forests was complemented by the expansion of activities by local operators. Construction of new railroads all over the South opened previously untapped forest resources for exploitation. The whole process was facilitated by the general rise of lumber prices between 1899 and 1907. Lumber production during that period was clearly profitable for the average producer, especially if the stumpage had been obtained at low cost. The extent of the growth is evident in the amount of capital invested in Mississippi forest industries: $10,800,000 were invested in the existing 608 sawmills in 1899, compared to $39,455,000 in 1,647 mills only a decade later.[33]

At the turn of the century, Mississippi mill owners were typically using wholesalers to ship their product to northern markets. Gradually the larger operators established their own selling agencies or cooperated with other producers in marketing their lumber, aiming to gain the wholesale cost for themselves. Despite the wide establishment of manufacturer-wholesaler combinations, separate wholesalers continued to market much of the smaller operators' production in Mississippi. The expanding lumber industry in Mississippi employed labor on a previously unknown scale. By 1909, forest industries employed over 60 percent of full-time wage earners in the state. During the first three decades of the twentieth century, the labor supply was insufficient to meet the demand. Most of the workers employed by Mississippi lumber operations were locals, both black and white, but labor was also imported from the old timber districts of the Northeast and Great Lakes states.[34]

Logging and mill operations were, as a rule, located in sparsely populated regions far removed from urban centers. With the expansion of large-scale lumbering into unsettled areas in Mississippi, timber workers moved into isolated company towns, especially in the pine belt region. Constructing sawmills in the middle of the forest (but adjacent to railroads), both in the pine hills and in the Delta, could prove complicated, as the operator had to erect housing and provide other services for the workers. When the timber in a given locality had been exhausted, the cheaply built houses were dismantled and moved to a new location. Laborers had to submit to rigid discipline and economic exploitation. They were forced to use the company store and charged exorbitant prices in a manner similar to sharecroppers at plantation stores. Many timber workers, especially blacks, worked in conditions approximating peonage.[35] Strikes at Mississippi sawmills were generally unknown, and the typical wage for a common laborer in the beginning of the 1910s was a mere ten cents an hour, rising to twenty cents during the next decade. In contrast to timber workers, a forest surveyor's fee for a day's work in 1916 was five dollars plus expenses.[36]

During the heyday of lumbering in Mississippi, over one hundred thousand acres of timberlands, most of it in the pine region, were logged annually. Initially, only the best-quality timber had been marketable, and fairly heavy stands of younger timber remained in the logged areas. These areas could be logged over as soon as the development of tram-roads made it economically feasible to bring lower-grade timber to the market. The advent of the log skidder, common among larger operators after the turn of the century, changed that. Trees that were not hauled to the mill were broken off by the skidder, so little if any timber remained standing after the logging crews had done their work.[37]

Between 1907 and 1916, except for a short period in 1912–13, lumber industries barely broke even as prices remained little above production costs, and profits were generally derived from an increase in the value of stumpage. Attempts by some lumbermen to raise prices through decreased production usually failed, as the majority had acquired their timberlands and production facilities with borrowed money and had to operate full scale in order to meet payments. After the United States joined World War I in 1917, lumber operations in the South became more profitable, a trend that continued until the Great Depression.[38] Because of cheap labor, tax incentives, and depletion of old-growth stands in the western and northern parts of the United States, the southern forest industry remained strong during the twentieth century. At the same time, technical innovations, such as the development of lightweight chainsaws in the late 1940s, made logging easier than ever before.

## A Bottomland Bonanza

Although the main focus of the northern invasion of southern woodlands was directed toward the region's intact pine reserves, the vast cypress stands of southern bottomlands offered lumbering interests additional opportunities for quick gains. Toward the end of the nineteenth century, the shrinking acreage of high-quality cypress reflected the expansion of lumbering in Mississippi. More rational utilization of the state's forest reserves, however, called for scientific surveys of the existing forest.

Formal study of Mississippi forest conditions was begun in 1857 by Lewis Harper, who unfortunately omitted discussion of the Delta forests on the grounds that the region had not been completely explored. Other descriptions of Mississippi forests, including references to the Delta, were published in the nineteenth century by Eugene Hilgard, A. B. Hurt, and E. G. Wall.[39] In the early twentieth century, Ephraim Lowe was the first to present an organized taxonomic treatment of Mississippi flora, including a detailed description of the vegetation found in the Delta.[40]

The first survey of importance in Mississippi, however, was carried out under the supervision of Dr. Charles Sprague Sargent for the purposes of the 1880 federal census. Sargent, Arnold Professor of Arboriculture at Harvard University, had been appointed special agent by the super-intendent of the census to survey all forest areas in the United States, a landmark effort in the rational development of the nation's natural re-sources. Unfortunately, many estimates in Sargent's classic *Report on the Forests of North America* remained little more than educated guesses, and the first exhaustive survey of southern bottomland hardwood stands had to wait for half a century. Sargent's general account of the Mississippi forest resources noted that the state's bottomland forests remained "al-most intact, although the most accessible cypress, which has long been cut in the Yazoo delta and the valley of the Pearl river to supply the New Orleans market," had become scarce.[41]

Surveys of the bottomland forest resources in the South for Sargent's report were performed by a German immigrant, Dr. Charles Mohr, who showed great concern for the future of the region's cypress stands. Describing the forest situation in the bottomlands of Alabama, Mohr reported that "[i]n 1831 Mr. Vaughn found these cypress swamps untouched by the ax. At present their resources are so diminished by the inroads made upon them during the last twelve years that, with a prospect of a rapidly-increasing demand for cypress lumber in the near future, he judges that they will be completely exhausted during the next ten years. This opinion is shared by all mill-owners here, who believe that in less than that time their business must come to an end."[42] According to Mohr, "[t]he large number of [baldcypress] logs harvested shows clearly with what activity the destruction of these treasures of the forest is being pushed; and the reports, as of heavy thunder, caused by the fall of the mighty trees, resounding at short intervals from near and far, speak of its rapid progress."[43] The situation in the Delta resembled that in Alabama: hardwood reserves were practically untapped while cypress stands were diminishing rapidly. Cypress "was once found in the greatest abundance in the [Delta] region, and immense quantities of cypress lumber" had been "furnished by the lower parts of Issaquena and Washington and the western parts of Warren and Yazoo counties." Mohr underlined the connection between the accessibility and exploitation of cypress resources. By 1880, the most valuable timber had "disappeared from the immediate neighborhood of the low river banks easily accessible at seasons of high water during every winter and spring." Consequently, only stands removed from the banks of the water courses and "only accessible to the raftsman during exceptionally high stages of water" now supplied good cypress.[44]

In the central and upper Delta, "more or less limited areas of undisturbed cypress forest" were still found, and cypress brakes remote from streams and surrounded "with a mire of forest swamp impassable to wagons" retained "their best timber" in 1880. The most extensive cypress groves were found "along Steele's bayou, between Deer creek and the Sunflower river, in Washington county; between that stream and the lower course of Bogue Phalia, and between the Mississippi river and Black creek above Greenville." There was furthermore "a very large body of cypress inclosing the 'California brake', upon the Little Sunflower, in the counties of Bolivar and Coahoma, extending through Tallahatchie county to the Yazoo River." Mohr could not help noticing the rapid change occurring in the ownership of the Delta lands after the war, and observed that "since swamp and overflowed lands have become the property of the state, planters have added many of these cypress tracts to their estates by purchase." Furthermore, many brakes had been "acquired by companies and [the] most extensive of these groves of cypress . . . [were] already in the hands of capitalists."[45] The early conservationist seriously doubted the approach taken by Mississippi in the disposal of its bottomland acreage:

> The cutting of these cypress forests is not wisely regulated under the ownership of the state. These lands have been thrown into the market at 50 cents an acre with the condition of settlement. Beneficial as such a law might prove in the disposal of lands fit for cultivation, it results, in the case of timber-land unfit for the plow, in the reckless destruction of one of the surest sources of public revenue. The state thus sells for 50 cents what on its face is worth to the purchaser hundreds of dollars, and which, when deprived of its value and rendered forever worthless, will be turned back to the state again.[46]

By 1880, Vicksburg had become "the center of a considerable lumber industry, depending for its supply of timber upon the cypress rafted down from the mouth of the Yazoo river." The first sawmill devoted to the manufacture of cypress lumber had been established in Vicksburg already in 1865. Prior to that, all Delta cypress timber was rafted down the Mississippi, mostly to Natchez and New Orleans, as still remained the case with most rafts. *Report on the Forests* furthermore noted the connection between growing cypress consumption and the emergence of a New South cotton kingdom in the Delta. Much of the lumber was used locally by planters and sharecroppers: "A second mill has lately been built at Vicksburg, and the combined annual capacity of the two is ten to twelve million feet. No manufactured lumber is shipped from here to farther south than Baton Rouge, nearly the whole production being consumed in the erection of small dwellings in the Mississippi and Yazoo bottoms. The logs received at these mills average 25 inches in diameter, with a length of from 30 to 70 feet."[47]

For most of the nineteenth century, only the timber accessible to streams subject to flooding was harvested, and cypress stands in the Delta were only "culled of their best timber wherever it could be obtained without the invest of capital, that is by simply floating the logs to the streams at times of freshet and overflow."[48] Antebellum methods in logging continued to be in use, and flooding remained essential for the successful harvest of the bottomland crop: "The river was extraordinarily high, the lowlands being overflowed to a depth of more than 10 feet. . . . No idle man was to be found on shore; everybody who could swing an ax, paddle a boat, or pilot a log was in the swamp engaged in felling and floating cypress timber. All the mill-hands worked in the swamps; fields and gardens were left untouched, and even clerks from the stores were sent to swamps as overseers."[49] As a rule, however, cypress lumbering was now carried out in a more organized manner. More or less professional logging crews, working under a foreman, cruised the cypress brakes, cut the timber, and prepared the floats for the market. The crew foreman, in turn, was closely supervised by the mill owner, who issued written instructions to his employees and might even pay a personal visit to the lumbering site.[50]

The infringement on the public domain by loggers, however, continued after the Civil War, and was not merely restricted to cutting on government land. The expansion of the levee system during the nineteenth century gradually began to hamper floating activities. As it was impossible for the levee districts in the Delta to keep watch over the whole levee line, it became a not uncommon practice for logging crews to dismantle parts of levees during low water in order to float their cut to the markets. After the floating was completed, the levee was hastily repaired.[51]

After the Civil War, the destination for much of the Delta cypress timber continued to be the first Mississippi sawmill in Natchez, owned and operated by the Andrew Brown Lumber Company. The preserved correspondence between the management of the lumber company and its field workers offers valuable insights into the hard work of logging in the Delta.[52] In February 1866, the logging crew foreman William I. Hindman, while asking for instructions on how to pay taxes on the company's Delta timberlands, informed Brown that he was patiently waiting for the water to rise in the Black Bayou in order to float his timber to the Natchez mill. Hindman could not attend to other lumbering activities, as the felled cypress logs needed constant attention—they could be easily lost with a sudden rise in the water level.[53] Later in the year, Hindman began to experience labor problems and informed his employer in November that his crew "dew not like me be case i croud them." Still, Hindman could not afford to estrange his men, as hands were hard to find in the Delta "with out going to Vixburg."[54] A few weeks later, he was "compel[le]d to quit work in the Swamp own the account of the heavy Ranins" and requested that the company send him "seventy five dollars to pay my men of an[d] charg[e] the same to A[.] Brown."[55] To his employer, Hindman gave this excuse-filled explanation:

> I am sorry to say to you that I have quit work in the Swamp own the account of the watter being two deap[.] I can[']t get men to work in the watter an[d] if I co[u]ld[,] it wood not pay own the account of the disadvantage of getting to work[.] I tho[ugh]t I had better stop the expenc[e.] It is hard own you I [k]now an[d] for my part it hurt me as bad as aney thing co[u]ld happin to me but ther[e] is now posabel chan[c]e to dew aney thing fo[r] we have had the largest Rane that ever I saw fal[l] sins I bin in the Swamp an[d] thar[e] is now chan[c]e of the watter run[n]ing down this season[.] If I had my timber an[d] dam reddy I co[u]ld be a floatin[g] now[.] I have cut down 1.75 one hundred an[d] seventy five tears of timber and b[u]ilt my camps[.]
>
> You may [k]now that I made al[l] of the efford that I co[u]ld to hur[r]y a long but I am more convinst that Baker Brake is the best and the Shores[t] plase to get timber out and I examend al[l] of the timber that is a bov[e] the dam and I think we can cut it without paying aney thing but you may be dishartend with this consem and I mite be two but s[t]ill I am willing to try it agane but I will try it in the summer[.] I want to cut for fifte[e]n hundred tears of timber.[56]

The conditions remained difficult in the bottomlands, and on December 30, Hindman informed Brown that he had his "men cutting timber in two and half feet watter an[d] it was sow cold they co[u]ld not stand it." By February 11, 1867, he was "not dewing any thing," confessing that "I am short of money to gow own an[d] I am a fraid I will have to abandon the buisness[.] The arrangement I had for hands fail[e]d an[d] it liavis me in a bad fix[.] At present i dew not now what I will dew." In late February, flood waters from the Mississippi entered the Sunflower River system, and the water level in the tributaries rose fast. Lacking hands, Hindman requested Brown to forward one hundred dollars, though "[f]ifty will dew if you are scarse of money," for hiring help, as he was cutting "a lit[t]le Sipress as well as ash" just by himself.[57]

By mid-March, Hindman was on board the "S[t]e[a]mer Cairo Bell[e]," on his "way to Quiver for the California timber with a good prospect of sucksess," and hoped to "be abel to get som[e] timber out of the Brak[e] becides that that now laing in the bayou." While traveling to the logging site, Hindman had learned that there was "a party going to steal" some previously felled timber below Holland Landing, now afloat, and "left word for them to let it lay." By early May, he had succeeded in getting some timber out of the woods, but again the hapless logger experienced labor troubles: "What hands I had quit me and I dew not [k]now what to dew[.] Ther[e] is a plenty of watter to work and i here the watter is rising a gane a bove[.] It is a falling at this plase[.] The Bayou is in a horabel condition to get timber down but if I can get men I will gow back a gane[.] I am very near worn out[.] I never worked harder in my life."[58] Whether Hindman was working hard and providing Brown's sawmill with prime cypress is not clear, but by the end of 1867 the Delta logger and his Natchez employer had a severe—and final—falling out. Hindman complained that Brown did "not intend to settle with," or even write to him, while the mill owner closed Hindman's account in his books with a debt of $616.18, noting that this amount had been drawn by Hindman "and in no wise accounted for, he being unprincipled and worthles[s] there is no hope of recovering the same."[59]

In the following decade, Brown's company changed ownership and became known as the Learned Lumber Company, after Rufus Learned, Brown's son-in-law.[60] The supervision exercised by the new principal owner over his bottomland workers, however, remained close. The extensive correspondence between Learned and his trusted foreman, Henry Wilson, between 1879 and 1882 provides a most illustrative chronicle of bottomland logging at the time. In April 1879, Wilson was preparing to float timber to Natchez and asked for "about 15 hands (white prefer[r]ed) so as to rush matters straight along."[61] The following year, Wilson asked

Learned to send up "at once 4 or 5 hands" and ended his letter begging for at least "some one, white or black," as he was getting behind in the operations with "chills & rain forever." [62] Similar requests for labor continued through the early 1880s.

At the request of his employer, Wilson had cruised the Hollywood Brake "as much as water and mud would permit." He noted that the lumber was fine and accessible, and there were enough sloughs and bayous for transportation purposes. Wilson was certain of getting at least eight thousand cypress trees from the brake. While confessing that no one could "form any conclusions as regards the am[oun]t of timber without a thorough knowledge of the [logging] lines," Wilson thought there was "a sufficient queantity to keep" Learned's mills "going for the next eight or ten years." [63] In January 1882, rapidly rising waters forced Wilson out of his camp. [64] In February, Wilson again pleaded for "Ten or 15 men[,] 3 or 4 white men" and specified his order: "Black good hands if you can possibly get them." [65] In the end of April, Wilson was at last satisfied with the size of his labor force and informed Learned that while "the colered men you sent are good hands," he "had knou use" for the white workers. [66] Only a few days later, Wilson was "getting a long very well" in his Coles Creek logging area: he had two tows ready and asked Learned to send up a tugboat as the water level was "falling fast in the Swamp." [67] In August 1882, Wilson and his crew continued their work in the bottomlands. During the dry season, their main occupation was deadening the trees. Still, the crew foreman was "on the Go all day," and hoped to have a horse, as he had to visit the nearby towns "for Medizine" and could not "Stand the walking." [68]

Without doubt, the job of a foreman was a demanding one: in addition to satisfying the needs of his employer, he had to maintain a positive working relationship with his loggers under grueling conditions. Furthermore, there was always the question of maintaining a sufficient stock of food and other supplies at the remote logging sites. In October, 1866, a purchase from a St. Louis merchant for a Delta logging crew included four blankets, three barrels of flour, two barrels of pork, one barrel of beef, one bushel each of white beans and dried apples, augmented by twenty pounds of soap and a box of "good common tobaco." [69] Thirteen years later, the lists of supplies for a crew remained similar: Irish potatoes, onions, grits, flour, molasses, rice, dried apples, red pepper, candles, coffee, tobacco, and blankets.[70] Sometimes the loggers had to put up with second-rate supplies. "The b[arre]ll of Pork you sent last is simply bad, it smells awfully, but we are trying to put up with it," complained Wilson to Learned in 1879.[71]

In the harsh swamp conditions, substances such as tobacco were highly valued by the workers. Great care had to be taken by the foreman to secure a continuous supply and to divide it equally among the men. When ordering buckets of tobacco, Wilson on more than one occasion specifically asked that it be proportioned in equal shares, as he possessed "no means of weighing it to the hands" in the field.[72] Once Wilson even asked his employer to "get out whiskey Licenc[e]s," which "only" cost $12.50. The reason was clear: "For the men are all Grum[b]ling about the water— They say that they want whiskey."[73]

Technological advances during the last years of the nineteenth century revolutionized cypress logging. It would have been difficult for William Hindman and Henry Wilson, wading hip-deep in the swamps with their crews and floating their cut, to imagine the rapid mechanization of the trade in the near future. Southern float logging was largely replaced in the 1890s by the pull boat method: a stationary engine, mounted on a barge, skidded the logs over the soft surface of the bottomlands for distances up to two thousand feet. A dredge boat worked in advance of the pull boat and opened canals in the swamps, giving access to cypress brakes in areas where natural channels were lacking. Around the turn of the century, the overhead-cableway skidding method came into use in connection with railroads, and opened previously inaccessible cypress stands for logging. In this method, the cypress logs were brought to the railroad track by a carriage traveling over a cable suspended between two trees some six hundred feet apart. Heavier machinery and other developments soon enabled longer distances to be covered.[74]

In 1915, a U.S. Department of Agriculture bulletin noted that much of the cypress lands, acquired by nineteenth-century speculators for between 25 cents and $1 per acre, commanded prices from $70 to $125 for the standing cypress alone. Because of the large investment required for

logging, cypress lumbering had become a highly concentrated industry after the abandonment of float logging. The cost for logging railroads built
on pilings over swamps ranged from $9,000 to $12,000 per mile, while
more temporary logging spurs built on mud piles or sawdust beds cost
from $1,000 to $2,000 per mile. Cypress timber also entailed high storage costs, as the bulk of the stock had to be air-dried from eight months to
two years on lumber yards. As a rule, kiln drying was used only for shin-
gles and lath. In the mid-1910s, the mill-run price of cypress in Louisiana
averaged $23.50 per thousand feet; the average stumpage value was es-
timated around $7.50, while logging and manufacturing costs amounted
to $13, creating a net profit of some $3 per thousand feet for the operator.
While there were "various forms of waste in logging" in certain cypress
regions, utilization was "complete, in the most literal sense of the term,
among the Mississippi River operators." Stumps were cut low, top logs
were also taken, and "not a living cypress tree" remained after the logging, as "practically everything left by the axe" had been broken by the
overhead skidder.[75]

Cutover cypress lands of the lower Mississippi Valley were considered
mostly unproductive in the 1910s, being held only for their potential value
for agriculture after drainage and clearing. Their speculative value averaged from $2 to $2.50 per acre, with considerable variation by elevation.
The average cost of drainage was estimated at $17 per acre, but the diffi-
culty of clearing land covered with stumps increased expenses drastically.
With the operator's profit set from $15 to $25 per acre, market values for
reclaimed swamp lands ranged from $60 to over $100 per acre. On the
more productive soils for agriculture, such as those of the Delta, growing
timber as an investment had little to offer.[76]

The depletion of Northern timber reserves following the Civil War was
not limited to conifers such as the white pine. Beginning in the late 1870s,
scarcity of high-quality hardwood in the North turned the lumbermen's
attention to America's last great hardwood reservoir, in the Lower Mississippi Valley, where an estimated 80 percent of the nation's hardwood
stock stood. The scarcity of wood in the central prairies, which were undergoing rapid settlement, acted as an additional stimulus for lumbering

in the South and led to the emergence of a continental system of timber transportation.[77] By this time, lumbering had depleted much of the Delta's most accessible cypress stock, but other species of bottomland trees were still found in abundance, despite the inroads made in the forests by agricultural clearing. As Mohr observed, "covered by these splendid forests of hard woods," the Delta possessed "a wealth of timber of the most valuable kind and in surprising variety." Besides, the beneficial effects of the Civil War on forest growth could be detected in the landscape, as "[m]ost of the clearings made in this region before the outbreak of the war, by the planters settled lower down, have since been abandoned and are again densely covered with the young growth of the trees of which the forest was originally composed."[78]

According to responses to an official 1875 circular, nearly one-third of the border county of Holmes was covered by "heavy swamp forests" still sustaining large quantities of cypress. In the core counties of the Delta, about 80 percent of Leflore and Washington were still heavily timbered with bottomland hardwood forest capable of yielding an average of 40 to 70 cords of wood per acre. Bolivar County was "reported as one dense forest, with a few clearings, including immense quantities of cypress, cottonwood, ash, the gums, and all the oaks, except live oak." Tunica County claimed to possess at least fifty thousand acres of cypress in addition to other kinds of timber, the latter still remaining "valueless, for want of enterprising capital."[79] In the beginning of the 1880s, it was generally believed that "not one acre in fifty over this whole region of hard-wood forest has yet been stripped of its tree covering."[80]

In 1880, the highest concentration of sawmills in the core region of the Delta was found in Bolivar County with eight mills in operation. Leflore and Washington followed with three mills each. Of the border counties, Panola maintained nine sawmills; Holmes, five; and Carroll and Talla-hatchie, four each.[81] The situation was summed up in the Sargent report: "The industries, however, which depend upon the hard-wood forests for material are still in their infancy in Mississippi, and are capable of enormous development."[82]

It was not until the 1880s that Northern lumbering operations began to show serious interest in the hardwood forests along the Lower Mississippi, and the city of Memphis on the northern edge of the Delta began its rapid rise toward the status of "The Hardwood Capital of the World." At the time, Delta residents replying to a federal forest questionnaire could boast at possessing "the largest supply of fine timber and the best country in the world" knowing there was nothing "but the hand of man to destroy them" for the needs of cultivation.[83] Nevertheless, the hands of the late-nineteenth-century Delta settlers could prove exceedingly destructive, as explained by Charles Mohr:

> During the last few years, however, the country has been entered again for cultivation by a class of small farmers, who from being farm hands have now risen to the position of independent landholders. It is astonishing to see the utter disregard of these settlers for the forest wealth of the country, which in a short time could not fail to be of great commercial value. On the shores of Indian bayou [in Sunflower County] may be seen clearings with hundreds of the finest black walnuts among the deadened trees, while many of the noblest specimens of this valuable timber are felled for fence rails or trifling purposes. The amount of oak and hickory timber destroyed here annually is amazing.[84]

Mohr furthermore noted that "[m]uch of the destruction of the timber can be traced to wasteful methods practiced by the negroes." A great amount of potential wealth was about to be wasted by freedmen on the threshold of a new, prosperous era for lumbering—"[u]nder present methods any one having rented a plantation will, for the most trifling wants, cut down a tree, regardless of size, and without any effort to preserve for future use the parts not immediately wanted, so that the next quarter of a century will probably see the entire destruction of vast quantities of timber stored in the whole of this great territory."[85]

Messrs. Edward J. and Alphonse Bobet of New Orleans were among the early hardwood operators in the Lower Mississippi Valley. During the 1860s and 1870s, the Bobet brothers purchased oak staves from all over the region and shipped them to European markets via Liverpool, England. Work conditions in the bottomlands were as hard for those making staves as for those logging cypress, as described by foreman W. B. Clement in May, 1873: "[It is] verry warm and rodds verry mudy[;] teams miring down every day and river fawling so fast[,] water will be gone before i can

an epiphyte growing upon bottomland trees. Already in 1808, Henry Ker noted the many uses of it: "On this moss Spanish horses subsist the year round. The inhabitants use it for mattrasses, &c.; there is also much of it exported for the purpose of stuffing saddles, chairs, &c. after it is cured in a proper manner. It is found only on low ground, and near water courses. It has a taste not unpleasant, and is full of juice. Large quantities of it are taken to the northern states."[92] Similarly, Christian Schultz commented that "[t]his singular and equally useful flying vegetable, . . . when prepared and cured, makes cheap mattrasses, equally pleasant and elastic with those made from horse-hair."[93] In 1837, in addition to requesting three thousand cords of wood from the manager of her bottomland plantation in Rodney, Jefferson County, a landlady urged him to "[t]ry to get any moss."[94] The moss was gathered by men known as "swampers" who navigated the swamps during floods in small boats. Wages were relatively low, so the swampers were typically people who could not find more profitable farm labor. Moss was most often shipped to New Orleans, the center of the "moss-ginning" industry with six factories in 1880, where it was dried, ginned, and sold for insulation and other purposes all over the country. Moss-ginning was a big business in the late-nineteenth-century South: in just one year, some thirty-five hundred bales of rough moss weighing ten million pounds and valued at $315,000 were processed in New Orleans alone.[95]

During the last years of the nineteenth century, the rapid extension of the railroad system into the interior Delta was imperative to the enormous growth of hardwood industry. In the words of William Faulkner, "it was as though the train . . . had brought with it into the doomed wilderness, even before the actual axe, the shadow and portent of the new mill not even finished yet and the rails and ties which were not even laid."[96] The first railroad in the region, the Louisville, New Orleans & Texas Railroad, had been "granted a heavy subsidy in lands" in 1881. By the turn of the century, much of the two million acres had been sold, either to large investors or small planters. Both were, however, opening it up for cultivation "with great rapidity." Indicative was "a recent sale of 150,000 acres" of "average" Delta timberlands "by the Yazoo & Mississippi Railroad Company to Messrs. George T. Houston & Company, lumber dealers in

Chicago," reported at "approximately $1,000,000." The lands were ex-
pected to "be thrown upon the market as farm lands" as soon as they were
"denuded of the marketable timber."[97] Whereas the early plantations in
the Delta had been carved from old-growth forests, the later ones were
founded on cut-over lands where merchantable timber had been removed
by sawmill interests.

In September 1882, Geo. S. Irving, a Vicksburg "Cotton Factor, Commission Merchant and Dealer in Groceries, Wines and Liquors," wrote a
long letter to Rufus Learned, the Natchez lumberman. Irving complained
that Delta timber was becoming inaccessible, as "the Memphis & Vicksburg Rail Road has & Is Seizing all the timber that come out of Sunflower
River Except the Carroll[,] Campbell & Merrigold timber[.] Campbell
Is a Stock holder In the Memphis & Vicksburg RR & Is In the Ring."
The angry merchant noted that this was "Capital & Monopoly against
the Poor Labor of the Yazoo." If nonrailroad interests were to "Submitt
to this," Learned and other lumbermen might as well "Sell out or Burn
down" their mills, as they would "only Be able to Get Such Timber as will
Be Gotten out By the Larger Concerns. . . . [A]s to the Men that Started
this Scheme," noted Irving, "thay are Regular Tricksters."[98]

By the early 1880s, the clearing of land titles and "[t]he Grant of the
State of the Swamp Lands" in the Delta had, in the words of the lumber
merchant, done "a Big Thing for Speculators." For the time, many new
owners refused to sell "any Lands to any one" and temporarily "Killed
the Swamp Country," as raftsmen could not get hold of new timber reserves.[99] Irving continued that

> You May Make The Timber Men Well Win Every Suit as all the Timber Seized
> By the Rail Road was Chop[p]ed Before the Road Bought the Land[.] The State
> In Some Instances May heave Claims But the Road has None[.] Mr Wilson with
> his Millions Will Pay for all this. . . . [T]he Present Superintendent I Believe to
> Be a Gentleman & he Is Real[l]y Sick of heaving Such Buisiness Forced upon
> him[.] I Think he Is Disposed to Do Right In the Matter[.] [T]he others had
> a Different Idea as thay Made Statements to Gentlemen here that thay had
> Several Thousand Trees Coming out of the Swamp & Now you See Into the
> Little Game[.] It has Put The Timber Men In Such a State that thay will Do

Nothing In the Way of Offering Thaier Timber[.] I hope the Matter May Be Settled[.] I Will Do all I can to Bring about a Just Compromise But Will Stick to the Raftsmen Should the Rail Road contend for the Timber.[100]

The situation described by Irving proved to be a temporary one, and soon the hardwood timber of the Delta was being cut down at an unprecedented rate. The growing utilization of bottomland resources was reflected in the expansion of the local economy, largely driven by lumbering activities. In 1894, Greenwood possessed, in addition to "two good restaurants" and "three flourishing banks," a lumber mill, a barrel factory, a hardwood factory, and a barrel head and stave factory. Among items wanted by the townspeople, besides an opera house, remained a "furniture and spoke factory."[101] Another expanding Delta town, Greenville, boasted "several" sawmills in 1901, and extended "a cordial invitation" to "capitalists seeking location for any kind of industrial enterprises, either in cotton or wooden manufactures."[102] The invitation was accepted not only by established businessmen but also by many aspiring entrepreneurs.

LeRoy Barry Allen's family left "the fast-eroding hills of Calhoun County" in the early 1890s for the town of Greenwood in Leflore County, where his father realized his longtime dream of becoming "a real live, honest to goodness Delta merchant." After a fire destroyed Allen's Greenwood store, the family relocated up the Yazoo in Philipp, Tallahatchie County, in the mid-1890s. Philipp, named after an official of the Schlitz Brewing Company, was a sawmill town, and Allen became the manager of the company commissary. A subsidiary of the Milwaukee brewing company known as the Delta Cooperage Company had purchased thousands of acres of surrounding bottomland hardwoods with the intent of manufacturing oak staves for beer kegs. In addition to a large band sawmill, Philipp therefore sustained a stave factory. A narrow-gauged railway hauled sawlogs from the company forest to the mill. During high water, additional timber was floated in rafts down the Tallahatchie to Philipp. The brewing company's experiment of manufacturing staves from Delta hardwoods, however, proved unsuccessful as beer stored in the containers became discolored. Consequently the stave factory was soon closed and the sawmill and timberlands sold to the John O'Brien Land & Lumber Company of Chicago.[103] Other considerable processors of Delta hardwood timber on the eastern edge of the floodplain included the Lamb-Fish Company at Charleston in Tallahatchie County and the Carrier Lumber Manufacturing Company at Sardis in Panola County. Between 1900 and 1929, the latter maintained the Sardis and Delta Railroad to transport its bottomland cut to the mill. Much of the hardwood timber east of the Delta in northeastern Mississippi had already been exhausted by 1900, and the scale of lumber operations in the hill region was consequently limited.[104]

Despite all the agricultural and lumbering fervor in the bottomlands, turn-of-the-century boosters could still claim that "probably not more than 30 to 35 per cent of the lands of the fertile Yazoo-Mississippi Delta" were in cultivation. They further declared that "[e]ven between parallel lines of railways not more than fifteen or twenty miles apart, there lies almost unexplored territory of the most fertile land to be found on the face of the globe, and peculiarly adapted to cotton raising." The heavily timbered lands, however, were "being rapidly brought to the front, large tracts of land having recently been sold to investors and sawmill men, and many large mills of the latest type are being erected to cut into a seemingly inexhaustible wealth of timber." The hardwoods were utilized especially for "box and barrel material, large quantities of which are sent to Europe." For example, the Memphis-based Anderson-Tully Company in 1899 acquired a box plant in Vicksburg and greatly expanded its operations, purchasing extensive timberlands in the region between 1916 and 1924.[105]

In 1896, Frank Hamilton ran a hardwood logging operation adjacent to the Mississippi River on Concordia Island in Bolivar County, where he had "six to eight log wagons going all the time; eight-wheeled wagons, the wheels boxed in to keep them from bogging up in the mud, and from eight to ten oxen to a wagon. Many of the logs were so large they couldn't get but one on a wagon." The methods used for lumber transportation on the Mississippi had not changed much during the nineteenth century, as described by Hamilton's wife, Mary:

They hauled the logs to the river bank and rolled them off in the river; then the men would float them up against the raft and fasten them together with poles. Gum and oak they called "sinkers"; these they would pin between cottonwoods, ash, and cypress, "floaters," to keep them from sinking. A large rope was tied every fifty feet with wire cables to trees along the bank or to posts driven in the ground. When they got enough logs rafted together to send to Greenville, they would send for tugboats, bunch the rafts together, and float them there to be sold.

Rough weather, rapid rises in the water level, and caving river banks could result in the breakup of a raft and severe economic loss. Once Hamilton and his partner "lost more than a hundred big fine logs." The original owners of the logs had their names "stamped in the end of every log, but there were people living on the Mississippi River then who made their living stealing timber like that, towing the logs to the bank, sawing the ends off and stamping their own names instead." [106]

In the fall of 1896, the Hamiltons moved to a timber camp about a mile from present-day Tutwiler, Tallahatchie County, as "there was a big stave works opening up . . . where he could get work." Much of the interior Delta had not yet been cleared for agricultural purposes, and the area "was all just a wilderness of woods and canebrakes. . . . Timber of all kinds stood so close together as almost to shut out all daylight; tall cane, blackberry vines, and a tangled mass of all kinds of vines wove around and over it all." Initially Frank Hamilton and his associates were making staves by themselves: "He bought the timber and paid so much a tree for all he used. If they split one and ruined it, they had to pay just the same. Their staves were inspected and taken up and paid for in the woods once a month." [107]

The flood of 1897 caused the loggers to abandon their Tutwiler camp and seek refuge from the rising waters on old Indian mounds. After the flood had subsided, they relocated to a new camp near present-day Parchman, Sunflower County, where Frank Hamilton began to work as a foreman for a European stave company. The company employed hundreds of workers, both Americans and Eastern European immigrants, or "Slavonians," in their ethnically segregated camps. Hamilton now "worked by the month for a good salary and a commission on all staves got out by the Americans" he was supervising. Still, Hamilton "was among the Slavs a good deal. They worked in gangs, six to a gang, and for his lunch each man carried half a loaf of bread and a pint of whiskey. Our boys told Frank they would be glad to adopt the Slav style of working if they could adopt their style of lunches." As the company "paid so much for each tree, the more staves they got out of a tree, the better for the company and for the men. It saved not only money but labor, for throwing the large trees was the hardest part of the work." Mary Hamilton admired the "Slavonian" workmanship, as "[t]heir staves were hewed smooth as glass, and they used every part of the tree but the bark and the smallest limbs." The good quality of their staves did not help the Eastern European loggers in the long run; after the area had been cleared in 1898, the "Slavonians" were deported for violating immigration laws.[108]

The lifestyle of the Delta logging camps was rough: "The country was filling up fast with people, mostly timber men, and almost all of them drank in those days, so besides the saloons in the little towns starting up, there were blind tigers all over the woods." Race relations in railroad and logging camps often became strained: "At Webb [in Tallahatchie County] they have a killing or a lynching on dull days between paydays and fights or some kind of amusement like that every Saturday night." Despite occasional drinking bouts, Frank Hamilton continued to find work and was able save enough money to become—albeit temporarily—an independent farmer in Sunflower County.[109]

Unfortunately, the available estimates for hardwood stands in the southern United States remain inadequate until the first Southern Forest Survey of 1932. The earlier federal surveys, such as the 1910 investigation by the Bureau of Corporations, were carried out at a time when hardwoods "were regarded as having comparatively little value, and satisfactory estimates could not be secured." Detailed study of bottomland lumbering is furthermore complicated by the natural fact that, unlike pine and other softwoods, hardwoods typically grow in highly mixed stands. The very heterogeneity of the bottomland hardwood forest led to the evolution of a highly specialized industry for the utilization of its different tree species. In the first decades of the twentieth century, baldcypress was used in naval construction and sweetgum in the rapidly expanding veneer and plywood industries. The more valuable species of oak constituted material for flooring and for the construction of truck and car bodies. Packing boxes were made from cottonwood, elm, and pecan lumber, while overcup oak, tupelo, and hackberry were utilized in barrel making. Ash and persimmon found even more specialized uses. As manufacturing and marketing of hardwood products centered on highly diversified products, the hardwood industry became much less concentrated than the commercial operations utilizing pine. The Lamb-Fish Lumber Company in Tallahatchie County was an unusually large operator. In 1915, it boasted a bottomland hardwood tract of seventy-four thousand acres, "the largest hardwood mill in the world" with an average daily capacity of 150,000 feet of one-inch lumber, and two tennis courts and a nine-hole golf course for its white-collar employees. Still, typical hardwood company holdings in the Delta were not comparable to the ones worked on by the largest softwood operators in the southern pine belt, and individual holdings of hardwood stands and the output by typical hardwood mills averaged much smaller amounts than was the case for pine and other softwoods. For example, in 1920, less than thirty companies in the alluvial Mississippi Valley could report an annual lumber cut exceeding ten million board feet.[110]

During the first five years of the twentieth century, the hardwood cut in the states of Ohio, Indiana, and Illinois plummeted by nearly 50 percent. With the exhaustion of hardwood reserves in this traditional center for hardwood lumbering, cutting aimed at meeting the growing needs of the vehicle, furniture, and other hardwood-consuming industries shifted to the Appalachian states, which furnished almost half of the country's hardwood by 1906.[111] The hardwood reserves of the Delta similarly attracted the attention of lumbermen and were heavily utilized at the time. After the turn of the century, mechanized equipment became common in larger hardwood operations. At the same time, the so-called Eight-Wheel Wagon, invented by the Lindsay Brothers of Laurel, Mississippi, made the loading and transportation of logs more efficient. The Memphis lumber industry now rivaled cotton as that city's foremost business, and a local editor could support the claim that "cotton could be removed entirely from the market and lumber alone would support the city."[112] The industry's importance to the local economy was evidenced by the chartering of the Lumbermen's Club of Memphis on December 19, 1898. In 1905 the city's position as the premier hardwood producing center in the United States was obvious by the presence of over forty manufacturers and wholesale dealers of hardwood lumber in the city, which employed over six thousand people and received over seventy-five million feet of hardwood logs by river transportation.[113]

The growing economic importance of hardwood manufacturing in the first decades of the twentieth century was also demonstrated by an attempt to establish a monopoly in the field. Three prominent lumber organizations, the Red Gum Association, the Oak Manufacturers Association, and the Hardwood Manufacturers Association of the United States, merged in 1917 and became the American Hardwood Manufacturers Association, headquartered in Memphis. In 1919 a suit was brought against 333 members of the new organization charging them with illegal business practices and restraint of trade. The case went to the United States Supreme Court and resulted in the association's disbanding.[114]

During the first decade of the twentieth century, Delta hardwoods had clearly shifted "from the position of an encumbrance to that of a considerable asset." A tract of 6,240 acres of mixed hardwoods in Yazoo County had been sold in 1897 for less than $10,000 at an average price of $1.50 per acre. Ten years later, a $100,000 offer was supposedly made for it but refused.[115] In 1908, H. S. Trager from New Orleans contacted Charles B. Allen of the Nanachehaw Plantation in Warren County. Trager had learned from the *New Orleans Times Democrat* that Allen seemed to be "in the Poplar timber business," and offered for sale some 1,500 acres of bottomland and upland hardwoods along the Mississippi in southern Wilkinson County containing an estimated 7,600,000 feet of poplar, hickory, white oak, and ash in the stump. At the "rediciously low" price of $10,500, this "very finest lot of Hardwood to be seen anywhere" was also "the cheapest lot of timber ever put on the Southern Market." Trager indeed seems to have been hard-pressed for money, as he was prepared, in addition to the usual provisions, to grant Allen as many years as desired "to get the same cut in."[116]

By the beginning of the 1920s, the output of hardwood timber in the Appalachian region was rapidly declining, and the remaining hardwood reserves in the Lower Mississippi Valley became an increasingly important supplier for hardwood industries elsewhere in the eastern United States.[117] Accordingly, "the last of the great hardwood regions" was "well on its way toward total exploitation." The supply of mature hardwood timber had already been largely exhausted in the northernmost part of the region, including the Delta. Close to Memphis, the exhaustion had reached a point where it was "profitable to return to cut-over areas for trees that were formerly regarded as too small to log and for less valuable species, such as tupelo and water gum," which at the time of the first logging were unmerchantable but now commanded "a ready sale."[118]

## Inside the Hardwood Industry

By 1920, all timber found in the Delta bottomlands had been subjected to cutting, as there now was "a steadily increasing interest in the utilization of smaller trees, inferior trees and logs, and species formerly rejected." In the first years of the twentieth century, the most heavily logged hardwood species had been oak and sweetgum, which were in great demand by the veneer industry. The rapid depletion of oak and gum stands of adequate size and quality was, however, beginning to result in the cutting of secondary species, such as sycamore and tupelo. The exhaustion of old-growth forests in the northern part of the Mississippi Alluvial Valley was also reflected by the change in the size of operations: small, portable mills had begun to replace large ones that were "finding themselves forced either to buy logs in order to continue operation or to move down river into southern Mississippi and Louisiana" where "a reasonable prospect of a 20 to 25 years' supply of material" existed.[119] Still, forest industries continued to rank second only to agriculture in the Delta economy, being important especially in the southern portion of the floodplain. While plantation owners continued to own most of the forested land, the principal supplies of merchantable timber were now firmly held by lumber companies and timber operators.[120]

The Chicago-based Paepcke Leicht Lumber Company was among the most influential hardwood operators in the early-twentieth-century Delta, and the archives of the company's Greenville plant provide illuminating insights into the cutthroat business of bottomland lumbering. In addition to business correspondence, the company records contain several professional estimates and maps of hardwood stands in the region. The company was founded in 1893 by the German-born Hermann Paepcke. Paepcke arrived in Chicago in 1881 and entered the lumber business with a small planing mill. The ambitious immigrant soon bought out his partner, added box manufacturing, and aggressively expanded his business. Paepcke accurately anticipated the coming shortage of hardwood lumber in the North and purchased the Wolverine Lumber Company of Cairo, Illinois, in 1892. Around the turn of the century, the company acquired vast holdings along the Lower Mississippi. In 1898, Paepcke bought the sawmill and logging equipment owned by J. H. Leavenworth of Greenville, largely because the purchase included the company's twenty-five thousand acres of standing cottonwood and sweetgum timber. At the time, top grade cottonwood was utilized for wagon boxes and the manufacture of buggies and burial caskets, while top grade sweetgum was in demand by furniture makers. Lower grades of both species were used in the construction of boxes and other packing materials. By 1909, Paepcke's business ventures included sawmills, veneer mills, and box plants in St. Louis, Cairo, Greenville, Helena, and Blytheville, Arkansas, with a branch office and lumberyard in Memphis. Among other holdings were two box companies in Chicago, the Blytheville, Leachville & Arkansas Southern Railroad, a fleet of barges, and the steamer *Hazel Rice*. At the time Paepcke's businesses operated on some 125,000 acres of timberlands and employed about six thousand people.[121]

During the 1910s and early 1920s, operations at the company's Greenville plant were supervised by C. Fred Berry, who was born in 1864 in Pennsylvania. Berry had worked as the manager of the Cairo plant since the turn of the century and arrived in the Delta in 1910. He soon rose from the position of Greenville plant manager to vice president of the whole Paepcke Leicht Lumber Company.[122] Berry's extensive correspondence with the company's board and other executives on the acquisition of Delta timberlands reveal a shrewd and sometimes ruthless businessman, albeit one with a certain sense of humor.

Between 1915 and 1919, the Paepcke Leicht Lumber Company was engaged in a prolonged competition between several operators for the cypress and hardwood timber growing on J. W. Johnson's Panther Burn Plantation in Sharkey and Washington counties. In the fall of 1915, Memphis surveyor G. W. Calhoun estimated that the five cypress brakes on the property covered close to one thousand acres and supported almost twenty-seven million feet of merchantable cypress. Cypress in two of the brakes was found to be very thick, long-bodied, free from low limbs, and of good quality. Unfortunately, a full quarter of the timber in these two brakes was dead, "caused by a ditch having been dug which keeps most of the water drained out of the brakes." Emphasizing the negative changes in local hydrology for cypress, Calhoun continued that "[i]t is plain to see, from marks on the trees, that water used to stand from one to four feet deep most of the year." In addition to the brakes, the approximately fifteen hundred acres of surrounding bottomlands supported about two hundred thousand feet of fair-quality ash and a full million feet of small red oak, overcup oak, and sweetgum. Another estimate on the property by J. P. Soper echoed Calhoun's figures at 28,195,000 feet of cypress and 850,000 feet of hardwoods.[123]

On December 7, 1915, C. Fred Berry contacted R. B. McMahon, a Greenville merchant, who held the option on the Panther Burn timber. Berry had consulted with the company's assistant treasurer in Chicago, R. L. McClelland, and informed McMahon that, "on account of the area being considerably less than we at first understood, and the amount of timber being less than one-half of what we first supposed it to be, the first proposition made to us by you [amount unknown] could not be considered." The company hoped McMahon would "take this proposition up again with Mr. Johnson with the view of securing a modification of the price." In his reply, McMahon replied that the planter had "not indicated that he wished to make any other proposition, other than the one submitted." Despite McMahon's wishes that the matter could be taken up in the near future, the cypress at Panther Burn remained unsold for the time being.[124]

The stalemate ended in May 1919, after the death of J. W. Johnson, when Hermann Paepcke learned in Chicago "that both the Darnell Lumber Company and the James E. Stark Lumber Company are investigating this property with a view of purchasing it, and that they have made offers of $375,000 and $450,000 respectively." Paepcke contacted Berry in Greenville and wondered whether these offers could be verified. Berry immediately held a meeting with B. O. McGee, "one of the new owners of the Panther-Burn Plantation." Another of the property's new stakeholders was the former U.S. Senator and powerful Delta planter LeRoy Percy. Berry informed Paepcke that additional holdings could be obtained, as "Mr. McGee states that they have about 4500 acres of other timber land, largely gum, although there is some ash and elm scattered through the tract. This timber they have never had estimated, but he advises that they have made arrangements with Spain & Co. of Memphis to cruise this timber." The owners "would desire to retain the land," but "would give ample time" for tree removal while having "a certain acreage turned back each year." McGee furthermore welcomed the company's estimators to the property, and G. W. Calhoun was again dispatched to the bottom-lands.[125]

Meanwhile, Hermann Paepcke had received information that "so far the owners of the Pantherburn property have not actually received any bids for the reason that they have declined to talk price with anyone until after an examination of the hardwood has been made by their inspectors." Thus the reports of existing bids seemed to be "nothing but idle gossip. While timberlands no doubt are looked after pretty closely, I hardly believe that any sane lumberman should be anxious to buy the timber at [such] prices." In his reply, Berry agreed with Paepcke: "I was talking with a gentleman this morning who said that he had a conversation with Mr. McGee on Saturday, and Mr. McGee said that they had not yet decided on a price to ask for the property, but that they expected to get at least $500,000.00." Berry still believed "that somebody is going to buy this timber and pay too much for it, for the reason that the lumber market at the present time has started a big boom on timber values, and while no sane man is going to pay an unusual price, I fear that there will be some insane man do some bidding." As an illustration of the booming timber prices, Berry gave an example of "a party [that] last week bought a small piece of oak, estimated to cut between one million and one and a half million feet, about $2/5$ red oak and $1/3$ white oak, and paid $25.00 per M stumpage, borrowed the money at 6% to do it, and gave a premium of 6% to get the loan. It will cost them $12.00 to put this timber to the railroad track. This may be sanity, but I doubt it."[126]

By early June, Calhoun had produced a new estimate and map of the Panther Burn Plantation timberlands, adding five hardwood stands to his 1915 cypress survey. The stands covered 4,130 acres altogether and added fifteen million feet of merchantable bottomland timber to the prospective deal. Dominated by sweetgum with almost ten million feet of timber, the tracts also contained over two million feet of cypress scattered among the hardwoods. Overcup oak, elm, red oak, cottonwood, ash, and miscellaneous hardwoods covered the rest of the forested area on the plantation. Calhoun noted that "most of the timbered area on Panther Burn other than the [cypress] brakes is fairly cut over land. The best Oak and Ash, most of the Cottonwood[, and] also some of the best Gum have been cut for plantation purposes." In addition, over five hundred acres of forested land on the plantation had just been devastated by a tornado. The "Cyclone Portion" was not included in the official estimate, but had probably supported around three million feet of hardwoods, "of which two million feet will be a total loss."[127]

The Calhoun map of the Panther Burn timber also showed an adjoining tract of some five hundred acres in Sharkey County already the property of the Paepcke Leicht Lumber Company. Purchased in 1918, the tract was hit only a year later by the same half-mile-wide tornado that ravaged the Panther Burn timberlands. Surveyor J. N. Hall outlined the tornado's path on his map of the area and wrote: "I can not describe the destruction wrought." He regretted to inform the company that "[a]ll of your best timber is down," except for some small strips containing not more than fifty acres.[128]

To minimize insect and other damage, the fallen timber had to be removed quickly. Soon after receiving Hall's report, the company entered into a contract with L. E. Smith, who agreed to "cut, log and remove [by July 1, 1920] the blown down and damaged timber" from the tract. All timber down to fifteen inches in diameter was to be removed, and "all logs to be cut in 12, 14 and 16 foot lengths." Smith had been working for one W. A. Morgan who "had a note against this man to the amount of

$700.00, secured by a mortgage on his logging outfit," but the company quickly took "over this mortgage in order to get the logger to come with us." Smith was to receive $14 per thousand feet for all timber delivered to the railroad at Nitta Yuma. In September 1919, the company assisted Smith "in buying some additional logging outfit" for the job, including a yoke of oxen and a wagon. The change of employers did not change Smith's fortunes; in September 1920, he was adjudicated bankrupt with a balance due to the Paepcke Leicht Lumber Company of $306.88.[129]

In a letter dated June 4, 1919, C. Fred Berry informed Hermann Paepcke on the developments regarding the Panther Burn property and reported on the activities of the Greenville plant. While wet weather had "delayed all logging operations very much, [with] some at a complete standstill," the sawmill had cut 813,354 feet and the veneer mill 689,083 feet of timber during May 1919. Shipments had been "rather light owing to largely to lack of business" but, in the last couple of days, several carloads filled with packing materials such as "Egg cases" and "Poultry boxes" had left the Greenville plant. The remaining orders for "2 cars [of] Cracker boxes" and "1 [car of] Peters Cartridge [boxes]" were to "move tomorrow and next day."[130]

A few days later, Berry was prompted by W. R. Satterfield, the company's new assistant treasurer in Chicago, to obtain more information on the Panther Burn timber and contact "Senator Percy and ask for a copy of Spain's estimate, or that you be permitted to see it." Berry did this, but Percy "did not offer to give us the amount of the [Spain] estimate, but we judge from the tone of his conversation that it was somewhat below their expectations." Percy also informed the Paepcke Leicht management "that they expected to sell this property at a public sale in the fall, but qualified it with the statement that if they had a satisfactory offer in the mean time they would sell the property." Speculating on the company's chances of getting the timber, Berry wrote to headquarters that Percy, McGee, and the other owners would hardly "give any preference to the Darnell-Love Lbr.Co., as their relations have been somewhat strained for some time."[131]

Berry also had other business to attend to, as on June 12, 1919, at 4 P.M., an accident that proved fatal occurred at the box shop of the company's Greenville plant. John James, a forty-year-old black employee who had worked at the plant for some years with an average hourly wage of 27½ cents, was severely injured while operating the ripsaw: "While [James was] assisting to put on a belt with a stick, the stick caught in [the] pulley and struck [him] in abdomen." The foreman, Charles W. Ford, "smelt the belt burning" and "went under the shop and found" James "sitting on the ground a few feet away [from the belt] holding his hands to his stomach." He "asked John what was wrong, and he said a stick hit him, and that it hurt pretty badly." Ford then told James "to go out and get some fresh air and not work until he felt better." After half an hour, however, it became evident that James was "pretty sick and wanted to go to the Doctor." According to Ford, James saw the company's physician, Dr. Payne, but refused to enter the local hospital and "insisted on being taken home."[132]

John James died two days later at home, cared for by his wife, Francis James. His death alarmed the Paepcke Leicht management, as "under the laws of Mississippi a suit can be filed for injuries received." Satterfield advised Berry "to make an early settlement [with the deceased's wife], judging that it might be well to go to the extent of $1000.00 or $1500.00, if necessary." Mr. Watson, the company's lawyer, furthermore encouraged Berry to settle the case as soon as possible, "even if necessary to pay $1000.00 or more" and suggested that anything under a thousand dollars "should be considered a good settlement." On June 20, Mrs. James came to Berry's office accompanied by a friend and was offered $500, $425 to be paid directly to the widow and $75 to the undertaker. The next day, a relieved Berry was able to inform Satterfield that "as the woman seemed to be entirely satisfied with the settlement which we made, we were very much gratified to have the matter closed up." Having "no reason to fear that there will be any further difficulty in reference to this case," Berry and his colleagues were "very much relieved to have the matter off of our hands and mind" and could again concentrate on the issue of the Panther Burn timber.[133]

Satterfield had begun to doubt the soundness of the whole Panther Burn venture, "but in view of the fact that we have had it estimated and been somewhat active in the matter, it is Mr. McClelland's idea [McClelland was now treasurer] that we might still stay in the game so to speak." The company could now "either make arrangements with other prospective buyers to stay out of the market, or assist them in buying; and thereby derive a modicum of profit." Satterfield strongly objected to any idea of an auction of the Panther Burn timber, as "[t]his manner of selling timber is unique to say the least, and I think we could well say to the prospective buyers that there was no use of those interested, standing around and raising each other's bid." The best solution would be to "make an arrangement with the buyers that would enable us to get part of the Gum at a reasonable price and enable them to get the other timber at a reasonable price." Thus it would be necessary for the company's Greenville representatives "to keep rather active in the matter, in the way of communicating with the other buyers in a diplomatic way, and also by conferring with the owners from time to time, about matters of more or less consequence."[134]

In a few days, Berry "succeeded in securing a copy of the cruise made by Spain & Co . . . , Mr. B. O. McGee having given this estimate to Mr. James W. Berry [C. Fred Berry's son, also employed by the Paepcke Leicht Co.]." The new estimate was almost identical with the one made by Calhoun "on the gum, red oak, elm, ash, cottonwood and miscellaneous." In the new estimate, the last category had been broken down to pecan at 224,000, maple at 22,000, and sycamore at 70,000 feet of merchantable timber. However, there was quite a difference in the amount of overcup oak, as Spain estimated it at only 569,000 against Calhoun's full one million feet; "Calhoun also estimated the cypress which was scattered through the hardwood," something Spain had not done, "as the owners erroneously told them that they had an estimate on the cypress, meaning the cypress breaks, but not the cypress scattered through the hard wood."[135]

By July 15, Berry was getting anxious about the headquarters' lethargic approach: "These people are asking us if we want to submit any proposition and we have jollied along with them about as long as we can, and we must either make up our minds to submit them a proposition or tell them that we are not interested." Only the next day, he had to wire Chicago "that the Panther Burn timber had been sold to Darnell-Love. This had been rumored for two or three days, but we saw nothing authentic until it was published in yesterday's Memphis paper." The rumors also placed the price at half a million dollars, a figure confirmed by Berry's conversation with McGee a couple of weeks later.[136]

According to Berry, McGee asked him: "What was the matter with you people [that] you did not let us hear from you on the Panther Burn Timber, and let someone else beat you to it." McGee then made an acidic remark—"Guess you people thought you were the only people big enough to swing this deal, and wanted us to come to you"—and told Berry that the Panther Burn people had received "Five Hundred Thousand Dollars, One Hundred Thousand dollars cash, One Hundred Thousand Dollars in two years, One Hundred Thousand Dollars in three years, One Hundred Thousand Dollars in four years and One Hundred Thousand Dollars in five years, and all timber to be removed in six years from date of sale." Berry judged this information correct, and figured it made "high priced stumpage for Darnell-Love. . . . The interest charge [on the amount] will be $60,000.00. There is 15 million feet of gum and other hardwoods, which at $5.00 per thousand stumpage would be a big price. This would make $75,000.00 for that item, leaving $485,000.00 for the 25 million feet of cypress, or $19.40 stumpage. Logging will be at present time at least $8.00, Loading $3.00 and freight to [the Darnell-Love mill in] Leland about $2.00 or about $33.00 logs at mill." The ill-fated Panther Burn venture was closed in the company's records with Berry's rhetorical question: "What do you think of it?"[137]

The Paepcke Leicht Lumber Company proved more successful with a simultaneous operation nearby involving some thirteen hundred acres of eastern Washington County timberlands that had originally belonged to the Hampton dynasty. In 1915, the tract was estimated to support close to eight and a half million feet of merchantable timber. In addition to four million feet of "Excellent" gum and one million feet of medium-sized cypress, there were 1.7 million feet of red oak of "medium size and excellent quality," 1.5 million feet of "medium quality" overcup oak, and close to three hundred thousand feet of ash. The property was offered for sale "either for land and timber, or for the timber alone." On August 1, 1916, the company purchased the merchantable timber on the tract for $20,415, after discounting the original estimate by 50 percent. The standing timber on the property was to be removed by December 31, 1921, and the land turned back to the owners.[138]

Some nearby timberlands had been purchased by an independent hardwood operator, W. W. Gary, who had "secured a right of way from Percy to his timber," and was expected "to put in a railroad and log this timber for sale." Gary was also interested in purchasing the adjoining tracts. In addition to the Hampton timber, these included "one piece of 280 acres owned by a man named Towner; one piece of 320 acres owned by a man named Gage, and 80 acres owned by a negro named Jones." As to Paepcke Leicht's latest acquisition, C. Fred Berry was happy to inform the company board "that this man Gary had offered the owner of the Hampton timber, Mr. Stoner, $1.50 per acre advance over our price, but he was not able to do anything with Mr. Stoner, as he was tied up on the option to us." Gary then contacted Berry and "stated that he would like to make an arrangement to log off" the Paepcke Leicht timber in addition to his own. Berry was anxious to know more about Gary and learned that he already had made discreet inquiries whether Paepcke Leicht "might be disposed to re-sell our Hampton timber at $1.50 an acre advance. This would give us a quick profit of about $2000.00," Berry noted, "but of course we do not care to consider it." Berry furthermore found out that Gary, "a comparatively young man," lived in Sunflower County and was "said to be a very good logger." Berry did "not favor giving him a logging job" on the Paepcke Leicht tract, as the company already had its own team of loggers in the locality.[139]

The independent operator with his own railroad could, however, be utilized by the company in the loading and moving of the their logs to the main railroad line. The company was furthermore interested in the timber on Gary's land. After meeting Gary, seeing "some of the gum which he is cutting," and noticing its "very fine quality," Berry attempted to make a contract with Gary for a purchase of one million feet at the price of nine dollars per thousand feet. Berry was also able to offer the company board a personal character judgment on Gary: "We are quite well pleased with what we have seen of this man. He seems to be very frank and quite a hustler, and pushing his operations with vigor. He has invested his money in his logging outfit, and naturally, wants to secure a profit." This Gary undoubtedly was able to do in November 1916 by declining

Paepcke Leicht's offer and making instead a more lucrative deal with the Cincinnati-based Korn-Conkling Company.[140]

In September 1919, the executors for the owners of the Hampton tract, the Memphis-based insurance company Thos. Wellford & Sons, had been contacted by "some parties who are interested in the land," presumably for agricultural purposes, and wanted to know when the land was ready to be returned. Blaming bad weather conditions, Berry informed the owners that "none of this land" had been "cut off entirely and [it was not] ready to turn back." Still, "[t]he greater part of it . . . had the largest saw mill timber cut off of it," while some tie timber remained scattered over the land. By September 1921, over four million feet of timber had been removed from the stumpage, and the company's account on the tract showed a profit of $4,601.00. Berry's son James now surveyed the tract and estimated that some six hundred thousand feet of timber still remained on the land. As the logging rights were to expire in a few months and the stumpage was small or even "not worth logging," the company attempted to "dispose of the remaining timber to Mr. W. W. Gary, as it is located on his spur, and close to his mill." In what Berry described as "a good deal" for the company, Gary decided "to take his chances on removing" the remaining timber and paid $1,750.00 for it, "payable in three notes six months after date, with 6% interest."[141]

In February 1922, Thos. Wellford & Sons again contacted Paepcke Leicht, asking "if you will not kindly give us, in a general way, some idea of the amount and character of timber remaining" on the tract, as the timber rights had expired on December 31, 1921. They furthermore "beg[ged] to advise that we are endeavoring to interest a local cooperage company in the remaining timber on this tract." Berry had to inform the executors that the forest had been exploited to the fullest: "We finished cutting practically all the timber of any value before our time expired. The remaining timber consists of only some scrub oak, which we consider worthless as logs but may be of very little value as a tie proposition." The company was now ready to turn back the land, as they only had "a few logs scattered along the track which runs into this timber."[142]

LeRoy Percy was not the only Delta luminary to do business with the Paepcke Leicht Lumber Company. On January 10, 1913, the former governor of Nebraska, George L. Sheldon, together with his wife Rose, sold

all the merchantable timber on their three Washington County plantations to the company for twenty-two thousand dollars in cash. The Shel-
dons granted ten years for the removal of timber from their Whitehall and
Ashland plantations while Loudon was to be logged within three years.
The company was given the right to establish logging camps and "build a
tram way, log road or railway over and across said plantations [Ashland
and Whitehall]" and "accept from the timber herein conveyed, all persimmon, mulberry, pecan and all oak thereon under 14 inches in diameter at
the stump, save and except suck [such] oak as may be used for cross ties
by the grantee for the purpose of laying said railway, log road, or tram
way, when other timbers are not conveniently available." The company
could furthermore operate the railroad as long as it wished, "provided
in so doing grantee shall haul the farm products and Plantation supplies
of us, the grantors, or our vendees of either of said plantations, free of
charge."[143]

The company had estimated the twelve million feet of "the larger,
choicest timber" on the property to be worth $151,300. Close to five
million feet of red oak, valued at twenty dollars per thousand feet, four
million feet of gum at five dollars per thousand feet, and one million feet
of white oak at twenty dollars per thousand feet constituted the most
valuable stumpage. Fully two-thirds of the one and a half million feet
of cypress among the Sheldon timber was deemed "Questionable" and
was not included in the calculations. The three plantations furthermore
supported some four hundred thousand feet of cottonwood and three
hundred thousand feet of ash. The logging of the Sheldon timber was,
however, postponed year after year as the company had other timber
rights expiring annually. By June 1920, only 116,164 feet of timber had
been removed from the land, and C. Fred Berry was getting concerned as
the deadline of January 10, 1923, approached.[144]

"There is no doubt in our minds whatever but that an extension on
this timber would be necessary," Berry wrote the Chicago office, as "[w]e
hardly dare take the chance of removing 6 million feet in 1921 and 6 million feet in 1922. Furthermore, this amount would probably be more than
we could handle at the Greenville plant, considering other timber which
we must also remove." His "idea was to accidentally meet the Governor

and feel him out on the idea of the extension, and owing to the financial conditions existing now, we feel that this would appeal to him." Berry planned to make Sheldon "an offer covering 3 years, 4 years and 5 years, and making the 5 year period the most attractive." Consequently the company succeeded in getting a four-year extension to the contract by paying Sheldon an additional eight thousand dollars. The former governor's financial situation was, however, worsening rapidly and began to threaten the whole logging operation. In March 1921, Berry contacted Satterfield in Chicago with information he hoped to "be of value" to the assistant treasurer: there was evidently a fifty-thousand-dollar mortgage against Sheldon's land, "covering 3500 acres, more or less, held by Alliance Trust Co., Memphis, Tenn." A five-thousand-dollar mortgage payment was due on February 1, 1922, with annual payments of the same sum until paid in full. Besides, there were "separate notes for the interest at 7 %[,] and a note for 3500.00 covering the first year[']s interest [that] was due Feb. 1st, 1921," although Berry could not "learn if this was paid when due." A desperate Sheldon had "told Jim [James W. Berry?] sometime ago, he could not pay his taxes, intended to let the tax sale take place and take his chances on making redemption before it was too late to do so."[145]

On May 27, 1921, Satterfield turned down Sheldon's offer to sell to the company for fifteen thousand dollars the land on which the stumpage stood. The company was "not interested in acquiring any more land" as there was no conceivable use for it in the future. "In short," wrote Satterfield, "we would not be interested in buying your land even at $5.00 per acre and the only reason that we considered the purchase at all was to protect our timber rights." The company instead optioned to pay the mortgage holders fifteen hundred dollars in exchange for a lien waiver, "so far as the same affected the title to our timber and the right to cut and remove same as evidenced" by Sheldon's deeds. The company's correspondence with the Hon. George L. Sheldon concluded with regrets

that you have become so involved and your affairs so tangled that financial loss is threatened. All of the officers of the Company have nothing but the kindest of feeling toward you personally. . . .

I wish to thank you in behalf of the Company for your co-operation and friendship displayed in this matter and sincerely hope that you will find some way to save your equity in this magnificent piece of property.[146]

The logging of section 16 in Township 21 North, Range 7 West, the Bolivar County School Section, presented the Paepcke Leicht Lumber Company with complications of a different kind. The stumpage on the section, estimated at one million feet of cottonwood, 150,000 feet of elm and hackberry, and 300,000 feet of oak, gum, and ash, was purchased in July 1918 for four thousand dollars. The entry of the United States into World War I had greatly increased the demand for cottonwood lumber, which was used as packing material for military supplies. In light of a recent telegram from Chicago stating that "we must continue to operate mill buy cottonwood logs best price can secure," C. Fred Berry was undoubtedly happy to inform the company's accounting department "that the cottonwood alone would be worth the purchase price."[147]

The company's contracted logger for the job, R. C. Little, was eager to start work on the section and claimed to be able to "log out about 200,000' per week, until he has the job completed." In a few days, however, Little and his crew ran "against a snag in logging the Bolivar County School timber": the owner of the adjoining sections on the west and north sides "would not permit them to come out on their roads." The troublesome landowner, Mr. Allan Gray of Evansville, Indiana, had issued "positive orders" to his local manager "not to allow any one to come on this road [into the school section], even in automobile." In a subsequent meeting with Gray's manager, James W. Berry and R. C. Little "offered to keep the roads in good shape and reimburse them for any damage which might be done to the roads and bridges," but were unconditionally turned down. In a letter to the Chicago office, C. Fred Berry reported additional complexities in the case: "We understand that Mr. Gray is in a sanitarium, and have been informed that he is suffering from temporary aberation of the mind."[148]

After some legal consultation, the company decided "to proceed to work and continue to use the road until stopped legally by Mr. Gray or his representative." C. Fred Berry also made it clear to Gray's local representatives "that they were delaying and interfering with War material and they should take this under consideration in making this movement." According to Berry, the patriotic element should have given the neighbor "a sober second thought" and made him "hesitate to stop us when it comes to a show down." When county officials deemed the roads into the school section public property, Gray still refused to back down and decided instead "to stir up trouble over the boundary line." Claiming that the boundary in the northwest corner of the section was incorrectly located, he demanded a new survey and warned the company "not to cross our boundaries there, for any purposes whatsoever, unless you wish to incur the consequences of doing so." Berry's report to the Chicago office on the survey problems also touched upon broader issues of regional identity: "Mr. James W. Berry says that this [Gray's] outfit is a lot of 'Damned Yankees', from which you will see that Mr. James has become quite a thoroughbred Southerner." [149]

Eventually the boundaries were resurveyed and logging in the section commenced in earnest. By November 1, 1920, work on the tract had been completed, with close to 1.2 million feet of timber cut and delivered to the railroad track at Dahomy. The school section did not contribute to the war effort nearly as much as was originally hoped for. Instead of one million feet of cottonwood, only 244,000 feet was removed from the tract, and much more of the cut, 664,019 feet, consisted of gum. Legal and surveying expenses added $287 to the original cost of timber, giving the Paepcke Leicht Lumber Company an average stumpage price of $3.83 per thousand feet. Logging and transportation costs for the timber were considerably greater at $16.29 per thousand feet. [150]

Between 1906 and 1922, the Greenville office of the Paepcke Leicht Lumber Company made many individual logging contracts for the removal of "all merchantable timber" from various bottomland hardwood stands in the western Delta, especially in Bolivar, Washington, and Sharkey counties. Still, the majority of the contracted operations during that time were made to obtain cottonwood, a species characteristic of recent alluvium. Most of the company's cottonwood stumpage therefore grew on the banks of the Mississippi. In a typical contract, made in 1909 between the company and "Fisher Johnson of Beulah, Mississippi," Johnson agreed to deliver three hundred thousand feet of cottonwood "under

good single splice, accessible to tow-boat, at the head of Beulah Lake."
He furthermore agreed "to commence at once, and cut all this timber
down, and let lay with tops on to dry out, for at least 30 days, then to log
up and prepare same for float." Johnson was paid $7.50 per thousand
feet and advanced $1.00 per same "to enable him cut and prepare tim-
ber for float." Spikes necessary for rafting were furnished free, "except
freight charges; same to be counted as delivered in rafts, and charge [John-
son] 3c each for any difference in spikes furnished and those returned."
The cottonwood operations were not restricted to the Yazoo-Mississippi
floodplain, and much of the company's timber was obtained from the
Arkansas side of the river.[151]

The creator of the Paepcke Leicht Lumber Company, Hermann Paepcke, died in 1922 and was succeeded by his son, Walter, and C. Fred Berry
retired the next year. In 1928 a considerable reorganization took place in
the business as the Paepcke Leicht Lumber Company became a part of
the new Chicago Mill and Lumber Company. Other participants in the
merger included the Arkansas Oak Flooring Company and three Memphis companies with extensive hardwood holdings: R. J. Darnell, Inc.;
Hudson Hardwood Flooring Company; and Penrod-Jurden Company.
Around the same time, the company acquired more timberlands around
Greenville. New holdings were obtained also in Louisiana as the company
purchased the sawmill and timberlands of the Kurz Brothers Company
of Tallulah in Madison Parish, where James W. Berry became the plant
manager. The Chicago Mill and Lumber Company was to continue the
tough entrepreneurial traditions of its predecessor.[152]

A 1920 report by the Forest Service concluded that, because of the transformation of the cutover areas into agricultural lands by drainage and
clearing, "the cut in the bottom-land region of the lower Mississippi Valley can not be maintained from second growth to the same extent as has
been the case in the Northeastern and Central States. Once the present
stand of timber on these bottom lands is gone, the hardwood supply of
the country will be permanently reduced."[153]

The earliest attempts at conserving southern forest reserves concentrated on the prevention of widespread forest fires, and legislation to that

effect was passed soon after Mississippi had attained statehood. The 1828 "Act to Prevent Damages which may happen by the Firing of Woods, Marshes, and Prairies" made wood burners liable for damages caused to private property and set fines from fifty to five hundred dollars, while slaves could "receive thirty-nine lashes on his or her bare back" for offending the law.[154] Throughout the nineteenth-century South, such legislation remained practically unenforced, and it was "extremely difficult to convict the hunters, lazy negroes, and fishermen who set the fires."[155]

The problem presented by human-induced forest fires in the bottomland hardwood forests was limited when compared to the upland pine forests.[156] Still, G. W. Featherstonhaugh was able to document many such fires while traveling in the Arkansas bottomlands across the Mississippi from the Yazoo-Mississippi Delta: "After travelling some distance through the forest, we got upon an extensive bottom, where we again found the country on fire, the leaves and twigs all burnt up, and every thing as black as soot." Upon leaving their campsite, they

> perceived that our fire was creeping through the leaves, and that, if not extinguished, it might produce a serious conflagration. Thinking it right to leave Nature as clean as we found her, we spent about a quarter of an hour bringing pails of water from the stream until the fire was out. Many careless persons do not take so much trouble; they kindle a fire, and then leave it unextinguished; the consequence of which frequently is, that many thousands of acres are burnt over, the mast upon which the deer and bears would have fed is destroyed, the buildings of the farmer destroyed, his fences burnt down, and his corn-fields injured. The hunters, too, sometimes, with the intention of driving the game to a particular quarter, will purposely fire the country in various places, indifferent to the devastation and inconvenience they cause; and all this merely to get a few deer with greater dispatch than they would do by going a little farther into the country. It is vain to remonstrate with these men; they live by getting deer, and as they look upon the farmer as an intruder, have little or no sympathy for him.[157]

Governmental pleas for the conservation of America's forest resources were made as early as the mid-1860s.[158] Early appeals were not necessarily limited to forest fire prevention, as evidenced by the evolving forest survey system—and the sour comments made in 1885 by Nathaniel H. Egleston, chief of the Division of Forestry at the U.S. Department of Agriculture. Egleston lamented that the lumberman "is concerned with no crops except those of the forest. His aim and interest are to level the trees and convert them into lumber as soon as possible."[159] Similarly, special agent A. B. Hurt had asserted a year earlier that "Mississippi has done little to preserve its bountiful supply of valuable woods, and nothing whatever to replete with a new growth the forests already destroyed."[160]

Despite much evidence to the contrary, a few people came to realize the economic potential for forest regeneration in the South even during the era of "cut out and get out." Scientific forest management in the region was initiated in 1892 by Gifford Pinchot on the Biltmore estate near Asheville, North Carolina. Pinchot's pioneering efforts were followed by the German Carl A. Schenck, who established the South's first forest school, Biltmore, in 1898. Before the school closed in 1913, some 350 foresters had received technical forestry training at Biltmore.[161]

The turn of the century witnessed the emergence of legislation to promote forest fire protection and forest regeneration in several southern states. Louisiana created the first state forestry agency for forest fire protection in 1904, and Alabama adopted a general forest administration law in 1907. The Mississippi Department of Forestry was founded only in 1926, and during the heyday of lumbering in the state, the vast majority of operators had no faith whatsoever in reforestation practices. Cutover lands, especially in the pine region, showed no prospects for immediate returns but continued to be taxed. The result was an urgent need to dispose of them, usually to prospective farmers. Large-scale advertising campaigns conducted by the Mississippi Land and Development Association and other organizations close to lumbering interests lured hundreds of people from the North to pursue their dream farm in the barren, cutover pinelands of southern Mississippi. Land companies appeared and speculation ran high on the land market during the first part of the 1910s. The poor Ultisol soils of the pine region, however, did not yield the promised six crops a year, and most stump-land settlers had to leave their farms and return to more familiar surroundings after exhausting their savings. Gradually the idea of using cutover pine forests for agricultural purposes was abandoned, and some Mississippi lumber companies even began to show interest in reforestation programs.[162]

Efforts at reforestation in the South had already been made by some influential lumbermen, most notably by Henry E. Hardtner, who came to be called the "Father of Forestry" in Louisiana. As the president of Louisiana's Urania Lumber Company, he attended Theodore Roosevelt's Conference of the Governors on conservation issues in 1908, authored the state's pioneering Reforestation Act of 1910, and entered the company's lands under it in 1913. According to the act, timberlands under contract with the state would not be taxed until the timber was cut. Hardtner's lead was soon followed by other lumbermen with the counsel of professional foresters such as W. W. Ashe, Austin Cary, and H. H. Chapman. The Great Southern Lumber Company of Bogalusa, Louisiana, began planting programs on cutover lands in 1920, and by 1929 some twenty-three thousand acres had been reforested.[163]

State and regional forestry associations had also appeared, and the most influential of these, the Southern Pine Association, was formed in 1914. Two years later, the association held the first Southern Forestry Conference to promote state legislation for forestry. The next year, it sponsored the Cutover Land Conference of the South to utilize the millions of acres of cutover land. By 1925, practices to make lands more productive had been started on eighty-two properties totaling 4.7 million acres. Wildfires still presented a great problem for commercial forests, so in 1927 the American Forestry Association launched a three-year education project. Teams of so-called Dixie Crusaders crisscrossed the backwoods of Florida, Georgia, Mississippi, and South Carolina, distributing forest fire prevention messages.[164]

The greatest incentive for sustained-yield and other modern forestry practices in the South was, of course, economic. The rapid expansion of the pulp and paper industry had, for the first time in southern history, made it economical to utilize small trees taken from the "second," regenerated forest. Since 1891, there had been a pulp and paper mill in Hartsville, South Carolina, and the use of southern pine for paper products got underway during the early twentieth century, thanks to the sulfate process developed in Germany in 1884. This new technology made it possible to produce bleached pulp from the highly resinous pines and surpassed the older, more expensive caustic soda or sulfite methods for pulp production.[165] The Southern Forest Survey of 1932 applauded the comeback of pine and provided encouragement for the continuing expansion of the industry.

Although the emphasis of southern forestry work since its beginning was on pine habitats, hardwood operators also began to recognize the possibilities for sustained yield during the mid-1920s. For example, the Hardwood Manufacturers Institute, based in Memphis since 1924, employed a forester to advise timber holders in methods of selective cutting.[166] In 1925, the Thistlethwaite Lumber Company of Opelousas, Louisiana, put eleven thousand acres of logged hardwood land under reforestation contract with the state. The contract was signed during the administration of State Forester V. H. Sonderegger, who was among the first southern foresters to show interest in hardwood management. As the first such contract in Louisiana, it placed almost half of the Thistlethwaite family's twenty-four thousand acres under scientific management. Interest in hardwood management among landowners, however, remained low, and silvicultural practices were employed at only one other Louisiana hardwood locality, the ten-thousand-acre Brewster-Neinstedt tract in Palmetto, before World War II.[167]

The Thistlethwaite project was supervised for three years by John A. Putnam, who was later to play a crucial role in the development of bottomland hardwood forestry in the Delta and the whole South. Putnam was born in Michigan, raised in Iowa, and studied for some three years at the University of Michigan prior his arrival in Louisiana in 1923. After the disastrous 1927 flood, he was hired by the Southern Forest Experiment Station to conduct a survey of Louisiana's hardwood lumber industry. Putnam cooperated closely with the state's Division of Forestry, represented by G. H. Lentz, a professor on sabbatical leave from the New York College of Forestry at Syracuse. In addition to examining the effects of the 1927 flood on the hardwood timber supply, Lentz and Putnam studied the growth characteristics of hardwood species. They pointed out that hardwoods grew at a remarkable rate and showed great potential for management. The greatest problem for hardwood forestry in the 1920s remained that of intermediate cutting, as only the biggest logs of high quality could be commercially utilized. All sizes of southern pine could be used in pulp and paper production, but existing technology did not allow for the use of immature hardwoods in that industry until the 1940s. Putnam nevertheless continued his study of the hardwoods and was largely responsible for hardwood counts of the Southern Forest Survey of 1932, which covered the bottomland hardwood forests of the Lower Mississippi Valley from Missouri to Louisiana. Reports by Putnam and the other surveyors provided the first scientific evaluation of the extent and condition of the Delta bottomland hardwood forests after decades of intensive human-induced transformation.[168]

Now the land lay open from the cradling hills on the east to a rampart of levee on the west, standing horseman-tall with cotton for the world's looms—the rich black land, imponderable and vast, fecund up to the very doorsteps of the Negroes who worked it and of the white men who owned it; . . .—the land across which there came now no scream of the panther but instead the long hooting of locomotives: trains of incredible length and drawn by a single engine, since there was no gradient anywhere and no elevation save those raised by forgotten aboriginal hands as refuges from the yearly water and used by their Indian successors to sepulchre their fathers' bones, and all that remained of that old time were the Indian names on the little towns . . .

—WILLIAM FAULKNER, "Delta Autumn," *Go Down, Moses*

CHAPTER SEVEN

# A Transformed Landscape

DURING THE NINETEENTH CENTURY, a growing number of Americans recognized the forest as the basis of industrialization, agricultural expansion, and material advancement. This soon resulted in a significant diminution of the area occupied by forests in North America. At the most conservative estimate, some 153 million acres of forest had been cleared for agriculture by 1860, and at least another 11 million acres by the lumber industry, mining, and urban spread; a couple of hundred years after the colonists' arrival in the early seventeenth century, about one-quarter of the original forests of the eastern United States had disappeared.[1]

Toward the end of the nineteenth century, the quickening destruction of the southern forests due to the booming lumber industry resulted in the vast deforestation and erosion that ultimately led to the adoption of sustained-yield forestry practices in the twentieth century. By 1919, the onslaught across southern timber stands had reduced the primeval forest by approximately 40 percent: only some 178 million acres of the original 300 million remained.[2] Lumbering and clearing for agricultural purposes had played the greatest part in the reduction of forested area and left a lasting imprint on the southern landscape. In addition to its massive assault on pine forests, the southern lumber industry concentrated much of its cutting efforts on hardwood stands of high commercial value, and mature bottomland forests were especially affected.

## The Nature of the Delta in 1932 and Beyond

It is a difficult task to estimate the amount and condition of hardwood forests supported by the Delta bottomlands before the Federal Forest Survey authorized in 1928 by Congress through the McSweeney-McNary Forest Research Act. Until the late nineteenth century, "hardwoods were often considered only an encumbrance on the agricultural lands of the lower Mississippi," and no comprehensive surveys were conducted. Consequently, a 1913 report by the U.S. Department of Commerce and Labor noticed with regret that little effort had been made to determine the amount of hardwoods in the South, and the quantity of even the more valuable hardwoods remained much less known than that of various southern pines.[3] Still, the only American hardwood reserve of importance in the beginning of the 1920s was located in the lower Mississippi Valley, and many industries dependent upon the hardwoods were "in great need of accurate information as to the extent of existing stands and what they can count on for the future."[4] Faced with a decreasing supply of hardwood timber of high quality, the lumber industry and the federal government for the first time showed genuine interest in assessing the region's forest reserves.

In addition to making an inventory of the existing supply of timber, the 1932 Federal Forest Survey in the South aimed to determine the growth and drain rates of American forests in connection with known and projected trends in the use of forest products and attempted to create a basis for policy formulations in the future. The actual survey was conducted by the regional forest experiment stations of the U.S. Forest Service. Work in the state of Mississippi and the Delta was carried out by the personnel of the Southern Forest Experiment Station, based in New Orleans. The station had divided each surveyed state into territorial units of four to twelve million acres, conforming as well as possible to the natural boundaries of major forest types. Not surprisingly, the Yazoo-Mississippi floodplain constituted the Bottom-land Hardwood Unit in Mississippi. The Delta, estimated at some 4.4 million acres for the purposes of the survey, was inspected by "three-man crews, each of which consisted of an expert timber cruiser and two assistants." The surveyors gridironed the whole Delta regardless of land ownership or occupancy and altogether sampled 2,301 quarter-acre plots.[5]

The surveyors found that close to 60 percent of the Delta had been transformed into an agricultural landscape. The overwhelming majority of this acreage was in cotton cultivation, and the clearing of land for agricultural reasons continued. The highest proportion of land in cultivation was located in the counties of Sunflower, Coahoma, Bolivar, and Leflore in the central part of the floodplain. In addition to the creation of crop and pasture lands, the emergence of modern infrastructure had made significant inroads into the original forest: some 4 percent of the land was classified as "other areas," consisting not only of waterways but also of levees, roads, railroads, and towns. Still, as table 3 shows, some 1.7 million acres, or almost 40 percent of the Delta, could be classified as forest land. Most of the forested acreage was found in the Delta's southernmost counties of Issaquena, Sharkey, and Warren.[6] But what was the character of the Delta forest in 1932, after roughly a century of American settlement? According to the surveyors, an immense change had taken place in the landscape encountered by John James Audubon, Thomas Nuttall, Henry Ker, and other early-nineteenth-century travelers who unanimously had described the Delta as an almost unbroken expanse of old-growth forest and canebrakes teeming with diverse wildlife.

The Southern Forest Survey noted that, in the beginning of the 1930s, forest land in the Delta was confined chiefly to two areas: the backwater area in the lower part of the floodplain and the batture land between the Mississippi and the main levee line. In addition, some poorly drained interstream sites in the middle and upper parts of the Delta still retained some forest, though they were described as "badly deteriorated." The existing forest areas were generally found to have been "cut over one or more times for sawlogs or other products."[7] Except for the area subjected to backwater flooding, few unbroken areas of forest remained on the floodplain. As explained in Faulkner's "Delta Autumn," the Delta wilderness had retreated "southward through this inverted-apex, this V-shaped section of earth between hills and River until what was left of it seemed now to be gathered and for the time arrested in one tremendous density of brooding and inscrutable impenetrability at the ultimate funnelling tip."[8]

Old-growth forest—meaning practically uncut stands that maintained the characteristics of original mature forests—were found to occur on approximately 5 percent of the Delta's forested area. A remarkable stand of bottomland hardwoods was found by the surveyors inside the levee on Jackson Point in Coahoma County. Individual sweetgum trees on this tract measured close to 170 feet high and 60 inches in diameter at 4.5 feet above the ground.[9] In 1932, such primeval stands made up only about 2 percent of the total land area on the floodplain. Another 2 percent of the Delta was covered by so-called culled old-growth stands from which "an appreciable quantity of high-grade timber" had been removed. Half of the Delta's forested area was deemed second-growth forest, supporting young timber in varying sizes (one inch or more in diameter at 4.5 feet above the ground). These forests had developed after the original forest had been removed by clear-cutting, tornadoes, or fire, or after agricultural acreage had been abandoned. Some of the second-growth area could be utilized for lumbering purposes, as the trees in these stands ranged in size from recently established to immediately merchantable. An additional 14 percent of the Delta forest consisted of cutover second-growth stands that showed no commercial promise in the immediate future. Almost four hundred thousand acres, or 9 percent of the whole Delta acreage, consisted of forest land classified as "old-growth cut-over." This class made up over 20 percent of the "forested" area, and was described as formerly mature forest "from which practically all trees of merchantable size and quality had been removed, leaving too small a volume per acre to justify another logging operation in the immediate future."[10] (See tables 3 and 4.)

*Table 3.* Land Use in the Yazoo-Mississippi Delta, 1932

| Land Use Classes | Acreage | % of Area |
|---|---|---|
| *Forest* | (% of Forest Area) | |
| Old-growth | 92,800 (5.3%) | 2.1 |
| Culled Old-growth | 98,900 (5.7%) | 2.2 |
| Cutover Old-growth | 394,100 (22.7%) | 8.9 |
| Second-growth | 872,600 (50.2%) | 19.7 |
| Cutover Second-growth | 240,800 (13.9%) | 5.4 |
| Killed by Fire or Flood | 37,700 (2.2%) | 0.9 |
| Total Forest | 1,736,900 | 39.2 |
| *Nonforest* | | |
| *Agricultural* | | |
| In Cultivation | | |
|   Old Cropland | 2,192,100 | 49.6 |
|   New Cropland | 127,600 | 2.9 |
| Out of Cultivation | | |
|   Idle | 74,700 | 1.7 |
|   Abandoned | 46,100 | 1.0 |
| Pasture | 57,300 | 1.3 |
| *Other* | | |
| Levees, Roads, Towns, etc. | 185,700 | 4.2 |
| Total Nonforest | 2,683,500 | 60.7 |
| Total | 4,420,400 | 99.9* |

Source: Eldredge, *Preliminary Report on the Forest Survey of the Bottom-Land Hardwood Unit in Mississippi,* 3 (table 1), 5 (table 2).

*Numbers do not add up to an even 100 due to rounding.

*Table 4.* Distribution of the Forested Area in the Yazoo-Mississippi Delta by Condition and Topographic Location, 1932

| | Acreage by Location | | | Total | % of |
|---|---|---|---|---|---|
| Condition | Swamp* | Batture† | Bottoms‡ | Acreage | Forest Area |
| Old-growth | 4,500 | 30,200 | 58,100 | 92,800 | 5.3 |
| Culled Old-growth | 19,600 | 2,300 | 77,000 | 98,900 | 5.7 |
| Cutover Old-growth | 45,300 | 15,100 | 333,700 | 394,100 | 22.7 |
| Second-growth | 65,700 | 176,600 | 630,300 | 872,600 | 50.2 |
| Cutover Second-growth | 17,400 | 30,200 | 193,200 | 240,800 | 13.9 |
| Killed by Fire or Flood | 8,300 | 5,300 | 24,100 | 37,700 | 2.2 |
| Total Area | 160,800 | 259,700 | 1,316,400 | 1,736,900 | 100.0 |

Source: Eldredge, *Preliminary Report on the Forest Survey of the Bottom-Land Hardwood Unit in Mississippi,* 5 (table 2). Cf. Stover, *Forest Resources of the Delta Section of Mississippi,* 9 (table 2).
*Forested area usually under water for a large part of the year.
†Land lying between the river and the levee, and thereby subject to overflow. Also, similar land located along sections of river banks where no levee exists.
‡All forestland not described as swamp or batture.

Logging operations had customarily concentrated on timber stands of highest economic value. Consequently, half of the remaining old-growth forest was dominated by overcup oak and water hickory, two species that in 1932 were "in limited demand." Despite their dominance in the old-growth stands, forests of this type made up less than 13 percent of the total forested acreage in the Delta. Found on the poorly drained flats and along the edges of swamps, the overcup oak–water hickory forest type was characteristic of the lower part of the floodplain. Some 160,000 acres of forest land classified as swamp would only a century before have been dominated by mature cypress, but now a mere 6,000 acres of old-growth cypress stands were to be found in the Delta. Much of the batture area was covered by cottonwood and black willow, two species that grew rapidly and had therefore retained their commercial importance. Stands of mixed hardwoods made up the rest of the forested acreage in the Delta. Constituting roughly 70 percent of the forest land, the mixed hardwoods type had been severely decimated by logging. The extent of old-growth stands of this type was restricted to less than 4 percent of the total forest area. The late-nineteenth-century sweetgum promoter Richard Abbey probably could not have imagined that only fifty years after he began campaigning for the increased use of the abundant bottomland tree, "extensive areas of pure, virgin red gum" were considered "rare" in the fragmented forests of the Delta. Within a century, the Yazoo-Mississippi floodplain, consisting mostly of mature bottomland hardwood forest in the beginning of the 1830s, had largely been logged over. The predominantly agricultural landscape of the 1930s, characterized by cotton fields and cutovers, bore little resemblance to the Delta inhabited by Native Americans or even mid-nineteenth-century American settlers.[11]

The immense ecological transformation of the Yazoo-Mississippi Delta had not been limited to the reduction of forested acreage or the removal of mature trees: the native fauna of the bottomlands had similarly undergone significant changes. As significant as those changes had been, however, scientific surveys of bottomland nature for other than purely economic purposes began only after most of the habitat had disappeared. Historical fluctuations of the Delta animal populations are therefore even less understood than those of the commercially utilized forest. The Delta had formed an important part of the original range of several animals that had more or less disappeared by the beginning of the 1930s. The most striking examples of local—and even global—faunal extinction were provided by the red wolf, cougar, Carolina parakeet, and ivory-billed woodpecker.[12]

The classic theory of island biogeography by R. H. MacArthur and E. O. Wilson suggests that loss or extinction of area-dependent stenotopic species occurs as a result of a decrease in total forest area and forest patch size. As islands of forest decrease in size and become more distant from each other, the faunal extinction rates become higher and colonization rates decrease. Newly formed fragments of forest that previously composed larger islands tend to be supersaturated with species that cannot be sustained for long. A so-called faunal collapse will result, even without any further reduction in the forested area. Species with narrow habitat requirements are therefore subject to a greater than normal threat of extinction. As the habitats of such species decrease in size and become more patchily distributed, formerly continuous populations are fragmented into series of small ones having very limited interactions. Such small populations are much more likely to undergo extinction than is a single large population—simply because the number of individuals in such populations is smaller. Should a small population die out, the site is much less likely to be recolonized than would be a piece of suitable habitat embedded in a large and continuously inhabited range.[13] Almost all the vanished animals of the Delta shared one common characteristic: they originally inhabited a broad geographical range but were largely restricted to, or at least preferred, a particular habitat. In Mississippi, these species were found mainly in the mature bottomland hardwood forests of the Delta and other, smaller floodplains.

As a general rule, bird surveys provide the best faunal data for assessing the correlation between habitat fragmentation and population declines. Many such studies support the hypothesis that the number of forest species and the population densities of forest interior species decrease with cumulative losses of forest area. Naturally functioning bottomland hardwood forests of sufficient patch size are therefore of great ecological importance. Modern research has pointed out that any reduction in the size of such a complex ecological system may have a disproportionately large effect on the resident fauna. It may furthermore affect migratory visitors and many species of adjacent areas, such as nearby aquatic communities.[14]

Incidentally, the Chicago Mill and Lumber Company, the successor of the Paepcke Leicht Lumber Company, played a decisive role in the destruction of the largest remaining tract of old-growth bottomland hardwood forest in the Lower Mississippi Valley—and the disappearance of the ivory-billed woodpecker from the face of the earth. In 1926, the Singer Sewing Machine Company, as the sole owner of the so-called Singer Tract of some eighty thousand acres in Louisiana's Madison Parish, leased the area to the state as a game refuge, but reserved the right for commercial development. In 1937, while James T. Tanner was conducting the field work for his classic study of the ivory-billed woodpecker in the locality, the logging rights were sold to the Chicago Mill and Lumber Company and cutting in the area commenced at once. The National Audubon Society immediately began to lobby for the creation of an inviolate wildlife refuge in the area, receiving support for the plan from both state and federal sources. Even President Franklin D. Roosevelt and Louisiana Governor Sam Jones seemed sympathetic to the project.[15]

Negotiations involving the Singer Sewing Machine Company, the Chicago Mill and Lumber Company, the state of Louisiana, the U.S. Fish and Wildlife Service, and the Audubon Society, however, came to a halt in 1943. While even the head of the War Production Board had maintained that parts of the tract could well be excluded from the war production program, the Singer Sewing Machine Company passed the decision over to the Chicago Mill and Lumber Company, which stubbornly declined "any limitation whatever on their plan to complete cutting of timber in accordance with their contract rights." During the negotiations, the company's chairman of the board, James F. Griswold, reportedly boasted to the representatives of the Audubon Society, "We are just money grubbers. We are not concerned, as are you folks, with ethical considerations."[16]

In the end, the last confirmed nesting area of the largest American woodpecker was completely cut over for military supplies such as packing crates, tea chests, and plywood gasoline tanks for fighter planes. German prisoners of war performed much of the labor. According to the noted wildlife artist Don Eckleberry, who in April 1944 made the last unquestionable sighting of the species in the United States, the POWs working in the Singer Tract "were, as any Europeans would be, incredulous at the waste—only the best wood taken, the rest left in wreckage to rot."[17]

"Patriotic cutting" was not restricted to the Tensas bottomlands on the western side of the Mississippi; the few remaining old-growth stands in the Delta were similarly affected. The property of Allan Grey—the sworn enemy of the Paepcke Leicht Lumber Company executives in 1918—had escaped substantial cutting until now. According to a Delta birdwatcher, M. G. Vaiden, "[u]p to the second World War there was another great estate composed of 12,341 acres of virgin timber located nine miles south of Rosedale, Mississippi, Bolivar County, and known as the Allan Grey Estate. It was reduced to zero during the War for the small PT boats used so effectively against the Japs in the south Pacific." After the area was logged over, the six resident breeding pairs of ivory-billed woodpeckers observed by Vaiden permanently vanished from the area.[18] No confirmed sightings of the species have been made in Mississippi for decades, although in March 1987, an unseen bird responded to taped vocalizations of the ivory-billed woodpecker for twenty-eight minutes in a bottomland swamp along the lower Yazoo. Later extensive searches for the species in Mississippi and elsewhere have invariably failed, and the ivory-billed woodpecker must today be considered extinct.[19]

Based on the preceding documentation of humans' extreme transformation of the floodplain forest, it would not be difficult to conclude that human-induced habitat alteration has been the sole cause of most Delta extinctions. Extinction, however, is a complicated biological process, and numerous factors usually contribute to the disappearance of a species. In order to evaluate the ultimate importance of anthropogenic habitat alteration in the extinctions of Delta fauna, it is important to briefly examine the general characteristics of extinction and address other factors that might have played a part in the animals' disappearance.[20]

A population or a species will become extinct when its rate of mortality is continually greater than its rate of recruitment. Extinction is, of course, a natural part of the evolutionary process, and since the late Jurassic period innumerable species of animals have evolved and most of them have become extinct. While humans cannot have had any role whatever in the vast majority of these extinctions, there is evidence of human agency in almost every avian and mammal extinction since the late Pleistocene. Humans have greatly increased the rate of animal extinctions, and the average longevity of a bird or mammal species has consequently been dramatically reduced. In fact, many scientists maintain that during historical times there are no documented examples of continental bird species being extinguished by nonhuman agencies. In any case, "natural" extinctions of continental birds seem very rare.[21]

The extinction of a species can often be viewed as a two-stage process, and causes of extinction can be divided into two classes: "ultimate" and "proximate." Ultimate causes of extinction are the reasons that have led to a situation in which there is a small population, while proximate causes consist of the reasons why the last individuals of this population die. Numerous studies of extinctions have shown that after the population of a species has decreased below a certain level by ultimate causes such as habitat destruction, proximate causes will deliver the *coup de grâce* and result in extinction. The concept of the "minimum viable population" is often used in this context, referring to the critical population size below which the population is doomed to quick extinction.

The proximate causes are often divided into four classes: demographic stochasticity, genetic deterioration, social dysfunction, and extrinsic forces. Demographic stochasticity refers to the random variation in population variables such as sex ratio, birth and death rates, or the distribution of individuals among age classes. Random change in these variables, such as the occurrence of broods consisting of members of the same sex, is much more threatening to a small population. Small populations also experience genetic deterioration as a result of losing alleles by genetic drift and inbreeding. This loss may limit a population's ability to respond to changes in its environment through natural selection. It must furthermore be noted that different proximate causes often contribute to each other. For example, a skewed sex ratio typically increases the rate of inbreeding depression and vice versa.

Some species have characteristic social behavior that makes them more liable to extinction. Elaborate social behavior, such as herding, group mating displays, or group defense, can considerably assist certain ultimate extinction forces in the reduction of population size. A good example of this is the destruction of the North American bison by commercial hunting during the nineteenth century, a process made extremely easy by the herding behavior of the species. It must also be noted that once a population has declined to a certain level, there may simply be too few individuals left to stimulate or consummate social behavior. For example, certain species of colonial birds satiate their predators by synchronous breeding, and thus a decrease in the size of a breeding population typically results in increased mortality from predation. A species may furthermore require some sort of social facilitation in order for mating and offspring production to be vigorous. Females of some bird species have short estrous cycles during which fertilization must occur, while males can fertilize females only during certain periods. Thus asynchrony between the sexes can prevent mating completely when there are too few individuals in a population.

A small population can furthermore be threatened by fluctuations in the biotic and abiotic environment. Extrinsic forces, such as temporal variation in habitat parameters and enemy population sizes, can produce major population declines. A greater probability then exists that extrinsic forces like epizootic diseases or adverse weather will extirpate a small population.[22]

Even a superficial examination reveals that at least one human activity, in addition to habitat alteration, contributed greatly to most animal extinctions on the Yazoo-Mississippi floodplain. Hunting of the abundant wildlife of the Delta bottomlands provided the early settlers with much-needed supplemental protein, but it also served as a diversion from everyday toil in the forests and fields. Early successional patterns and free-roaming livestock near plantations attracted both herbivores and predators and increased encounters between settlers and wildlife.[23] Much of the hunting was carried out in order to control the species harmful to the settlers' fields and livestock. Consequently, the shooting of agricultural pests, such as the Carolina parakeet, and various large predators, such as cougar and bear, was common in the frontier environment of the nineteenth-century Delta. The excitement of big game pursuit could, however, make antebellum hunters forget their original goal of protecting their agricultural acreage: "But before we had left the field [after the hunt], the horses, dogs, and bears, together with the fires, had destroyed more corn in a few hours, than the poor bear and her cubs had, during the whole of their visits."[24]

During the nineteenth century, the planters of the Delta came to emulate European aristocracy in their hunting practices, as seen proper for men of status and power. Hunting enabled the sportsman to appreciate the God-given order in nature—and in human society. Whereas blacks and poor whites hunted out of necessity, planters throughout the South maintained that they hunted solely for sport and amusement. Hunting practices could be wasteful even among the nonplanter class, as described by Mary Hamilton: "We had fried venison steak, roast apples, and good coffee with rich sweet cream. Next day I asked Bob what he did with the rest of the venison. 'Left it in the woods,' he said. 'We never save any but hindquarters. Our women won't cook anything else.' "[25]

Delta hunters typically began their career with squirrels, raccoons, and opossums, graduating through geese, ducks, turkeys, and deer to bobcat, bear, and cougar. Hunting, especially that of deer and bear, became a social activity. Hunting dogs were prized possessions and their feats were described at length in family letters.[26] Many a Delta planter—and a Faulkner character such as Major de Spain—would have subscribed to Benjamin Grubb Humphreys' claim that "I was never happier than when in wild pursuit, following my dogs through the cane brakes after a bear."[27]

Descriptions of deer hunts abound in the correspondence and reminiscences of nineteenth-century Delta residents.[28] Rowland Chambers of Satartia, Yazoo County, recorded in his 1858 diary many such accounts. Stalking deer in the Delta bottomlands proved tiring for the 51-year-old dentist, but on September 22, he was lucky enough to kill "3 fawns and one squirrel."[29] A deer hunt in the cane thickets of the Delta bottomlands could easily turn into a bear hunt: in September 1857, William Worthington and a hunting companion with their pack of four hounds came across a female bear with two cubs. "Madam Bruin" was finally brought down with the seventh shot while the cubs "made their absence."[30] Black bears must have been very common in the antebellum period; in addition to threatening the ample supply of free-roaming swine, they attacked hog pens.[31] Consequently, an overseer of a Delta plantation reportedly killed fourteen bears during one fall in the 1850s.[32]

During the nineteenth century, Wade Hampton III evolved into the ultimate planter-sportsman of the region. Already in antebellum times, the heir of the southern planting dynasty spent much of his time chasing deer and bear in the canebrakes of the Delta. In the course of a few days in April 1855, Hampton's hunting party killed five bear and a deer, and the following November, a week-long hunt by him and another prominent Delta planter, Dr. Orville M. Blanton, yielded ten bear with the loss of seven dogs.[33] In 1857, Hampton had "a large English party now with him [on the Wild Woods Plantation in Washington County], Lord Althorpe and his friend, . . . Cap[tain] Tower of the guards, and Mr and Mrs Portman." The male European guests were "all hunters" but initially "killed but few Bear, and no deer at all" while their more experienced host had earlier "killed two [bear] and caught a cub" in just one day. Captain Tower "who was in every action in Crimea" turned out to be "a very nice fellow" with great regret for "not being able to stay here longer." On November 8, 1857, the guests finally found some sport as Hampton "took them bear-hunting and we killed *four*." This particular hunt turned rather exciting as "Lord Althorp (or as Sam calls him "Lord") . . . literally had his clothes torn off. I had to furnish him with my *drawers,* so as to enable him to come home decently."[34] During Reconstruction, the temporarily fallen Delta aristocrat found contentment in the bottomlands; in the middle of severe financial crises, Hampton reported in 1866 to "have killed 5 bear, one panther, and one wild turkey" in his last three hunts, and two hunts in 1875 produced three bear.[35] With his fortunes reversed after 1876, Wade Hampton III attained fame not only as a prominent politician and businessman but as one of the nation's foremost hunters.

In comparison to deer and bear, cougars and wolves were encountered less often since the commencement of European settlement.[36] The hides of cougars shot at the Doro plantation in Bolivar County were used to replace the cane seats of chairs. In frontier style, even "[t]he children had their little panther chairs."[37] Toward the end of the nineteenth century, the populations of all large game species plummeted. In 1893 Washington County planter and businessman Clive Metcalfe complained in a plantation ledger entry that his brother "came and we went hunting [but] did not start a single thing. The bear have all disappeared from this part of the country."[38]

At the turn of the century, however, the remaining Delta canebrakes still supported a few black bear and served as the setting for the most famous bear hunt ever. For five days in November of 1902, President Theodore Roosevelt and his hunting party camped on the banks of the Little Sunflower in Sharkey County. The nation's best-known trophy hunter had never before participated in a southern black bear hunt on horseback with hounds, a pastime warmly recommended to him by Wade Hampton III, and wanted to have the experience while it still was possible. The hunt was arranged by Stuyvesant Fish, the president of the Illinois Central Railroad, with prominent Delta planters and politicians as members of the hunting party. At Roosevelt's request, John McIlhenny, a former

Rough Rider, Louisiana conservationist, and Tabasco tycoon, joined the company of John M. Parker, E. C. Mangum, George M. Helm, Huger Lee Foote, and LeRoy Percy. Fifty-six-year-old former slave Holt Collier of Greenville served as the chief hunting guide. Collier had killed his first black bear in Washington County at the tender age of ten, and during his fifty-two years as a bear hunter reportedly slew over three thousand, most of them in the bottomlands of the Delta.

On the first day of the hunt, Collier positioned the President on a stand with Foote and promised to drive a bear to the spot with his hounds. After several hours of hard work, Collier and his assistants succeeded in doing this—only to learn that the hunters had left their stand for a late lunch. Desperate, Collier then caught the exhausted animal with his lariat and presented it to Roosevelt to be shot. As a sportsman of the highest class, the president refused to kill an animal under restraint, and the already wounded 235-pound male bear was finally knifed to death by Parker and Collier. During the rest of the hunt, Roosevelt was unable to get a shot at another bear and had to return to Washington without his trophy. To the President's considerable irritation, the incident with the tied bear became widely publicized. On November 16, 1902, Clifford Berryman's editorial cartoon for the *Washington Post* depicted the scene in a humorous light, with the old male bear drawn as a cute little cub. Berryman and other cartoonists subsequently associated Roosevelt with the cub and gave birth to the term "Teddy Bear." The name was also adopted by makers and marketers of stuffed toy animals, eventually creating a lasting cultural icon and international industry.[39] Teddy Roosevelt's Delta excursion evidently prompted other northerners to follow suit: wealthy tourists continued to hunt down the few surviving black bear in the floodplain's fragmented forests until the 1920s.[40]

In addition to mammals, various species of bottomland birds were commonly shot for food, but also for sport, practice, and even curio items. The crest and upper mandible of an ivory-billed woodpecker could be used as a watch charm or ornament for the hunter's shot pouch. During his travels on the Mississippi, Audubon frequently noticed "that on a steam-boat's reaching what we call a *wooding-place,* the *strangers* were very apt to pay a quarter of a dollar for two or three heads of this Woodpecker."[41] The shooting also targeted visiting avifauna, such as migrating waterfowl and the now-extinct passenger pigeon. In the 1840s, enormous pigeon flocks were still encountered along the lower Mississippi.[42] Christian Schultz's description of an early-nineteenth-century pigeon hunt leaves no doubt about the destructiveness of antebellum hunters:

> During the remainder of the day we met with nothing worthy of notice, till towards sunset, when we were crossed by innumerable flocks of pigeons, which passed so near us that we were able to distinguish their eyes. We had fine sport for about an hour; but although well provided with pigeon-shot, we could not kill more than one or two at a time. Every one thought the guns were spelled or bewitched, as each of us had frequently shot five and six ducks at a time, and I had even brought down fourteen pigeons from a tree at a single shot. I attributed it to the nearness of the object, which prevented the shot from being sufficiently scattered; and in order to convince them, I took a pair of pistols, and put thirty or forty shot over the ball, with which I brought down two at the first shot; but after charging it with an ordinary load, it was easy to kill from three to seven at every fire.[43]

In the antebellum Delta, pigeons were also harvested from their nightly roosts, using lighted pine knots to blind the birds and then beating them down with long poles.[44]

The pursuit of bottomland game was not restricted to birds and mammals; alligators were commonly hunted on the interior streams of the Delta. During the 1850s, large-scale commercial hunting of the species for shoes, belts, and purses commenced all over the South, but "the most detested loafer in the animal kingdom" was still found on the Yazoo, "though the steamers disturb him much, and send him to take his siestas in the more retired bayous, and the secluded swamps and ponds."[45] The value of hides skyrocketed in the following decades, and alligators were hunted to extinction over much of their range. By the mid–twentieth century, protection under state legislation had become requisite for the species' survival in the United States.

Although hunting can be the ultimate cause of the demise of a species, it is highly probable that habitat destruction in the form of land clearing and logging acted as the ultimate cause for most, if not all, animal extinctions in the modern Delta. Forest fragmentation led to the species' initial decline, and various proximate causes, especially hunting, then extermi- nated the isolated and sedentary populations. The rapid disappearance during the late nineteenth century of species ecologically as dissimilar as the ivory-billed woodpecker and the cougar indicates a profound change in the Delta bottomland hardwood forest complex.[46]

The human-induced transformation of the Delta landscape was by no means completed by the early 1930s. Instead, economic activities on the floodplain were to undergo substantial change throughout the rest of the century, with considerable effects on the region's human inhabitants— and the remaining forests. Between 1929 and 1954, acreage under cotton cultivation in the eastern United States decreased dramatically, from 43 million to 17 million acres. During the same time period, production de- creased only slightly, from 14.6 million to 12.8 million bales, as the aver- age yield per acre had increased from 164 pounds in 1929 to 341 pounds in 1954. The reduction in cotton acreage, however, had far-reaching con- sequences for the tenant farmer. The new tenancy-based plantation sys- tem created during the Reconstruction came to an end between 1935 and the mid-1950s.[47]

The Great Depression had focused public attention on the plight of southern tenants. Ironically, from the beginning, federal farm programs heavily subsidized planters while largely ignoring the predominantly black workers. New Deal programs administered by the Agricultural Ad- justment Administration (AAA) further strengthened the position of the landholder and contributed to the eviction of thousands of tenants. All federal subsidies for acreage reduction and related payments were chan- neled through large landowners and, despite AAA contracts forbidding the displacement of tenants, the funds were invested in labor-saving ma- chinery. Much of that federal money came to the Delta; already in 1934, 44 percent of all AAA payments in excess of $10,000 were received by ten Delta counties. Between 1933 and 1935, the Delta and Pine Land Company alone was the recipient of some $320,000 in federal benefit payments, while the Parchman prison farm was given $75,000 in 1933 alone.[48]

During the 1930s, technological innovations led to the emergence of the so-called neoplantations, which were more capital-intensive but required less labor than earlier plantations. By the beginning of the 1940s, cotton could be planted and cultivated almost without manual labor, which was now needed only for hoeing and harvesting. Soon the advent of mechanical cotton harvesters and new herbicides turned the remaining sharecroppers into part-time wage laborers. The cotton plants were now also sprayed with calcium cyanide in order to defoliate them; unlike humans, the mechanical harvesters could not find the white bolls among the green leaves. The application of new defoliants, herbicides, and pesticides, combined with efficient spraying techniques, greatly increased yields from the 1950s onward. The new chemicals were expensive to use, were customarily overapplied, and had harmful effects on the native flora and fauna, not to mention the workers in the fields. Dramatic changes also took place in cotton ginning. Because the mechanically harvested cotton contained more moisture and trash than the hand-picked variety, specialized machinery had to be developed to clean and dry the product during the ginning. At the same time, the harvest period was compressed into some six weeks, as the new high-capacity gins could process more than ten bales of cotton per hour. Ironically, the neoplantations came to resemble their antebellum predecessors in certain distinctive ways. The few remaining workers again lived in grouped housing, while the owner recentralized his power over the work force and equipment. The decline in the number of tenant houses maintained by the Delta and Pine Land Company illustrates the rapid changes in bottomland agriculture; in just a few years during the mid-1940s, close to 40 percent of the company's 850 tenant houses were vacated. Because of the permanently reduced need for labor, the human population of the Delta has decreased steadily since the 1940s. Today the number of people inhabiting the Delta's ten core counties has dwindled down to the levels of the 1910 census.[49]

After World War II, bigger farms, commercial fertilizers, pesticides, and an expanding market for agricultural commodities resulted in increased profits, and new land for agriculture was in high demand. New drainage techniques became crucial to the transformation of the remaining forested lowlands into agricultural acreage. Although cotton was still grown on many plantations along the lower Mississippi, including in the Delta, most southern neoplantations concentrated on soybeans, grain, sorghum, peanuts, and cattle. Since the 1950s, a growing number of plantations also adopted rice culture, pecan growing, and even catfish farming as the backbone of their agricultural operations. For example, between 1965 and 1976, the town of Belzoni in Humphreys County evolved into the self-proclaimed "Catfish Capital of the World." The Chicago Mill and Lumber Company also diversified its business base during the 1960s. In 1965 the company changed ownership and was reorganized, with the general offices moving from Chicago to Greenville. The company soon entered into agribusiness in Louisiana's East Carroll, Franklin, and Tensas parishes with rice and soybeans. By 1980, the venture had turned into one of the largest row-crop farms in the country with thirty thousand acres under cultivation. Still, Hermann Paepcke's legacy had not been forgotten, as the company retained about two hundred thousand acres of timbered bottomland and employed some seven hundred people in its three sawmills and two box factories.[50]

By the mid-1970s, the levee system along the main stem of the Mississippi consisted of some twenty-two hundred miles of levees, with over fifteen hundred additional miles of authorized levees along the tributaries. The construction and installation of this extensive flood control system exerted a two-fold effect on land clearing in the bottomlands. Direct clearing of the remaining bottomland hardwood forests resulted in project areas, while induced clearing took place because of economic externalities. Construction or even expectations of the construction of a flood control measure provided incentives for landowners to convert previously flood-threatened land, occupied by hardwood forests, into farmland.[51]

Soybeans (*Glycine max*) became the major crop in many such areas along the Mississippi, and the desire to convert the remaining areas of bottomland hardwood forest into productive row-crop fields was largely inspired by the cultivation of this new, profitable crop. Between 1930 and 1982, the acreage devoted to its cultivation in the United States increased from one million to seventy-one million acres. Introduced in North America as early as 1804, soybeans came to be planted more widely during the early twentieth century, initially not for oil but for fodder. The demand for vegetable oils and high-protein livestock feeds increased during World

War II, and the cultivation of soybeans spread rapidly through the South. At the same time, developments in processing technology made soybean oil suitable for human consumption. Fertile bottomland areas along the Lower Mississippi were especially affected by the expansion of soybean cultivation: tenfold increases in soybean acreage were not uncommon in the region between 1950 and 1970. In Mississippi, the amount of land in soybeans increased from mere twelve thousand acres in 1930 to over two million in 1970, most of it in the Delta. By the 1970s, most counties and parishes along the Mississippi between southeastern Missouri and northeastern Louisiana devoted over half of their total harvested cropland to soybeans. During this time, the Delta and Pine Land Company evolved into one of the largest soybean planting seed suppliers in the United States.[52]

Not all floral introductions to the South were as economically successful as soybeans. Probably the most famous example of failure is the introduction of the Asian kudzu (*Pueraria lobata*), a weedy vine capable of incredible growth—more than a foot per day. It was originally introduced to the South during the late nineteenth century as a shade plant. Since the fast-growing kudzu is a legume, capable of adding atmospheric nitrogen to depleted soils, and showed great promise as livestock forage, it came to be viewed by many as the crop of the future. Extensive planting commenced during the 1930s to control erosion on hillsides and fallow fields throughout the South, and was heavily subsidized by the U.S. Department of Agriculture. The use of kudzu as fodder, however, proved problematic, as regular grazing tended to kill the plant. But when kudzu escaped into the woods, it created an eerie landscape and became a danger to native flora—and commercial timberlands—by shading out its competitors.[53]

As a result of the onset of the Great Depression and the virtual disappearance of old-growth forests, Mississippi lumber production in the mid-1930s plummeted to 1880 levels. Many landowners were unable to pay property taxes and forfeited their titles. In the backwater area county of Issaquena, some 35 percent of the land reverted to the state in 1934. The Chicago Mill and Lumber Company's Greenville plant also shut down for three years during the onset of the Depression. At the same time, however, the depletion of mature pine forests in Mississippi contributed to the continuing exploitation of Delta hardwoods. In 1938, 165 mills, most of them small operations, produced eighty-one million board feet of lumber from the remaining Delta forests. The annual cut continued to greatly exceed the growth, and by the early 1940s all the best timber had been removed and the land cut over two or more times.[54]

After World War II, the hardwood industry encountered new competition from synthetic materials and products such as nylon carpets, which quickly began to replace wood products. Consequently, the residential hardwood flooring market suffered heavily after 1960. Hardwood flooring sales in Memphis plummeted from 1.2 billion to 96 million feet between 1955 and 1975, and by the beginning of the 1980s, the former "Hardwood Capital of the World" supported only one hardwood sawmill.[55]

In 1950, 37 percent of the land area in the Yazoo-Mississippi Delta was classified as forested. Between 1935 and 1950, the decrease in the amount of forested area in the Delta was limited, while in the alluvial lands of Arkansas and Louisiana deforestation proceeded rapidly. The rates of clearing differed because, unlike the bottomlands of Arkansas and northeastern Louisiana, the Delta had largely been logged over by the beginning of World War I. As explained in the preceding chapters, lumber companies had sponsored Delta drainage districts during the early twentieth century, harboring hopes that the creation of drainage on cutover lands would facilitate their sale. Substantial acreage was sold but the existing drainage often proved inadequate for agricultural purposes. The development of sustained-yield forestry largely ended the sales of cutover lands, and after World War II the lumber companies commonly began to restock their remaining bottomland holdings with certain fast-growing hardwoods. Many of the new forestry practices originated from the Southern Hardwoods Laboratory in Stoneville, Washington County, founded in 1936 through the efforts of John A. Putnam. Still, there was room for further agricultural development in the Delta: between 1945 and 1959, over three hundred thousand acres were cleared, mostly in the lower portion of the floodplain. By the beginning of the 1960s, less than 32 percent of the Delta could be classified as forestland.[56]

Encouraged by the continuous development of levee and drainage systems and aided by favorable climatic conditions during the 1960s, agricultural expansion proceeded in the backwater area of the lower Delta. Between 1957 and 1977, another three hundred thousand acres of poorly drained soils were cleared for agricultural purposes in Sharkey, Issaquena, Humphreys, Yazoo, Washington, and Warren counties. The Mississippi flood of 1973 unexpectedly inundated some 640,000 acres in the Delta's backwater area and once again demonstrated the risks associated with bottomland farming. Despite ambitious engineering projects, flooding has remained a recurring problem for the lowermost Delta. Federal agencies are still at odds today over the issue of how to best manage this area that supports most of the remaining bottomland habitat in the Delta. In September 2000, the U.S. Army Corps of Engineers released its elaborate plan for further drainage development in the Yazoo backwater area: the Yazoo Pumps Project. The U.S. Fish and Wildlife Service vehemently opposes the plan, claiming it contains "environmental features that are neither viable nor sustainable" while excluding "floodplain restoration as an explicitly stated project purpose."[57]

The intensively cultivated lands of the central and upper Delta have faced another challenge. Since the 1930s, the regulation of the Mississippi and Yazoo River systems through damming and leveeing generally prevented overflow and sedimentation. Consequently, erosional processes replaced depositional ones and, by the 1980s, erosion on the former floodplain progressed at a rate of 1.5 to 13 tons per acre per year.[58] For obvious reasons, activities to manage the flows of water in the Delta had originally focused on flood control and drainage, and little attention was paid to the region's groundwater supply. The floodplain's groundwater is stored in the Mississippi Delta Alluvial Aquifer, located in a shallow layer of sand and gravel between layers of other materials, mainly clay. The water-bearing layer may extend from fifty to two hundred feet in depth. The layer is recharged from the Mississippi in the west and the hillside streams in the east, with only minor recharge occurring from direct infiltration of rainfall or local waterways.[59]

Much of the reclamation effort along the lower Mississippi after World

War II, prompted by growing demands for soybeans and rice, was carried out by private interests, especially large-scale agribusiness operators. Cultivation of soybeans and rice and, more recently, farming of catfish, requires vast amounts of supplemental water. Consequently, the amount of irrigated land in Mississippi grew from 5,056 to 430,901 acres between 1950 and 1982, much of it in the Delta. Ironically, the formerly overflowed lands of the Delta today face problems of water availability: aquifers have experienced drawdowns as a result of extensive pumping of groundwater, and under present conditions, within a few decades there may not be sufficient water for the projected use. To attain the most effective combination of irrigation and drainage for the flat Delta land, modern agriculturists have furthermore been obliged to rebuild their fields. Since the 1950s, land grading, locally known as "land forming," became an inseparable part of the regional agriculture.[60]

The two most important factors affecting the maintenance of natural habitats in the South and the Yazoo-Mississippi Delta are still the patterns of land ownership and land use. Farmers and forest industries continue to emphasize the economic productivity of their lands, usually managing them without any long-term goals of protecting biological diversity. Since the 1930s, roughly a third of all the cleared land in the South has reverted to forest, though fragmented and immature. Large areas of southern upland forest have been converted to even-aged pine plantations, and these biological deserts today dominate the landscape of many states. Much of the increase in forested area has come from the abandonment of marginal farmland, a process largely restricted to upland regions. Consequently, while the amount of forested land continued to decline in the Delta during the 1930s and 1940s, the abandonment of farmland increased it elsewhere in the state. By 1960, the annual growth of softwoods in Mississippi exceeded the drain by 10 percent while the hardwood supply continued to dwindle.[61]

The remaining bottomland hardwood forests in the southern United States continue to undergo compositional changes and rapid reduction in area. Much of the surviving forest land is being managed for timber production or recreation in ways that reduce its viability as natural habitat; forests are also still being converted to farmland, but they are being used for suburban development, as well. Most of the mature bottomland hardwood forest in the South, Mississippi, and the Delta seems to have been permanently lost: old-growth stands of baldcypress were typically four hundred to six hundred years old at the time of European settlement, but today few individuals over two hundred years old remain. The fifteen thousand acres of the Congaree Swamp National Monument in South Carolina contain the last significant stand of "virgin" floodplain forest in the South. Only the relatively small areas of bottomland hardwood forest habitat in public ownership today seem secure from economic exploitation.[62]

Passage of the Weeks Act of 1911 enabled the federal government to acquire land from individuals and corporations, and many owners of logged, burned, and abandoned forestlands were happy to pass these back into federal ownership. National Forests (NF) in the South consequently became a heterogeneous mixture of land in various stages of recovery from former use. During the 1930s, the federal government purchased vast areas of Mississippi cutovers and expanded the NF system into the state. By the late 1970s, the federal government managed just over one million acres of NFs in Mississippi. These lands, located mainly in the pine belt region, amounted to less than 4 percent of the total land area of the state. Other, considerably smaller areas of federal land were found in national monuments and wildlife refuges, military reservations, and areas administered by the Bureau of Land Management. Additionally, most of the watersheds of the reservoirs managed by the U.S. Army Corps of Engineers were in the public domain.[63]

The only national forest on the Yazoo-Mississippi floodplain, the Delta NF, was established in 1961 in the southern backwater area to promote scientific hardwood management. Founded largely on lands previously owned by the Singer Sewing Machine Company, the Delta NF today encompasses some sixty thousand acres of bottomland hardwood forest in Sharkey and Issaquena counties—including Teddy Roosevelt's 1902 hunting grounds. Most of the habitat on this nation's only bottomland hardwood NF is second or third growth, but the Green Ash–Overcup Oak–Sweetgum Research Natural Areas in the northern part of the forest contain some significant old-growth stands. Like the batture land between the main levee and the Mississippi, the Delta NF has for decades maintained the highest populations of deer and turkey in the state. These two areas today constitute the largest remaining unbroken patches of bottomland habitat in the Delta.[64]

The long-term protection and preservation of all levels of biological diversity in Mississippi has traditionally been maintained through federal ownership and, to a lesser extent, the various natural areas purchased by state and local governments. During the last four decades, federal agencies have been mandated by numerous laws and directives to maintain natural habitats. It has, however, proved difficult or even impossible to acquire every variety of habitat necessary to conserve biological diversity on federal or other public lands. Portions of existing public lands in the eastern U.S. were allowed to be designated as wilderness areas by the Eastern Wilderness Act of 1975. Such areas in Mississippi, established after 1978 and completely protected from development and active resource management, are the 5,050-acre Black Creek and the 940-acre Leaf Wilderness Areas. Located in Perry and Greene counties in the southernmost part of the state, they amount to a mere 0.3% of the total wilderness area in all southeastern states.[65]

Two National Wildlife Refuge (NWR) complexes currently exist on the Yazoo-Mississippi floodplain. They are managed by the U.S. Fish and Wildlife Service for the benefit of migrating and wintering waterfowl, but lately the management of the refuges has been expanded to include habitats that also protect other species of wildlife. Today the NWR complexes in the Delta offer glimpses of the bottomland hardwood forests of the past. The Yazoo NWR Complex in the southern part of the floodplain covers over 75,000 acres and consists of five refuges: the 13,021-acre Yazoo in Washington County, 38,601-acre Panther Swamp in Yazoo County, 2,418-acre Mathews Brake in Leflore County, and 15,572-acre Hillside and 7,381-acre Morgan Brake in Holmes County. Land acquisition for the refuges began in 1936 with the purchase of 2,166 acres under provisions of the Migratory Bird Treaty and Hunting Stamp Acts, but the first one established, the Yazoo, was staffed and activated only in 1956. The North Mississippi Refuges Complex, established in 1989, is responsible for the management of three refuges in the northern part of the Delta. The 9,691-acre Dahomey NWR was founded in 1990, largely on the lands originally owned by Allan Gray and clear-cut during the 1940s; 260 acres of the Bolivar County School Section, logged over by the Paepcke Leicht Lumber Company between 1918 and 1920, have recently been added to the refuge. The Tallahatchie NWR covers 4,083 acres in Tallahatchie and Grenada Counties, while the 2,069-acre Coldwater NWR is mostly situated in Quitman County.[66]

In wintertime, the Delta refuges support hundreds of thousands of geese and ducks, and they provide nesting habitat for innumerable herons, egrets, shorebirds, and passerines during the summer months. Extensive nest box programs have considerably increased wood duck populations on the refuges. Deer, river otter, and alligator populations are thriving over much of the two refuge complexes, and black bear sightings have been on the rise.

Wildlife habitat in the refuges is heavily managed. Impounded areas within the refuges are managed either as permanent water, so-called green-tree reservoirs, or moist-soil units. The green-tree reservoirs serve as seasonal wetlands and are intentionally flooded during the late fall and winter, while moist-soil units are managed by burning and mowing existing vegetation with the aim of providing an optimal source of food for wildlife. The absence of recurrent natural flooding has made it impossible to maintain bottomland hardwood forest habitat on the refuges without carefully planned human manipulation of water levels. For example, most of the eighty former catfish ponds in the Coldwater and Morgan Brake NWRs are today managed at different successional stages to maintain diverse waterfowl and shorebird habitat, while some are stocked for the purposes of public fishing. The Hillside NWR on the eastern edge of the Delta is essentially a silt collection site created by the U.S. Army Corps of Engineers in connection with the Hillside Floodway/Yazoo Basin Headwater Project and turned over to the Department of the Interior in 1975. Originally designed with a life span of five decades, the project is approaching its intended silt load at nearly twice the expected rate. This is largely due to excessive deforestation in the loess hills east of the refuge, where many of the proposed silt collection reservoirs were never completed. Much of the forest on the northern part of the refuge has been eliminated, and siltation is expanding southward with harmful effects to existing wildlife populations.

Local farmers have been included in the management efforts in the refuges. They can plant crops such as corn, wheat, rice, soybeans, and milo on refuge lands, and in return leave a quarter of the crops unharvested; these are utilized by wildlife as food. Abandoned fields are being reforested with bottomland hardwoods such as native oaks, sweetgum, and baldcypress to reclaim the original habitat, retard erosional processes, and serve as filters to improve water quality. Still, fishing in the Yazoo NWR remains prohibited because pesticide levels found in fish, including the notorious DDT, continue to exceed federal standards.

In addition to natural areas protected by the federal government, the state of Mississippi maintains nine Wildlife Management Areas (WMA) in the Delta, coordinating their use with activities on other public lands. The WMAs, like other protected areas, are concentrated in the southernmost Delta; only two WMAs are located north of Greenwood. Some corporate landowners in the Delta, such as the Anderson-Tully Company, have for decades maintained parts of their timberlands with wildlife management objectives and opened them for public use. Of the four state parks located in the Delta, the Leroy Percy State Park in Washington County features a wildlife preserve while the Winterville Mounds State Park, located north of Greenville, serves as a small monument to the indigenous Missisippian cultures. All public and other protected areas in the Delta are surrounded by vast areas of private land, most often intensively cultivated agricultural acreage. Activities on private lands considerably affect public lands, as not enough public funding exists for protection of entire watersheds or to establish corridors of natural habitat connecting natural areas to each other. Consequently, habitat fragmentation continues to present a great threat to the remaining biological diversity in the Yazoo-Mississippi Delta. However, habitat alteration as the cause for decline in Delta animal populations is not always directly related to the fragmentation of the forest. Two resident bird species of the Delta, the least tern (*Sterna antillarum athalassos*) and the wood stork (*Mycteria americana*), are today considered endangered, but they suffer mainly from the construction of water reservoirs and channelization of rivers. Human alteration of natural water cycles has affected their nesting success by destroying much of the tern's sandbar nesting habitat and the stork's feeding grounds.[67]

## This Delta, This Land

The Yazoo-Mississippi Delta can rightfully be described as an identifiable bioregion, and therefore an appropriate setting for the study of interaction between humans and nature in deep time. The floodplain has constituted a definable ecological complex for at least the last five thousand years. The prerequisites for the sustained existence of southern bottomland hardwood forest in the area—a meandering river regime and modern climatic conditions—have remained generally unchanged during the late Holocene, and enable a meaningful examination of the human-nature interaction on the floodplain.

As stated in the introductory chapter of this study, environmental history can contribute to the general study of human-nature interaction by phenomenologically identifying various social, economic, and ecological processes in the past and analytically separating relevant patterns from each other. Pattern descriptions pertaining to major events in the environmental history of the Yazoo-Mississippi Delta during the last five millennia are presented in table 5. This representation aims to identify some major phases in socioecological dynamics and offers criteria for the identification of environmental change in the Delta. It furthermore attempts to describe the different ways in which successive ecohistorical formations functioned.

Fernand Braudel's concept of geographical time (*histoire de la longue durée*) provides time scales for developments in both the natural and human spheres of Delta history. For example, one clearly identifiable natural formation, the late Holocene bottomland hardwood forest of the Delta, has acted as a setting for two major civilizations: the Native American and Euro-American cultural complexes. These two civilizations can conveniently be used to divide the environmental history of the Delta into two ecohistorical periods that illustrate socioecological change in a long historical perspective. This loose classification, however, does not exclude the construction of more detailed analyses within or at the interface of the two general ecohistorical periods; the environmental history of the Delta during the late Holocene can be further arranged into several distinct ecohistorical formations. For example, table 5 identifies seven temporal phases that can be used in analyzing human-induced environmental change in the Delta. The time period extending from five thousand years ago to the late seventeenth century encompasses the era of Native American civilization in the Delta. It is divided into three ecohistorical formations, the pre-Mississippian, Mississippian, and post-Mississippian, each exhibiting distinct features in its subsistence technologies and modes of production. Similarly, the four temporal formations under the Euro-American heading differ considerably from each other in many of their characteristics. The properties of human-induced environmental change on the floodplain accordingly vary between these formations. All the groupings displayed in the table undoubtedly exhibit considerable internal variation, and should not be interpreted as rigid classifications—the boundaries could well have been drawn somewhat differently.

Ecohistorical periodization can be augmented by the theory of ecological revolutions introduced by Carolyn Merchant to examine the changes in the land and life of post-Columbian New England.[68] An ecological revolution is a major transformation in human relations with nonhuman nature stemming from the changes, tensions, and contradictions that develop between the human modes of production and ecology of a society. In a similar manner, the changing relationship between the modes of production and the modes of both biological and social reproduction contribute to ecological revolutions. According to Merchant, two major transformations in the regional ecology of New England took place between 1600 and 1860. The so-called colonial ecological revolution occurred during the seventeenth century. It was externally created by the incoming European life-forms and resulted in the collapse of the Native American socioecological complex. The transformation was legitimated by cultural symbols that placed Europeans above wild nature, which included the indigenous people. In the process, the animistic belief system of Native Americans, based on symbolic exchanges between humans and nature, was replaced by the European image of nature as female and subservient to an all-powerful male God, whose orders the settlers were executing in North America. The so-called capitalist ecological revolution began to take place during the late eighteenth century and was more or less completed by the beginning of the Civil War. This second major transformation was internally generated. It introduced an economy of increased human labor and land management, legitimated by mechanistic science. During the second ecological revolution, human consciousness was seen as divided into a disembodied analytic mind and a romantic emotional sensibility.

Ecological revolutions also occurred in pre-Columbian times. Vastly differing subsistence systems and human modes of production had evolved in the Delta since the arrival of the first Native Americans. Originally based on hunting and gathering, the Native American subsistence systems in the Delta changed over time to include various horticultural practices. The amount of horticulture and trade practiced by the Native American societies did not show a continuous, linear increase, but fluctuated considerably. For example, during the late Mississippian era, Native American populations of the Delta were probably larger and more dependent on agriculture as their subsistence base than ever before or after. During this time period, the environmental impact of Native Americans on the bottomland forests accordingly attained significant levels. Consequently, true wilderness, meaning nature uninfluenced by human activities, hardly existed in the Yazoo-Mississippi Delta at the commencement of Euro-American settlement. Still, the Delta landscape of the eighteenth and early nineteenth centuries possessed many features of wilderness because the decimation of aboriginal populations by virgin soil epidemics in the sixteenth and seventeenth centuries had enabled forest vegetation to reclaim the areas affected by Native American agriculture.

Despite occasional increases in population and cultivated acreage, the relationship between the human and nonhuman worlds in the Delta apparently remained sustainable through the indigenous era. With the arrival of Europeans, Native American consciousness, which presumably viewed all life-forms as equal subjects, was supplanted by a vision in which humans were largely considered separate from the rest of the world, which was viewed as a pool of resource objects. At the same time, European alphanumeric literacy replaced the aural and oral transmission of tribal knowledge. During the eighteenth century, the growing presence of

*Table 5.* General Patterns of Human-Induced Environmental Change in the Yazoo-Mississippi Delta, 3000 B.C. to the Present

| Period | 3000 B.C.–1000 A.D. | 1000–1400 | 1400–1700 |
|---|---|---|---|
| Human Population Dynamics | Fluctuating | Increasing | Rapidly declining |
| Society | Preindustrial (Native American) | → | |
| Human View of Nature | Nature as self-active; reciprocity between humans and nature | | |
| Natural World in Human Thought | Active subjects | → | → |
| Space | Named places | → | → |
| Mode of Production | Kin-ordered and/or tributary | Tributary | Tributary (capitalist) |
| Primary Uses of the Floodplain | Hunting and gathering with increasing agriculture | Diversified maize agriculture, hunting and gathering | Hunting ground for meat and deerskins |
| Main Form of Labor | Communal and slave | → | → |
| Bounty of Forest for Local Use | Fish and game, nuts and acorns, wood for domestic use | | |
| Bounty of Forest for Supralocal Use | Various trade items | → | → |
| Landscape | Mature forest landscape with increasing number of settlements | Mostly mature forest landscape with agricultural areas | Mature forest landscape with scattered settlements |
| Faunal Changes | Increasing amount of edge species | | Increasing amount of forest interior species |
| Forest Condition | Mature | Mature with human-induced openings | Increasingly mature |
| Floodplain Hydrology | Natural | → | → |

Cf. Björn, *Kaikki irti metsästä,* 18, 227, 232; Merchant, *Ecological Revolutions,* 24.

| 1700–1820 | 1820–1880 | 1880–1940 | 1940– |
|---|---|---|---|
| Negligible | Increasing, tens of thousands | Increasing, hundreds of thousands | Decreasing, hundreds of thousands |
| → | Industrial | → | → |
| Nature as passive object; human domination and mastery of nature | | | → |
| Passive objects, commodities | | Scientific objects, natural resources | |
| → | Rectangular, mapped space as property | | → |
| Capitalist | → | → | → |
| → | Cotton monoculture, cypress lumbering | Cotton monoculture, hardwood lumbering | Various agribusiness monocultures |
| → | Slave, tenant | Tenant, wage labor | Wage labor |
| → | Wood and fodder for domestic use, fish and game | Wood for domestic use (fish and game) | Fish and game for recreational purposes |
| Deer skins | Cypress lumber | Hardwood lumber | Hardwood lumber from managed and planted stands |
| Mature forest landscape | Increasingly agricultural landscape | Predominantly agricultural landscape | |
| Human-induced extinctions of ungulates | Increasing amount of edge species, species introductions | Numerous human-induced extinctions, species introductions | Permanently reduced number of species |
| Mature | Mature with human-induced openings | Increasingly young or clear-cut | Predominantly young |
| → | Small-scale leveeing and drainage | Large-scale leveeing and drainage | Totally transformed |

colonial Europeans in the lower Mississippi Valley caused an enormous increase in the amount of Indian trade. Over time, changes in the Native American society and economy undermined the old religious order and resulted in the adoption of new value systems among selected elites.

The Euro-American period of subsistence-oriented agriculture in the Delta was largely nonexistent, as the development of the market and transportation systems during the early nineteenth century geared the local economy toward production for the capitalist marketplace from the very beginning. In contrast to the ecological revolutions Merchant describes in New England, the temporal boundaries between the colonial and capitalist ecological revolutions in the Delta were more blurred because of the delayed but extremely rapid settlement by people of European origin. The analytical and quantitative consciousness of Euro-Americans emphasized efficient management and increasing control over nature from the establishment of the first cotton field on the Delta bottomlands. Consequently, major human-induced environmental change on the floodplain was unavoidable.

Probably the most important question in assessing human-induced change in ecological systems is whether the systems affected retain their resilience. Human modification of the Delta environment up to the late nineteenth century generally remained comparable to that caused by natural occurrences within disturbance regimes, and the change stayed within bounds that enabled the recovery of the natural systems after disturbance. As illustrated by the successful forest regeneration on the agricultural fields abandoned by the Mississippian people and by the Civil War era planters, Delta forests retained their resilience well into the modern era. Similarly, deer herds decimated by the deerskin trade during the eighteenth century made a comeback within decades. On the other hand, certain individual species, such as the ivory-billed woodpecker, Carolina parakeet, black bear, and cougar, were practically erased from the native fauna of the Delta already during the nineteenth century.

The massive human efforts to restrain the rivers for the first time endangered the ability for recovery of the whole natural complex. From the late nineteenth century on, the system of protective levees on the floodplain expanded enormously. As more dependable levees for the protection of croplands and other property were established, it became evident that maximum yields could be obtained from the flat Delta lands only by providing adequate drainage. These developments furthermore facilitated the removal of the original forest cover. After the 1927 flood and abandonment of the "levees only" policy, the hydrological system of the floodplain was remade in a way that offered little chance for the regeneration of the bottomland hardwood forest. The immense human-induced change in the hydrology of the Mississippi and its tributaries since the nineteenth century has aimed to make floodplains safe for agriculture and human habitation. After centuries of hard work and enormous investments by local, state, and federal governments, it is customarily assumed today that almost any amount of high water can be safely transported through the Lower Mississippi Valley. Still, as pointed out by the foremost student of the history of flood control in the region, Robert W. Harrison, the "task has been difficult, costly to carry out, and is not yet completed."[69]

Artificial levees and other devices have changed the character of the rivers: because of the smoother leveed channel, the rivers run faster and hurry flood crests along to the sea, ceasing much of the river's land-building activity. Constriction of the rivers has in fact forced floods to rise higher: the levees keep normal rises in the river level from topping the banks, making floods rarer but insuring that when floods do occur they are more catastrophic for the fields and settlements behind the levees. The common assumption that the days of the Mississippi overflows are over is unrealistic, and the potential for major flood damage still exists in the region. Despite the immense transformation of the hydrological system, the inclination of people to farm and live on floodplains continues to pose hazards to property and human life.[70]

Human inhabitants along the Mississippi have not succeeded in liberating themselves from the limits set by the natural world. In many places along the lower Mississippi and its tributaries—including the Yazoo-Mississippi Delta—it might have been wiser, from an economic as well as an environmental point of view, to regulate development on the floodplain rather than remake the original ecological complex. The human exploitation of the Delta's natural productivity has often turned into what William Faulkner described as "a mad and pointless merry-go-round . . . :

the timber which had to be logged and sold in order to deforest the land in order to convert the soil to raising cotton in order to sell the cotton in order to make the land valuable enough to be worth spending money raising dykes to keep the River off of it."[71]

The immense change in the bottomlands of the Yazoo-Mississippi Delta during the last three hundred years was first and foremost driven by economic interests. Industrialization and commercial development in Europe and North America created burgeoning markets for products that could be obtained by utilizing the bottomland hardwood forest complex to its fullest. Consequently the Delta encountered by Euro-Americans was indeed "too rich for anything else, too rich and strong to have remained wilderness."[72] Even before the arrival of the European settlers, the shift from hunting and gathering to agriculture and trade among the Delta's native peoples had led to a general increase of human influence on the natural environment. Still, it was the region's inclusion in the greater European economy that revolutionized the way the floodplain was exploited by humans. For the first time, local economic development became dictated by the supralocal needs of the world economy. The basic productivity of the floodplain ecological complex remained the same, but now it was the capitalist world system that ultimately mandated how the Delta was utilized as a pool of natural resources. During the nineteenth century, agricultural and lumbering activities expanded enormously compared with Native American practices. The first commercial products of the Delta in the industrial age, cotton and cypress, were later replaced by other crops and tree species, while land use on the fertile floodplain progressively intensified. As a result, great personal fortunes were made, always based on the biological productivity of the land. For most of the people involved in the transformation of the Delta bottomlands, however, economic gain and social mobility remained severely limited. Despite its distinctively American features, the environmental history of the bioregion known as the Delta is but one example of the processes idiosyncratically duplicated on a global scale.

Thus the factors that contributed to the successful utilization of the natural resources in the Delta are not unique to Mississippi, southern, or American history: exploitation of disadvantaged people and the natural environment in a globalizing culture geared at continuous economic growth is an unavoidable theme in modern history. Today we are fast realizing Faulkner's 1942 prophecy that, despite given our chance and "warning and foreknowledge, too," the natural world we devastate will be "the consequence and signature of [our] crime, and [our] punishment." Camping for the very last time in the vanishing forests of the Yazoo-Mississippi Delta, Uncle Ike McCaslin knew that "it was his land, although he had never owned a foot of it. He had never wanted to, not even after he saw plain its ultimate doom, watching it retreat year by year before the onslaught of the axe and saw and log-lines and then dynamite and tractor plows, because it belonged to no man. It belonged to all; they had only to use it well, humbly and with pride."[73]

# *Notes*

This book is a significantly revised version of my doctoral dissertation in North American Studies for the University of Helsinki, *Evolution of a Place*. Small portions of the text have been previously published in my articles "Down by the Riverside" and "'Home in the Big Forest.'"

## Chapter 1. Environmental History and the Yazoo-Mississippi Floodplain

1. Michael Williams, *Americans and Their Forests*, 3–4. For an overview of the gargantuan change in the American landscape after the European colonization, see Zelinsky, "Landscapes," 1289–95.

2. Among the many basins making up the Lower Mississippi Valley is the Yazoo-Mississippi Delta, which is actually more oval than deltoid in shape. The Yazoo-Mississippi Delta is usually called "the Mississippi Delta" or simply "the Delta" by the region's inhabitants. It must be noted, however, that the term "Mis-

sissippi Delta" in physical geography refers to the true delta of the Mississippi River at its mouth in Louisiana.

In this study, the term "Delta" is used interchangeably with "Yazoo-Mississippi Delta." The channel of the Mississippi River, from Memphis to Vicksburg, forms the western boundary of the Yazoo-Mississippi floodplain. The eastern boundary is defined by a series of bluffs that begin just below Memphis and run south to Greenwood and thence southwesterly along the Yazoo River, which meets the Mississippi just above Vicksburg. The area is approximately two hundred miles long and seventy miles across at its widest point, encompassing some seven thousand square miles of alluvial floodplain. For more information on the regional geography, see chapter 2.

3. Eldredge, *Preliminary Report on the Forest Survey of the Bottom-Land Hardwood Unit in Mississippi,* 2–5.

4. Cowdrey, *This Land, This South,* remains the only general work on the subject. Excellent smaller-scale studies include Silver, *New Face on the Countryside;* Kirby, *Poquosin;* Stewart, *"What Nature Suffers to Groe";* Davis, *Where There Are Mountains.*

5. Brandfon, *Cotton Kingdom of the New South;* Cobb, *Most Southern Place on Earth;* Woods, *Development Arrested;* Willis, *Forgotten Time;* Otto, *Final Frontiers.* Brandfon's and Willis's books are based upon their doctoral dissertations. See Brandfon, "Planters of the New South"; Willis, "On the New South Frontier." A highly recommendable, if somewhat dated, introduction to the Delta history is Frank E. Smith, *Yazoo River.*

6. While Fickle's *Mississippi Forests and Forestry* is indispensable for the study of exploitation of the pine forests of southern and central Mississippi, it is rather cursory with its treatment of the state's hardwood forests. Vileisis's *Discovering the Unknown Landscape* provides an excellent overview on the subject on a national scale but obviously must omit many important locations. Harrison's *Levee Districts and Levee Building in Mississippi* and *Alluvial Empire* remain the definitive monographs on their subject. See also his *Flood Control in the Mississippi Alluvial Valley* and "Clearing Land in the Mississippi Alluvial Valley." For a summary of his argument regarding the Delta, consult Harrison and Mooney, *Flood Control and Water Management in the Yazoo-Mississippi Delta.*

7. The following discussion on methodology and historiography of environmental history is based upon Myllyntaus and Saikku, "Environmental History" 1–20; Myllyntaus, "Environment in Explaining History," 141–59. See also Crosby, "Past and Present of Environmental History," 1177–89; Hughes, "Whither Environmental History," 1–3; John R. McNeill, "Observations on the Nature and

Culture of Environmental History," 5–43; Richard White, "American Environmental History," 297–335.

8. Worster, "Doing Environmental History," 290–91. See also Haila, *Vihreään aikaan,* 7–17; Haila and Levins, *Humanity and Nature,* 182–83.

9. The roundtable panel in *Journal of American History* 76 (March 1990): 1087–1147, remains probably the best introduction to the field. Articles include Worster, "Transformations of the Earth," 1087–1106; and "Seeing Beyond Culture," 1142–47; Crosby, "An Enthusiastic Second," 1107–10; Richard White, "Environmental History, Ecology, and Meaning," 1111–16; Merchant, "Gender and Environmental History," 1117–21; Cronon, "Modes of Prophecy and Production," 1122–31; Pyne, "Firestick History," 1132–41. See also Bailes, "Critical Issues in Environmental History," 1–21.

10. Myllyntaus, "Environment in Explaining History," 145–51.

11. Worster, "Doing Environmental History," 289–307; Massa, "Ympäristöhistoria tutkimuskohteena," 294–301; Myllyntaus, "Environment in Explaining History," 152–55. See also Worster, "History as Natural History," 1–19.

12. Christensen, "Landscape History and Ecological Change," 116–24. On the use of historical sources in ecological study, see Forman and Russell, "Evaluation of Historical Data in Ecology," 5–7; Peterken, "The Use of Records in Woodland Ecology," 81–87.

13. Massa, "Ympäristöhistoria tutkimuskohteena," 296. Classic examples of the ecological approach to anthropology include Wissler, *Relation of Nature to Man in Aboriginal North America;* Steward, *Theory of Culture Change;* Rappaport, *Pigs for the Ancestors;* Harris, *Cannibals and Kings.* See also Boyden, "Human Ecology and Biohistory," 31–34; Worster, "History as Natural History," 6–15.

14. Worster, "Doing Environmental History," 293. Good examples of this approach include Worster's *Dust Bowl* and *Rivers of Empire* and Cronon's *Changes in the Land* and *Nature's Metropolis.*

15. Worster, "Doing Environmental History," 293. For an example of a study with legislative emphasis, see McEvoy, *Fisherman's Problem.* The classic in this subfield of environmental historiography is Hays, *Conservation and the Gospel of Efficiency.* This book did not represent a radical departure from existing work in political history, but made the history of nature conservation an approved subject of academic study, a trend that has continued to attract the attention of American historians.

16. Starting with Nash's *Wilderness and the American Mind,* originally published in 1967, the intellectual component in American environmental historiog-

raphy has remained important. See also Nash, *Rights of Nature;* Merchant, *Death of Nature;* Worster, *Nature's Economy.*

Influential writings from the 1960s on the history of human ideas on the environment include Glacken's *Traces on the Rhodian Shore* and Lynn White Jr.'s "Historic Roots of Our Ecological Crisis," 1202–7. In his much-debated essay, White maintained that current large-scale environmental problems are to a great extent an outgrowth of the Judeo-Christian heritage, which has made it possible for humans to exploit nature with indifference to other organisms.

On the differences between native and colonial perceptions of nature and the use of natural resources, see Cronon, *Changes in the Land;* Polanyi, *Great Transformation.*

17. Worster, "History as Natural History," 5.

18. Myllyntaus, "Environment in Explaining History," 155. See also Opie, "Environmental History," 9–11.

19. Worster, "Nature and the Disorder of History," 2.

20. Cronon, "Uses of Environmental History," 10–12; idem, "Trouble with Wilderness," 7–25.

21. On the modes of explanation in environmental history, see Myllyntaus, "Environment in Explaining History," 155–59.

Such interplay between environmental and social factors is evident, for example, in the Irish "Great Famine" of the late 1840s and the Finnish *Nälkävuodet* ("Hunger Years") of the late 1860s. See Utterström, "Climatic Fluctuations and Population Problems in Early Modern History."

22. Haila, "Environmental Problems, Ecological Scales, and Social Deliberation," 68.

23. Yrjö Haila's guest lecture at the Renvall Institute, University of Helsinki, titled "Ekososiaalisen kompleksin käsite metsähistorian jäsentäjänä," on September 23, 1998, was most helpful in providing conceptual tools for addressing temporal and spatial problems in environmental history. Much of the following discussion on the culture-nature relationship is based on this lecture and his articles "Environmental Problems, Ecological Scales, and Social Deliberation," 65–87, and "Assessing Ecosystem Health across Spatial Scales," 81–102.

24. Haila, "Environmental Problems, Ecological Scales, and Social Deliberation," 76–77. See also Cronon, *Nature's Metropolis,* xix, 264–69; Pollan, *Second Nature.*

25. On the problematic definition of wilderness as a cultural construction of nature, see Cronon, "Trouble with Wilderness," 7–28; Haila, "'Wilderness' and the Multiple Layers of Environmental Thought," 129–47.

26. For critical examinations of ecological paradigms throughout history, consult Worster, *Nature's Economy;* Haila and Levins, *Humanity and Nature.*

27. Braudel, *The Mediterranean and the Mediterranean World,* 20–21. The original French notions of time in the famous introduction (*géographique, social,* and *individuel*) were later applied by Braudel as *histoire de la longue durée, conjonctures,* and *événements.* See Braudel, *Écrits sur l'histoire,* 11–13, 41–83. See also Haila and Levins, *Humanity and Nature,* 191; Massa, *Pohjoinen luonnonvalloitus,* 28–29.

28. Haila and Levins, *Humanity and Nature,* 190. The concept is borrowed from the Soviet ecologist S. V. Kirikov. Haila points out that sociocultural factors correspond with Braudel's temporal scales. Fundamental institutions such as religions unfold in geographical time, rules and customs may change during a conjuncture, while habits correspond to individual time. See Haila, "Environmental Problems, Ecological Scales, and Social Deliberation," 76–77.

29. Haila and Levins, *Humanity and Nature,* 199–201. See also Wolf, *Europe and the People without History;* Polanyi, *Great Transformation.*

30. The following discussion is based upon Haila and Levins, *Humanity and Nature,* 190–98. See also Haila, *Vihreään aikaan,* 124–41. The general ecohistorical periodization by Haila and Levins in many aspects resembles Stephen Boyden's classification of four distinct ecological and biosocial phases in human history. Boyden recognizes the primeval, early farming, early urban, and high-energy phases in environmental history. See Boyden, "Human Ecology and Biohistory," 36–39.

31. It must be pointed out that there is no general "developmental law" that requires human societies to evolve in this direction. A classic example of the opposite is the introduction of the horse to the Great Plains of North America by the Spanish in the seventeenth century, which encouraged some Indian nations to shift from a largely agricultural subsistence base to one dominated by horseback hunting of bison. On the concept of cultural development, see Henriksson, "Land Ownership and Cultural Development," 189–203.

32. The fur trade is a good example of a trading activity that can be used in defining an ecohistorical period, as it has affected vast areas and numerous societies. See Innis, *Fur Trade in Canada;* Åström, *Natur och byte,* 66–77.

33. The North Atlantic archipelagos of the Azores, Madeiras, and Canaries, or, Macaronesia, became the first Iberian colonies on the Atlantic and were turned into a pilot project for Europe's subsequent expansion to the Americas. Although the Canaries had been known to exist by the Romans, they were effectively brought under European influence only after their "discovery" by Lanzarote Mal-

ocello in 1336. Between that date and 1496, successive waves of European, mainly Spanish, colonists arrived and exterminated many of the aboriginal Guanches by warfare and introduced disease. After the disposal or enslavement of the indigenous inhabitants, the Canaries provided an ideal environment for Spanish settlement. Meanwhile, the Portuguese had concentrated their colonization efforts on the unpopulated Azores and Madeiras. Madeira as found by the Portuguese was completely forested, and among their first actions was an attack on the original forest. The land was cleared by ax and fire in order to obtain lumber for the European market and land for the cultivation of staple crops also aimed at the emerging European consumer. The rich, volcanic soil and subtropical climate of Macaronesia allowed efficient cultivation of both temperate and subtropical crops. Various crop species, such as wheat, sugar cane, vines, bananas, and tobacco, were soon introduced into the region. Sugar cane became the dominant crop during the early days of colonization. Intensive cultivation of sugar required great amounts of labor and water; use of local Guanche and imported African slave labor, along with the introduction of terracing and construction of irrigation channels, or *levadas,* solved these problems. On Macaronesian environmental history, see Crosby, *Ecological Imperialism,* 70–103; Saikku, "A extinção nas areas de expansão Europeia," 201–8.

34. For an overview of Innis's thesis, see the classic *Fur Trade in Canada.* Innis's staple theory has strongly influenced Finnish environmental historians, as evidenced in Björn, *Kaikki irti metsästä,* 16; Massa, *Pohjoinen luonnonvalloitus,* 19–20. For other impressive examples of the center-periphery juxtaposition, consult Cronon, *Nature's Metropolis;* Wallerstein, *Modern World-System.*

35. See also Crosby's *Columbian Exchange;* idem, "Biotic Change in Nineteenth-Century New Zealand"; idem, "Papua New Guinea, Its Demographic History and Infectious Diseases." For a summary of his argument, see his "Ecological Imperialism," 103–17.

36. Haila and Levins, *Humanity and Nature,* 208–11. The large-scale application of slash-and-burn agriculture in societies as dissimilar as preindustrial Finland and contemporary Brazil offers an example of the potential destructiveness of traditional subsistence methods.

37. Worster, "From Columbus to Rio: The Historical Roots of the Global Environmental Crisis," keynote address at the 4th Maple Leaf and Eagle Conference on North American Studies, University of Helsinki, September 7, 1992.

38. Haila and Levins, *Humanity and Nature,* 211–13.

39. Ibid., 213. The discussion on ecohistorical formations is based on ibid., 213–24.

40. Ibid., 213–24.

41. The term *Raubwirtschaft* ("robbery economy") refers to the ever-expanding human exploitation of natural resources. Ruthless exploitation of nature, with no concern for the future, is seen as characteristic of the species *Homo sapiens.* The term was introduced by the German economic geographer Friedrich, "Wesen und Geographische Verbreitung der Raubwirtschaft," 68–79, 92–95. See also Massa, "Ympäristöhistoria tutkimuskohteena," 298–300; and *Pohjoinen luonnonvalloitus,* 16–18, 59.

42. Odum, *Fundamentals of Ecology,* 9. A more recent textbook, Begon, Harper, and Townsend, *Ecology,* 851, defines ecosystem as "[a] holist concept of the plants, the animals habitually associated with them and all the physical and chemical components of the immediate environment or habitat which together form a recognizable self-contained entity." On the application of the ecosystem concept in environmental history, see Simmons, *Changing the Face of the Earth,* 11–19. Cf. the preface to the paperback edition of Richard White's *Land Use, Environment, and Social Change,* xvii–xix.

43. Worster, "History as Natural History," 13; idem, "Nature and the Disorder of History," 5–8; Haila, "Assessing Ecosystem Health across Spatial Scales," 83–87. See also Golley, *History of the Ecosystem Concept in Ecology;* Haila and Levins, *Humanity and Nature,* 70–74, 134–43; Worster, "Ecology of Order and Chaos."

Various editions of Odum's *Fundamentals of Ecology,* all committed to the mechanistic ideal, were the most widely used textbooks for ecology from the 1950s to the 1970s. On Odum and the development of the "energy-economic" model of ecology in general, see Worster, *Nature's Economy,* 291–315; Opie, *Nature's Nation,* 412–13.

44. Flores, "Place," 10.

45. Haila, *Vihreään aikaan,* 36–41.

46. Haila and Levins, *Humanity and Nature,* 184.

47. The following discussion on ecological complexes and their structure is based upon Haila and Levins, *Humanity and Nature,* 184–90. See also Forman and Godron, *Landscape Ecology,* 13–17; Urban, O'Neill, and Shugart, "Landscape Ecology," 119–25.

48. Haila and Levins, *Humanity and Nature,* 189–90.

49. Ibid., 72–74.

50. Flores, "Place," 6. Cf. Trimble, "Nature's Continent," 9–26.

51. Flores, "Place," 10–12. The applied equation for the notion of place has been proposed by geographer Tuan in *Space and Place,* 4–6.

52. Flores, "Place," 14. An excellent example of such a case study—and an early inspiration for the work at hand—is Richard White's *Land Use, Environment, and Social Change.*

## Chapter 2. A True Ecological Complex

1. Cohn, *Where I Was Born and Raised,* 12.

2. Bailey, *Description of the Ecoregions of the United States,* 1–2, appendix 2. See also Flores, "Place," 6–7.

3. Bailey, *Description of the Ecoregions of the United States,* 22–24; Cry, "Surface Waters of the Lower Mississippi River Region," 65–67; Harrison, *Alluvial Empire,* 18–21.

4. Kelley, "Some Aspects of the Geography of the Yazoo Basin, Mississippi"; Harrison, *Alluvial Empire,* 46–48; Harrison and Mooney, *Flood Control and Water Management in the Yazoo-Mississippi Delta,* 10; Cross and Wales, *Atlas of Mississippi,* 7.

5. Harrison and Mooney, *Flood Control and Water Management in the Yazoo-Mississippi Delta,* 10; Cry, "Surface Waters of the Lower Mississippi River Region," 68.

6. MacDonald, Frayer, and Clauser, *Documentation, Chronology, and Future Projections of Bottomland Hardwood Habitat Losses in the Lower Mississippi Alluvial Plain,* 7 (table 1.2).

7. Ibid., 11 (table 1.3).

8. Bailey, *Description of the Ecoregions of the United States,* 13, 22–24, appendixes 1 and 2.

9. Ibid. See also Muller and Willis, "Climatic Variability in the Lower Mississippi River Valley," 55–63. In the classic climate classification by Wladimir Köppen, Bailey's Subtropical Division (2300) lies within the *Cfa* (humid subtropical) climate, described as temperate, rainy, and having hot summers and no dry season. Eight or more months average above 50°F, and the coolest month is warmer than 32°F but colder than 65°F. For the Köppen classification, consult his *Die Klimate der Erde.*

10. Cox et al., *This Well-Wooded Land,* 2; Michael Williams, *Americans and Their Forests,* 3–4. Cf. Pyne, *Fire in America,* 75–76, 79–80.

11. Barnes, "Deciduous Forests of North America," 220–29; Christensen, "Vegetation of the Southeastern Coastal Plain," 336–40; Braun, *Deciduous Forests of Eastern North America,* 31–38, 290–91; Brockman, *Trees of North America,* 18–19; Sharitz and Mitsch, "Southern Floodplain Forests," 311–15; David Smith, "Forests of the United States," 9–19; Vankat, *Natural Vegetation of North America,* 132–37, 144–53.

12. Shantz and Zon, "Natural Vegetation"; Küchler, *Potential Natural Vegetation of the Coterminous United States.* See also Greller, "Deciduous Forest," 294 (figure 10.4).

13. Putnam, Furnival, and McKnight, *Management and Inventory of Southern Hardwoods,* 1 (figure 1); Sharitz and Mitsch, "Southern Floodplain Forests," 314–15.

14. J. R. Clark and J. Benforado, "Introduction," 1; Sharitz and Mitsch, "Southern Floodplain Forests," 314–16.

15. David Wm. Smith and Norwin E. Linnartz, "Southern Hardwood Region," 157–58; J. R. Clark and J. Benforado, "Introduction," 1. See also William H. Martin and Stephen C. Boyce, "Introduction," 16–18; Sharitz and Mitsch, "Southern Floodplain Forests," 316.

16. Chambers Diaries, no. 3 (January 1, 1858—December 31, 1858).

17. Ker, *Travels through the Western Interior of the United States,* 32.

18. Thorne and Curry, *Cultural Resources Survey,* 4–6.

19. Paul A. Delcourt et al., "History, Evolution, and Organization of Vegetation and Human Culture," 48–60, is a good introduction to the evolutionary history of the southeastern flora. The following review on the evolution of the southern floodplain flora is largely based on this work, augmented by Barnes, "Deciduous Forests of North America," 229–32; Davis, "Quaternary History and the Stability of Forest Communities," 137–53; Paul A. Delcourt and Hazel R. Delcourt, *Long-Term Forest Dynamics of the Temperate Zone,* 85–106; Webb, "The Past 11,000 Years of Vegetational Change in Eastern North America," 501–6; Webb, "Eastern North America," 385–409. See also Hazel R. Delcourt and Paul A. Delcourt, "Blufflands," 385–400; Paul A. Delcourt et al., "Quaternary Vegetation History of the Mississippi Embayment," 111–32; Royall, Delcourt, and Delcourt, "Late Quaternary Paleoecology and Paleoenvironments of the Central Mississippi Alluvial Valley," 157–70.

20. Thorne and Curry, *Cultural Resources Survey,* 19, 37–40, 136–38; Holloway and Valastro, "Palynological Investigations along the Yazoo River," 193–200, 236–40 (appendix 1). Discussion on pollen is based on palynological investigations at the Yazoo Basin (33.6°N, 90.1°W) and Nonconnah Creek, Tenn., (35.1°N, 89.9°W) in Holloway and Valastro, "Palynological Investigations along the Yazoo River," and in Paul A. Delcourt et al., "Quaternary Vegetation History of the Mississippi Embayment," 127–29.

21. Paul A. Delcourt et al., "Quaternary Vegetation History of the Mississippi Embayment," 112–13; Harrison, *Alluvial Empire,* 23–24; Thorne and Curry, *Cultural Resources Survey,* 7, 276–79 (appendix 4).

22. MacDonald, Frayer, and Clauser, *Documentation, Chronology, and Future*

*Projections of Bottomland Hardwood Habitat Losses in the Lower Mississippi Alluvial Plain,* 21–22; Pittman, *Present State of the European Settlements on the Mississippi,* 33.

23. Odum, "Summary," 433–34.

24. Braun, *Deciduous Forests of Eastern North America,* 291–92; David Wm. Smith and Norwin E. Linnartz, "Southern Hardwood Region," 156; Larson, *Aboriginal Subsistence Technology,* 20–22, 36, 51.

25. The terminology used in describing the physiography of southern floodplains is complex and varying. I have chosen to model my discussion after Putnam and Bull, *Trees of the Bottomlands of the Mississippi River Delta Region,* 4–10; Putnam, *Management of Bottomland Hardwoods,* 3–6; Putnam, Furnival, and McKnight, *Management and Inventory of Southern Hardwoods,* 4–7. For other classifications of the physiographic features of the floodplain, see Braun, *Deciduous Forests of Eastern North America,* 290–91; Sharitz and Mitsch, "Southern Floodplain Forests," 316–18. See also Harrison, *Alluvial Empire,* 31–37; David Wm. Smith and Norwin E. Linnartz, "Southern Hardwood Region," 154–55; Thorne and Curry, *Cultural Resources Survey,* 7–10. For a nineteenth-century description of the topography on the Yazoo floodplain, see Sargent, *Report on the Forests of North America,* 535.

For a generalized view of the physiographic features and vegetation along a major southeastern river, see David Wm. Smith and Norwin E. Linnartz, "Southern Hardwood Region," 155 (figure 5–2); Sharitz and Mitsch, "Southern Floodplain Forests," 317, 319, 332 (figures 2, 3, and 5).

26. Nuttall, *Journal of Travels into the Arkansas Territory,* 69, on January 9, 1819, at Island 66.

27. Lyell, *Second Visit to the United States of North America,* 214. However, see Paulding, "Mississippi," 337.

28. Harrison, *Alluvial Empire,* 33.

29. Williams, *Cat on a Hot Tin Roof,* 94.

30. Hurt, *Mississippi,* 20. On soils, see Cross and Wales, *Atlas of Mississippi,* 4–7, 82–84; Langsford and Thibodeaux, *Plantation Organization and Operation in the Yazoo-Mississippi Delta Area,* 8–9; William H. Martin and Stephen C. Boyce, "Introduction," 22–26, David Wm. Smith and Norwin E. Linnartz, "Southern Hardwood Region," 156–57.

31. Cuming, *Sketches of a Tour to the Western Country,* 350. Cf. Sinclair, "Studies of Soil Erosion in Mississippi," 533–40. On the forest vegetation of the loess hills, see Hazel R. Delcourt and Paul A. Delcourt, "Blufflands," 385–400; Caplenor, "Forest Composition on Loessal and Non-loessal Soils in West-Central Mississippi," 322–31.

32. Echternacht and Harris, "Fauna and Wildlife of the Southeastern United States," 81–82, 102–3, 107–110; Sclater, "On the General Geographical Distribution of the Members of the Class Aves," 130–45. There are some interesting patterns in the global distribution of certain flora and fauna that link temperate parts of eastern Asia with those of eastern North America. For example, the largest reptile found on the southern floodplains, the American alligator (*Alligator mississippiensis*), is closely related to the Chinese alligator (*Alligator sinensis*). See Echternacht and Harris, "Fauna and Wildlife of the Southeastern United States," 109–10.

33. Ker, *Travels through the Western Interior of the United States*, 31, on October 21, 1808, five miles south of New Madrid.

34. Dr. Charles Mohr's 1880 report on the forests of the Yazoo Delta, Mississippi, in Sargent, *Report on the Forests of North America*, 535.

The term biological diversity refers to the variety and the variability among living organisms and the ecological complexes in which they occur. Generally four different levels of biodiversity are recognized. These are genetic diversity in individual organisms, diversity within a species, diversity in communities or assemblages of species, and diversity of communities and populations across landscapes. See William H. Martin and Stephen C. Boyce, "Introduction," 1–2.

35. Hamilton, *Trials of the Earth*, 51–52.

36. Christensen, "Vegetation of the Southeastern Coastal Plain," 339–40; Braun, *Deciduous Forests of Eastern North America*, 293–95; J. R. Clark and J. Benforado, "Introduction," 1–2. On succession in general, see Begon, Harper, and Townsend, *Ecology*, 628–47. On the effects of flooding on the bottomland flora, see Barnes, "Deciduous Forests of North America," 318–19; Broadfoot and Williston, "Flooding Effects on Southern Forests," 584–87; Lugo, "Review of Early Literature on Forested Wetlands in the United States," 10–12; Muzika, Gladden, and Haddock, "Structural and Functional Aspects of Succession in Southeastern Floodplain Forests Following a Major Disturbance," 5–7; Putnam, Furnival, and McKnight, *Management and Inventory of Southern Hardwoods*, 132–59. For an idealized example on the relationship between floodplain topography, inundation, and habitat zonation, see Patrick et al., "Characteristics of Wetland Ecosystems of Southeastern Bottomland Hardwood Forests," 279 (figure 2).

37. Audubon, *Delineations of American Scenery and Character*, 46. The eerie atmosphere of the bottomlands at night startled even the seasoned outdoorsman: "How easy, I thought, would it be for the confused and agitated mind of a person bewildered in a swamp like this, to imagine in each of these luminous masses some wondrous and fearful being, the very sight of which might make the hair stand

erect on his head. The thought of being myself placed in such a predicament burst over my mind, and I hastened to join my companions"(47).

38. The following discussion on different plant communities of the bottomland hardwood forests is based upon Barnes, "Deciduous Forests of North America," 249–55; Braun, *Deciduous Forests of Eastern North America,* 290–95; Putnam, *Management of Bottomland Hardwoods,* 52–60; Putnam, Furnival, and McKnight, *Management and Inventory of Southern Hardwoods,* 6–7; Sharitz and Mitsch, "Southern Floodplain Forests," 328–47. See also Langdon et al., "Extent, Condition, Management, and Research Needs of Bottomland Hardwood-Cypress Forests in the Southeastern United States," 73–76.

Common and scientific names of the bottomland flora used in this study are based on Little, *Checklist of United States Trees.* For illustrations and short descriptions of the tree species mentioned in this chapter, consult Brockman, *Trees of North America.*

39. Nuttall, *Journal of Travels into the Arkansas Territory,* 66, on January 7, 1819, near the mouth of the St. Francis River.

40. Ibid.

41. De Puy van Buren, *Jottings of a Year's Sojourn in the South,* 43.

42. Ibid., 196. According to Ker, *Travels through the Western Interior of the United States,* 43, the moss cloaked "the wilderness with still greater gloom, hanging in weeping forms of mournful gray." See also Paulding, "Mississippi," 324.

43. Thorne and Curry, *Cultural Resources Survey,* 12–14, (figure 2), 43–49.

44. Hamilton, *Trials of the Earth,* 82.

45. Putnam, Furnival, and McKnight, *Management and Inventory of Southern Hardwoods,* 14–20; Sharitz and Mitsch, "Southern Floodplain Forests," 345–47; Crow, "Fire Ecology and Fire Management in the Forests of the Lower Mississippi River Valley," 75.

46. For different attempts to classify the bottomland hardwood forest flora, consult Braun, *Deciduous Forests of Eastern North America,* 291–95; Sharitz and Mitsch, "Southern Floodplain Forests," 312–14; Shelford, *Ecology of North America,* 89–119; David Wm. Smith and Norwin E. Linnartz, "Southern Hardwood Region," 160–61 (table 5–3).

47. Putnam, *Management of Bottomland Hardwoods,* 3–6. "Bitter pecan" is synonymous with "water hickory" (*Carya aquatica*).

48. The following discussion of the non-avian fauna of the bottomland hardwood forests is based on Echternacht and Harris, "Fauna and Wildlife of the Southeastern United States," 81–116; Kitchings and Walton, "Fauna of the North American Temperate Deciduous Forest," 350–58; Sharitz and Mitsch, "Southern

Floodplain Forests," 348–56; Wharton et al., "Fauna of Bottomland Hardwoods in Southeastern United States," 87–127; Wharton et al., *Ecology of Bottomland Hardwood Swamps of the Southeast,* 84–102.

49. Echternacht and Harris, "Fauna and Wildlife of the Southeastern United States," 114, define the "fauna" of a given area as the native animal species inhabiting that area, and the "wildlife" of an area as its fauna together with species introduced by humans. I have chosen to apply this distinction throughout the study, as it has obvious implications for assessing biodiversity in the bottomland hardwood forests.

50. Francis Baily on May 1, 1797, near the present-day town of Ashport, Tenn., in Baily, *Journal of a Tour in Unsettled Parts of North America,* 141.

51. In many historical accounts, the names "panther" and "mountain lion" have somewhat misleadingly been used for the cougar, which is the largest North American feline. The southeastern subspecies of the cougar are *Felis concolor cougar* and *F. c. corryi.*

The taxonomy of North American wolves is currently under revision. A special form of the southeastern wolf is often described as a separate species, known as the red wolf (*Canis rufus*). Some, however, believe it to be a hybrid between the wolf (*Canis lupus*) and the coyote (*Canis latrans*). In any case, the red wolf is currently endangered and under the threat of genetic swamping through hybridization with feral dogs and coyotes. See *Endangered Species of Mississippi,* "Florida Panther" and "Red Wolf."

52. On beaver as a natural disturbance factor, see Christensen, "Vegetation of the Southeastern Coastal Plain," 339.

53. Dickson, "Forest Bird Communities of the Bottomland Hardwoods," 67. See also Begon, Harper, and Townsend, *Ecology,* 819–25; Best, Stauffer, and Geier, "Evaluating the Effects of Habitat Alteration on Birds and Small Mammals Occupying Riparian Communities," 118–21; Shugart et al., "Relationship of Nongame Birds to Southern Forest Types and Successional Stages," 5–15; Stauffer and Best, "Habitat Selection by Birds of Riparian Communities," 1–15; Swift, Larson, and DeGraaf, "Relationship of Breeding Bird Density and Diversity to Habitat Variables in Forested Wetlands," 51–58.

The following discussion on the bird communities of bottomland hardwood forests is based on Dickson, "Seasonal Populations of Insectivorous Birds in a Mature Bottomland Hardwood Forest in South Louisiana," 261–68; "Bird Communities in Oak-Gum-Cypress Forests," 52–58; Hamel et al., *Bird-habitat Relationships on Southeastern Forest Lands,* 132–33; along with personal observations made in 1994, 1996, and 2003 in various National Wildlife Refuges in the

Delta. For illustrations and short descriptions of the bird species mentioned in the text, consult Robbins, Bruun, and Zim, *Birds of North America*.

54. Schultz, *Travels on an Inland Voyage,* 131.

55. In the 1850s, an all-white owl was captured in Bolivar County. Since the snowy owl (*Nyctea scandiaca*) does not range this far south, the bird in question was probably an albinistic individual of a more common species. See Jacobs Manuscript, 12.

56. On these three vanished species, see Saikku, "The Second Wave in the Southern Woods"; idem, "Extinction of the Carolina Parakeet"; idem, "'Home in the Big Forest.'"

57. Audubon, *Audubon's America,* 148–49; M. G. Vaiden to James Bond in 1963, as quoted in Jackson, "History of Ivory-billed Woodpeckers in Mississippi," 6. In 1963, Dr. James Bond worked for the Academy of Natural Sciences in Philadelphia. Bond was a well-known ornithologist who wrote the 1936 classic *Birds of the West Indies,* still in print. Ian Fleming, whose hobbies included ornithology, adopted Bond's name for his best-known literary character.

58. Audubon, *Audubon's America,* 149–50. On December 23, 1820, at the mouth of the Yazoo.

59. Sharitz and Mitsch, "Southern Floodplain Forests," 348–49; Wharton et al., "Fauna of Bottomland Hardwoods in Southeastern United States," 120–21.

60. Paul A. Delcourt et al., "History, Evolution, and Organization of Vegetation and Human Culture," 47–48.

## Chapter 3. *Enter* Homo sapiens

1. Richard White and William Cronon, "Ecological Change and Indian-White Relations," 417. For an example of a rather uncritical portrayal of the relationship between Indians and their natural environment, see Jacobs, "Indians as Ecologists and Other Environmental Themes in American Frontier History," 49, 62–64. For a good, if somewhat dated, literary survey of the historical studies dealing with the relationship between Indians and the natural world, see Richard White, "Native Americans and the Environment," 179–97. For a recent general study of the subject, see Krech, *Ecological Indian.*

2. Richard White and William Cronon, "Ecological Change and Indian-White Relations," 417.

3. Ibid.

4. Marshall, "Archaeological Provinces of Mississippi," 285–86, 288–89.

5. Larson's *Aboriginal Subsistence Technology* offers a good example of the interdisciplinary approach to the pre-Columbian history of North America. Butzer,

"Indian Legacy in the American Landscape," 29–33; Byrd and Neuman, "Archaeological Data Relative to Prehistoric Subsistence in the Lower Mississippi River Alluvial Valley," 9–16; Connaway and McGahey, *Archaeological Survey and Salvage in the Yazoo-Mississippi Delta and in Hinds County,* 7–12; Paul A. Delcourt et al., "History, Evolution, and Organization of Vegetation and Human Culture," 62–69; Bruce D. Smith, "Archaeology of the Southeastern United States," 1–53; Stephen Williams and Jeffrey P. Brain, *Excavations at the Lake George Site,* 386–91, 393–408; have served as the model for the following discussion of pre-Mississippian Native American subsistence technology in southeastern North America. On the adaptation of different cultigens, see also McAndrews, "Human Disturbance of North American Forests and Grasslands," 679–81.

The outline for the Native American cultural sequence in the Lower Mississippi Valley and the Delta has been compiled from Haag, "Prehistory of the Lower Mississippi Valley," 1–8; Johnson, "Prehistoric Mississippi," 9–20; Marshall, "Prehistory of Mississippi," 24–59; idem, *Indians of Mississippi.* There is considerable variation in the temporal placement of boundaries between various cultural phases by different authors.

6. Cf. Ballas, "Historical Geography and American Indian Development," 16–17; Butzer, "Indian Legacy in the American Landscape," 28–29; Paul A. Delcourt et al., "History, Evolution, and Organization of Vegetation and Human Culture," 62–63; Cowdrey, *This Land, This South,* 11–12; Flannery, *Eternal Frontier,* 173–185; Krech, *Ecological Indian,* 35; Thomas, "World as It Was," 35. Cf. "New Ways to the New World," 54.

7. Bruce D. Smith, "Archaeology of the Southeastern United States," 6–7.

8. Many archaeologists divide the pre-Columbian Native American cultural developments into three broad stages of the Paleo-Indian, Meso-Indian, and Neo-Indian Eras. The Meso-Indian Era more or less equals the Archaic period, while the Neo-Indian Era encompasses all of the later developments. Cf. Haag, "Prehistory of the Lower Mississippi Valley," 1; Stephen Williams and Jeffrey P. Brain, *Excavations at the Lake George Site,* 349.

9. Thorne and Curry, *Cultural Resources Survey,* 48.

10. Reasons for the extinction of North American megafauna such as the mastodon (*Mastodon americanus*) have not been fully explained, but eminent scientists have cited overhunting by Paleo-Indians as a contributing factor. Cf. Thomas, "World as It Was," 34; Edward O. Wilson, *Diversity of Life,* 234–37; Flannery, *Eternal Frontier,* 186–205.

11. Cowdrey, *This Land, This South,* 13; Ford, "Patterns of Prehistoric Food Production in North America," 341–43, 347–49; Hurt, *Indian Agriculture in*

*America,* 11–13. Other indigenous species cultivated and introduced beyond their natural range include maygrass (canary grass, *Phalaris caroliniana*), knotweed (*Polygonum erectum*), and little barley (*Hordeum pusillum*). Cf. Byrd and Neuman, "Archaeological Data Relative to Prehistoric Subsistence in the Lower Mississippi River Alluvial Valley," 11–16.

12. Bruce D. Smith, "Archaeology of the Southeastern United States," 48–50. Cf. the classic article by Sahlins, "Poor Man, Rich Man, Big-Man, Chief," 285–303.

13. Hereafter, all dates are A.D., unless otherwise noted.

14. Byrd and Neuman, "Archaeological Data Relative to Prehistoric Subsistence in the Lower Mississippi River Alluvial Valley," 16; and Bruce D. Smith, "Archaeology of the Southeastern United States," 61.

15. The following account of the general characteristics of Mississippian culture and subsistence economy in southeastern North America is based on Brain, "Late Prehistoric Settlement Patterning in the Yazoo Basin and Natchez Bluffs Regions," 331–65; Butzer, "Indian Legacy in the American Landscape," 33–37; Byrd and Neuman, "Archaeological Data Relative to Prehistoric Subsistence in the Lower Mississippi River Alluvial Valley," 16–18; Cowdrey, *This Land, This South,* 14–15; Cox et al., *This Well-wooded Land,* 39–40; Paul A. Delcourt et al., "History, Evolution, and Organization of Vegetation and Human Culture," 69–72; Doolittle, *Cultivated Landscapes of Native North America,* 21–193; Gibson, "Indians of Mississippi," 69–74; Herndon, "Indian Agriculture in the Southern Colonies," 283–97; Hurt, *Indian Agriculture in America,* 13–16, 27–33; Marshall, "Prehistory of Mississippi," 59–68; Bruce D. Smith, "Archaeology of the Southeastern United States," 53–63; Richard White and William Cronon, "Ecological Change and Indian-White Relations," 419–20; Stephen Williams and Jeffrey P. Brain, *Excavations at the Lake George Site,* 391–92, 408–19.

For an overview on the subject, see Doolittle, *Cultivated Landscapes of Native North America,* 124 (table 5.1), 126 (figure 5.1), 129–30 (figure 5.2., table 5.3), 140, 142 (table 5.6), 144 (figure 5.7), 163–64 (table 5.10, figure 5.15), 167.

16. The fascination of early European travelers with Indian mounds is evident, for example, in the 1801 description by James Hall, "Brief History of the Mississippi Territory," 565–67.

17. On Mississippian population concentrations, see, for example, Butzer, "Indian Legacy in the American Landscape," 34–36. On the Cahokia contact, see Stephen Williams and Jeffrey P. Brain, *Excavations at the Lake George Site,* 410–11.

18. On the correlation between pellagra and a maize diet, see William H. McNeill, *Plagues and Peoples,* 179.

19. Richard White and William Cronon, "Ecological Change and Indian-White Relations," 418–19; Cowdrey, *This Land, This South,* 16; Gibson, *Chikasaws,* 16; Krech, *Ecological Indian,* 212–13.

20. English translation by Bernard Shipp in 1881, as quoted in Harrison, *Alluvial Empire,* 106. On the various factors affecting the selection of Mississippian settlement sites, see Bruce D. Smith, "Variation in Mississippian Settlement Patterns," 479–98.

21. Cf. Antoine La Page Du Pratz's eighteenth-century account of Indian agriculture in the Louisiana bottomlands, quoted in Usner, *Indians, Settlers, and Slaves in a Frontier Exchange Economy,* 154.

22. On the concepts of slash-and-burn and swidden in relation to Native American agriculture, see Doolittle, *Cultivated Landscapes of Native North America,* 175–91.

23. Cf. Richard White, *Roots of Dependency,* 21.

24. It should be noted that this pattern of agriculture was confined to the broad natural levees of major rivers on the lower elevations of the Piedmont, the Cumberland Plateau, and the Mississippi Valley. The narrow levees of rivers and the constant threat of flooding on the Coastal Plain made it unattractive for agriculture. Thus the floodplain areas of the inner Coastal Plain were generally unoccupied by any permanent populations during the Mississippian period. This assertion is supported by all the maps made of the area prior to the early nineteenth century: Native American towns on these maps are always situated on or above the Fall Line or on the coast while the intervening area is shown without any occupation. See Larson, *Aboriginal Subsistence Technology,* 56, 58–59, 65, 221–22.

25. Byrd and Neuman, "Archaeological Data Relative to Prehistoric Subsistence in the Lower Mississippi River Alluvial Valley," 14–15 (table 3); Cowdrey, *This Land, This South,* 13–14; Bruce D. Smith, "Variation in Mississippian Settlement Patterns," 480–88.

26. Cf. Usner, *Indians, Settlers, and Slaves in a Frontier Exchange Economy,* 152; Richard White, *Roots of Dependency,* 10, 30–31. Featherstonhaugh, *Excursion Through the Slave States,* vol. 1, 347, claims that "there were still a great many elk and buffalo" in the St. Francis bottomlands on the western side of the Mississippi, "it being favourable to them from the great extent of the swamp, the luxuriance of the wild grass, and the absence of man." Richard White claims that

bison arrived to the region only after the introduction of European disease had resulted in great human depopulation and, consequently, significant relaxation of hunting pressure. See his *Roots of Dependency*, 11. Cf. Usner, *Indians, Settlers, and Slaves in a Frontier Exchange Economy*, 152. According to Usner, 83, French soldiers hunted bison near the mouth of the Yazoo River in the fall of 1738.

27. Belmont, "Appendix D: Faunal Remains," 457–60.

28. Cowdrey, *This Land, This South*, 14–15; Larson, *Aboriginal Subsistence Technology*, 38–47, 51–52; Thorne and Curry, *Cultural Resources Survey*, 72–84; Richard White and William Cronon, "Ecological Change and Indian-White Relations," 420. Cf. Pyne, *Fire in America*, 143–44. On the regulation of white-tailed deer populations by the Choctaw during historical times, see Richard White, *Roots of Dependency*, 8–12.

The competitive advantage of the longleaf pine in frequently burned woods is based on its thick bark, rapidly growing root system, and seeds germinating on exposed soil. For comparisons with the Indian use of fire in the Northeast, see, for example, Day, "Indian as an Ecological Factor in the Northeastern Forest," 334–39; Calvin Martin, "Fire and Forest Structures in the Aboriginal Eastern Forest," 23–26; Russell, "Indian-set Fires in the Forests of the Northeastern United States," 78–86.

29. Richard White, *Roots of Dependency*, 26–28.

30. Ibid., 7–10.

31. Nevertheless, some tribes, such as the Chickasaw, believed carnivores and eaters of carrion to be unclean and, consequently, taboo as food. See Gibson, *Chikasaws*, 16.

32. On the utilization of bivalves, see Barber, "Appendix D: Faunal Remains," 471–74.

33. According to Jackson, "Southeastern Pine Forest Ecosystem and Its Birds," 137, the enormous roosts of the passenger pigeon in the South were located in the oak-hickory forests of the uplands. James Hall, "Brief History of the Mississippi Territory," 567–69, describes an upland pigeon roost, utilized by the Choctaw in 1800. On rookeries, see Shelford, *Ecology of North America*, 111–13.

34. Catesby, *Catesby's Birds of Colonial America*, 88. See also Audubon, *Ornithological Biography*, 343; Alexander Wilson and Charles Lucian Bonaparte, *American Ornithology*, 135–36. The collections of the Milwaukee Public Museum, Wisconsin, contain a peace pipe ornamented with six bills and crests of the ivory-billed woodpecker. This pipe was made by the Iowa tribe of Oklahoma. See Nice and Nice, *Birds of Oklahoma*, 50.

35. Audubon, *Ornithological Biography*, 343.

36. Bailey, "Ivory-billed Woodpecker's Beak in an Indian Grave in Colorado," 164. Many Indian tribes believed certain animals to be useful in curing diseases. For example, the Wyoming Shoshone associated woodpeckers with the cure of venereal diseases.

37. Butzer, "Indian Legacy in the American Landscape," 36; Paul A. Delcourt et al., "History, Evolution, and Organization of Vegetation and Human Culture," 71; Richard White, *Roots of Dependency*, 29.

38. Consult the exhaustive Stephen Williams and Jeffrey P. Brain, *Excavations at the Lake George Site*. See also Phillips, *Archaeological Survey in the Lower Yazoo Basin*. The Lake George site is located southeast of the modern village of Holly Bluff in Section 11, Township 11N, Range 5W.

39. Stephen Williams and Jeffrey P. Brain, *Excavations at the Lake George Site*, 393.

40. Quoted in Herndon, "Indian Agriculture in the Southern Colonies," 284.

41. See Richard White, *Roots of Dependency*, 21–22.

42. For a discussion of the estimates on the total Indian population of the Americas at the time of the contact with Europeans, see, for example, Butzer, "Indian Legacy in the American Landscape," 45; Denevan, "Native American Populations in 1492," xviii–xxi; Krech, *Ecological Indian*, 93–94.

The estimates of precolonial Native American population within the Eastern Agricultural Complex vary enormously. Cf. Kroeber, *Cultural and Natural Areas of Native North America*, 147; Dobyns, *Their Number Become Thinned*, 41. Dobyns, 42, claims the pre-Columbian population of the greater Mississippi Valley was as high as 5,250,000 (2.53 persons per km$^2$).

43. Gibson, "Indians of Mississippi," 69. On major cultural areas of precontact North America, see, for example, Ballas, "Historical Geography and American Indian Development," 15–20. On southern Indians in general, see, for example, Hudson, *Southeastern Indians*; Swanton, *Indians of the Southeastern United States*; Usner, *American Indians in the Lower Mississippi Valley*.

44. The outline for the following discussion of the colonization of the Lower Mississippi Valley has been provided by Gibson, "Indians of Mississippi." See also Usner, *Indians, Settlers, and Slaves in a Frontier Exchange Economy*, 13–144; idem, *American Indians in the Lower Mississippi Valley*, 1–93.

45. Usner's *Indians, Settlers, and Slaves in a Frontier Exchange Economy* and *American Indians in the Lower Mississippi Valley* offer an excellent overview of the socioeconomic interaction between the colonial powers and Native Americans in the Lower Mississippi Valley.

46. Cf. Usner, *Indians, Settlers, and Slaves in a Frontier Exchange Economy*, 15.

47. Collot, *Journey in North America,* 45. See also Thorne and Curry, *Cultural Resources Survey,* 15–16.

48. On the initial contact between the French and Natchez, see Brain, "La Salle at the Natchez," 49–59. See also Brown, "Archaeological Study of Culture Contact and Change in the Natchez Bluffs Region," 176–93. On the annihilation of the Natchez, see Usner, *Indians, Settlers, and Slaves in a Frontier Exchange Economy,* 65–76; idem, *American Indians in the Lower Mississippi Valley,* 15–32.

49. Four of the original thirteen states (Virginia, North Carolina, South Carolina, and Georgia) had extensive areas of bottomland hardwood forest.

50. Usner, "American Indians of the East," 657–65, offers a good overview of the sociopolitical challenges faced by the eastern Indians.

51. By the 1850s, the westward-moving frontier ran from Minnesota (admitted in 1858) south through Iowa (1846), Nebraska (1867) and Kansas (1861) and then through Arkansas to Texas (annexed in 1845). Areas of settlement, however, had appeared on the West Coast, and the intervening territory was occupied during the next three decades. Officially the frontier came to an end in 1890; the unsettled area had been so broken up by isolated bodies of settlement that there could hardly be said to be a frontier line anymore. The event prompted historian Frederick Jackson Turner to present his famous thesis on the significance of the frontier for the development of American democracy. This also meant that by the late nineteenth century the last remaining patches of "virgin" bottomland hardwood forest in the Southeast were within the area of intense human settlement. See Turner, "Significance of the Frontier in American History."

52. Basil Hall, *Travels in North America in the Years 1827 and 1828,* 354–55. On the establishment of the American judicial system in the territory, see Wunder, "American Law and Order Comes to Mississippi Territory," 131–55.

53. Alexander Moultrie to Alexander McGillivray, February 19, 1790, in Whitaker, "South Carolina Yazoo Company," 393.

54. Seavoy, "Fletcher v. Peck," 439. See also Whitaker, "South Carolina Yazoo Company," 383–94.

55. The following account of the change among native communities after European contact is based upon Gibson, "Indians of Mississippi," 74–89; Usner, *Indians, Settlers, and Slaves in a Frontier Exchange Economy,* 165–74, 244–75; Richard White, *Roots of Dependency,* 34–146. For a critical examination of the source material on the La Salle expedition, consult Galloway, "Sources for the La Salle Expedition of 1682," 11–37.

56. Paul A. Delcourt et al., "History, Evolution, and Organization of Vegetation and Human Culture," 71; Stephen Williams and Jeffrey P. Brain, *Excava-*

*tions at the Lake George Site,* 414. For a nineteenth-century discussion of the epidemics, see Paulding, "Mississippi," 331.

57. Penicaut, "Annals of Louisiana," 61–62.

58. Richard White, *Roots of Dependency,* 10–12.

59. On the Tunica, consult the exhaustive Brain, *Tunica Archaeology.* On the initial contact with the Spanish, see ibid., 288–93. Interestingly, agricultural labor was performed by men at the Quizquiz settlement. See Stephen Williams and Jeffrey P. Brain, *Excavations at the Lake George Site,* 415. On the Tunica migrations, see ibid., 21–44, 293–314; Brain, "Late Prehistoric Settlement Patterning in the Yazoo Basin and Natchez Bluffs Regions," 358–61; Brain, "Tunica Archaeology," 44–48; Brain, "Tunica Triumph," 45–50. For Tunica demography, see Brain, *Tunica Archaeology,* 315–18. The Tunica population seems to have declined from about 5,000 to 1,200 between 1541 and 1699. In 1980, the overall tribal membership was officially listed as 200.

The Lower Delta Native American population estimate is quoted in Brain, *Tunica-Biloxi,* 24–25; and the estimate for the Choctaw and Chickasaw in Hale and Gibson, *Chickasaw,* 14. See also Richard White, *Roots of Dependency,* 5; Usner, *American Indians in the Lower Mississippi Valley,* 33–37, 49–51. Brain, *Tunica-Biloxi,* and Hale and Gibson, *Chickasaw,* are good introductory texts on the Tunica and Chickasaw. On the initial contact between the French and Chickasaw, see Stubbs, "Chickasaw Contact with the La Salle Expedition in 1682," 41–48.

60. Paul A. Delcourt et al., "History, Evolution, and Organization of Vegetation and Human Culture," 71; Ford, "Patterns of Prehistoric Food Production in North America," 347; and Marshall, *Indians of Mississippi,* 35–37. Peaches were also commonly grown by the Choctaw. See Bowman, "Early History and Archaeology of Yazoo County," 427; Nutt, "Nutt's Trip to the Chickasaw Country," 42. Nutt's 1805–6 diary contains a vast amount of information on the early-nineteenth-century Choctaw and Chickasaw culture under rapid change. On Chickasaw ethnohistory, see also Gibson, *Chikasaws,* 3–30. On the traditional Choctaw subsistence, see Campbell, "Choctaw Subsistence," 9–21; Richard White, *Roots of Dependency,* 1–33.

61. This definition of dependency is by Theotonio Dos Santos, as quoted in Caporaso, "Dependence, Dependency, and Power in the Global System," 23.

62. Richard White, *Roots of Dependency,* xv–xix. See also ibid., 315–23.

63. Much of the following discussion derives from Richard White, *Roots of Dependency,* 34–146, who presents a powerful and generally convincing thesis on the Choctaw subsistence economy's collapse into dependency. On the deerskin

trade in the Lower Mississippi Valley, see also Krech, *Ecological Indian*, 151–71; Usner, *American Indians in the Lower Mississippi Valley*, 62–67.

64. Nutt, "Nutt's Trip to the Chickasaw Country," 53.

65. Descriptive accounts of alcoholism among southern Native Americans include Nuttall, *Journal of Travels into the Arkansas Territory*, 65; Schultz, *Travels on an Inland Voyage*, 108.

66. Krech, *Ecological Indian*, 160.

67. Paulding, "Mississippi," 331–32.

68. On the adoption of the new form of agriculture among the Chickasaw in 1800, see James Hall, "Brief History of the Mississippi Territory," 542–43. See also Hodgson, *Remarks during a Journey through North America*, 269–83.

69. In addition to the Choctaw and Chickasaw, this nineteenth-century designation included the Cherokee, Creek, and Seminole. On these tribes, see the classic Cotterill, *Southern Indians*.

70. "Hints on the Subject of Indian Boundaries, suggested for Consideration," December 29, 1802, in Jefferson, *Writings*, 373–74. A few months later, Jefferson outlined a similar plan regarding Native Americans along the lower Mississippi in his letter to the then Governor of Indiana Territory, William Henry Harrison. See Thomas Jefferson to William Henry Harrison, February 27, 1803, as quoted in Robert M. Owens, "Jeffersonian Benevolence on the Ground," 417.

71. Kidwell, "The Mississippi Choctaws in the Nineteenth Century," 57–70; Wells, "Trail of Treaties," 45–55.

72. Richard White, *Roots of Dependency*, 146.

73. Flint, *Recollections of the Last Ten Years*, 282.

74. Nuttall, *Journal of Travels into the Arkansas Territory*, 66.

## Chapter 4. The Creation of a Cotton Kingdom

1. The following account of the general characteristics of antebellum agriculture in the South is based on Cowdrey, *This Land, This South*, 28–41, 67–80, 107–11; Cox et al., *This Well-Wooded Land*, 9, 38–39; Silver, *New Face on the Countryside*, 104–10, 139–88; Scheiber and Vatter, *American Economic History*, 43–44, 52–53, 133–34. See also Thomas Clark, "Agriculture," 5–7; Dethloff, "Agriculture," 11–16; Kirby, "Plantations," 26–27. Lewis Cecil Gray, *History of Agriculture in the Southern United States to 1860*; Moore, *Agriculture in Antebellum Mississippi*; idem, *Emergence of the Cotton Kingdom in the Old Southwest* are indispensable for the study of antebellum agriculture in the South and Mississippi, while good cartography is provided by Hilliard, *Atlas of Antebellum Southern Agriculture*.

2. Kirby, "Plantations," 27.

3. See the classic Cash, *Mind of the South*.

4. Moore, *Emergence of the Cotton Kingdom in the Old Southwest*, 97–115. On slave culture in general, see Genovese, *Roll, Jordan, Roll*; Levine, *Black Culture and Black Consciousness*.

Black folk narratives of the South, in the form of animal trickster tales and John and Old Marster stories, convey lively information on the slave and freedman experience. For examples from the Delta, see Dorson, "Negro Tales from Bolivar County," 104–16.

5. De Puy van Buren, *Jottings of a Year's Sojourn in the South*, 209.

6. On the average date of the close of cotton picking in the early 1880s, see "Report of the Statistician," 413. The average dates for Mississippi between 1883 and 1885 varied from November 23 to December 8.

7. Agelasto et al., "Cotton Situation," 372–76. On the development of cotton ginning, see also Aiken, "Cotton Gins," 568–69.

8. Ker, *Travels through the Western Interior of the United States*, 39–40, October 1808, along the lower Mississippi, possibly in the Delta.

9. Grim, "Maps of the Township and Range System," 89–93; Opie, *Nature's Nation*, 98–105.

10. Thorne and Curry, *Cultural Resources Survey*, 41.

11. Ibid., 41–49. The townships examined by Thorne and Curry were T16–21N, R1E; T20N, R2E; T16–19N, R1W; and T17–19N, R2W. On the problems associated with the use of land survey maps in reconstructing past landscapes, see Grim, "Maps of the Township and Range System," 91, 93.

12. "Descriptive Notes of Choctaw District, Ranges 5 & 6 West."

13. Collot, *Journey in North America*, 45. See also Bowman, "Early History and Archaeology of Yazoo County," 427.

14. Thorne and Curry, *Cultural Resources Survey*, 41–49, 194–200; Holloway and Valastro, "Palynological Investigations along the Yazoo River," 236–40 (appendix 1). Pigweed (*Amaranthus* spp.) was also among early Native American cultigens.

Schultz, *Travels on an Inland Voyage*, 108, describes settlements below the mouth of the St. Francis. These were "made upon a natural prairie; where, finding the land ready cleared, a few settlers, without any title or claim, have taken possession. Most of these people have come here since the cession of this country to the United States, expecting, as has heretofore been the case, to receive some encouragement from the government."

15. Collot, *Journey in North America*, 45.

16. Nuttall, *Journal of Travels into the Arkansas Territory*, 252, on January 27, 1820, on the Mississippi between Point Chicot, Arkansas, and Walnut Hills (later Vicksburg), Mississippi. Cf. Jacobs Manuscript, 100.

17. "Descriptive Notes of Choctaw District, Ranges 5 & 6 West."

18. Hazel R. Delcourt, "Presettlement Vegetation of the North of Red River Land District," 126–35.

19. Hutchins, *Historical Narrative and Topographical Description of Louisiana and West-Florida*, 54.

20. "Northern Portion of Warren County."

21. Nutt, "Nutt's Trip to the Chickasaw Country," 45–46.

22. Stephen Williams and Jeffrey P. Brain, *Excavations at the Lake George Site*, 392.

23. Bromme, *Mississippi*, 11.

24. Harrison, *Levee Districts and Levee Building in Mississippi*, 8–9; Kelley, "Levee Building and the Settlement of the Yazoo Basin," 287–88.

25. Jacobs Manuscript, 6–9.

26. Harrison, *Levee Districts and Levee Building in Mississippi*, 8–9; Kelley, "Levee Building and the Settlement of the Yazoo Basin," 287–88.

27. Harrison, *Levee Districts and Levee Building in Mississippi*, 8–9; Humphreys, "Autobiography," 241; Kelley, "Levee Building and the Settlement of the Yazoo Basin," 295.

28. Cauthen, *Family Letters of the Three Wade Hamptons*, xiii–xvi.

29. Charles Helme to Dear Brother, January 28, [1847], box 1, Helme Papers.

30. Paul C. Cameron to Thomas Ruffin, February 1, 1858, in Ruffin, *Papers*, 549.

31. Cauthen, *Family Letters of the Three Wade Hamptons*, xiv–xviii; Griffin to Dear Nefew [Wilkinson], March 20, 1853, and March 31, 1853, folder 1, Wilkinson Papers.

32. Olmstead, *Journeys and Explorations in the Cotton Kingdom*, 70.

33. Griffin to Dear Nefew [Wilkinson], March 20, 1853, and March 31, 1853, folder 1, Wilkinson Papers.

34. Ibid.

35. Griffin to Dear Nefew [Wilkinson], July 10, 1855, folder 1, Wilkinson Papers.

36. See, e.g., C. D. Bonney to C[alvin] Taylor, September 27, 1853, H-20/vol. 3, Taylor and Family Papers.

37. Griffin to Dear Nefew [Wilkinson], March 20, 1853, and March 31, 1853, folder 1, Wilkinson Papers. Cf. Affleck, "The Duties of an Overseer." For a good

example of the nature of events recorded, consult the 1860 plantation book for the Buena Vista Plantation near Brunswick Landing, Warren County, kept by overseer B. J. Britt, 2E522, Kiger Family Papers.

38. Charlie Davenport, an ex-slave from Adams County, in Rawick, *American Slave*, 34.

39. LeRoy Barry Allen Manuscript, 31.

40. Hamilton, *Trials of the Earth*, 55.

41. Hamp Kennedy, an ex-slave from Perry County in Rawick, *American Slave*, 86, 88.

42. Thomas Gale to My Dear Daughter [Anna Gale], April 13, 1868; November 1, 1868; in Gale, "Thomas Gale's New South Letters," 231.

43. Hesse-Wartegg, *Travels on the Lower Mississippi*, 88.

44. Cotton Plantation Account and Record Book, box 1, vol. 1, Panther Burn Plantation Records.

45. Paul C. Cameron to Thomas Ruffin, February 1, 1858, in Ruffin, *Papers*, 549.

46. Martha [Rebecca Blanton] to dear Parents [Dr. and Mrs. George Smith], March 21, 1857, Cooper transcripts of Blanton-Smith Letters.

47. Hamilton, *Trials of the Earth*, 208, 138.

48. De Puy van Buren, *Jottings of a Year's Sojourn in the South*, 52. See also Jacobs Manuscript, 107.

49. Bromme, *Mississippi*, 11.

50. De Puy van Buren, *Jottings of a Year's Sojourn in the South*, 52.

51. Amanda Worthington to Dear Brother [Robert Tilford], April 22, [1838 or 1839], box 1, folder 3, Worthington Family Papers. See also Martha [Rebecca Blanton] to dear Parents [Dr. and Mrs. George Smith], March 21, 1857, Cooper transcripts of Blanton-Smith Letters.

52. Charles Helme to Dear Brother, January 28, [1847], box 1, Helme Papers. Cf. Bowman, "Early History and Archaeology of Yazoo County," 433.

53. Humphreys, *Yellow Fever and the South*, 4–5; Otto, *Final Frontiers*, 41. See also Carter, *Lower Mississippi*, 300–306; Cohn, *Where I Was Born and Raised*, 120–41. For a report of four cases of yellow fever in Yazoo City, see Thomas Gale to My Dear Annie [Anna Gale], December 13, 1867, in Gale, "Thomas Gale's New South Letters," 230.

54. James H. Ruffin to Thomas Ruffin, April 30, 1833, in Ruffin, *Papers*, 77. Slavery was partly defended on the grounds that blacks were thought immune to malaria and many other diseases. See, e.g., Jacobs Manuscript, 7.

55. Wade Hampton III to Mary Fisher Hampton, November 2, 1857, in Cau-

then, *Family Letters of the Three Wade Hamptons,* 51. See also Sessions, "Diary," 243–44.

56. Martha [Rebecca Blanton] to dear Parents [Dr. and Mrs. George Smith], March 21, 1857, Cooper transcripts of Blanton-Smith Letters.

57. De Puy van Buren, *Jottings of a Year's Sojourn in the South,* 198.

58. Glenn, "Mississippi," 42–43.

59. De Puy van Buren, *Jottings of a Year's Sojourn in the South,* 49–51.

60. Wade Hampton II to My Dear Mary [Mary Fisher Hampton], November 17, 1855, in Cauthen, *Family Letters of the Three Wade Hamptons,* 40.

61. Cuming, *Sketches of a Tour to the Western Country,* 351–52. See also De Bow, "Agricultural Survey of Mississippi."

62. Price, "Mississippi Swamp," 55. Nonetheless, malarial aspects of bottom-land settlement still had to be downplayed by early-twentieth-century boosters. See [Tompkins], "Mississippi Alluvial Lands," 496.

63. Lewis, *Valley of Mississippi Illustrated,* 363.

64. *Statistics of the Population of the United States at the Tenth Census,* 67, 398. See also table 1. For pre–Civil War demographic studies of different Delta counties, see Criss and Hill, "Demographic Study of Bolivar County in 1850," 3–9; Clifton, "Demographic Study of Bolivar County in 1860," 10–20; Halsell, "Migration Into and Settlement of Leflore County," 219–33.

65. However, Jordan and Kaups, *American Backwoods Frontier,* 94–134, claim that the slash-and-burn techniques were largely imported to North America by Nordic settlers to the New Sweden Colony on the Delaware, especially by ethnic Finns.

66. Paulding, "Mississippi," 325–26.

67. Martha [Rebecca Blanton] to dear Parents [Dr. and Mrs. George Smith], March 21, 1857, Cooper transcripts of Blanton-Smith Letters.

68. Cf. Harrison, *Alluvial Empire,* 301.

69. Hamilton, *Trials of the Earth,* 168.

70. Cameron to Thomas Ruffin, March 21, 1857, in Ruffin, *Papers,* 550.

71. Jacobs Manuscript, 7.

72. "Improved" in U.S. Census usage refers to land in farms that was harvested, idle, plowable pasture, and fallowed or in crop failure. Until 1930, it also included land occupied by farm buildings.

73. Michael Williams, "Clearing the United States Forests," 14–15.

74. Faulkner, *Go Down, Moses,* 193.

75. Faulkner, *Requiem for a Nun,* 226–27.

76. *Seventh Census,* 447, 456, 458; *Agriculture of the United States in 1860,* 84–85. See also table 2.

On July 23, 1855, Alexander Hamilton Polk sold 2,880 acres, "lying South and West of Deer Creek" in southwestern Bolivar County, to Frances Anne Devereaux of New Orleans for $46,232.19, more than $16 per acre. See "Land Sale, 1855," in Polk Papers.

77. The following account of the general characteristics of postbellum cotton agriculture in the Delta and the South in general is based on Brandfon, *Cotton Kingdom of the New South;* Thomas Clark, "Agriculture," 6–7; Cobb, *Most Southern Place on Earth,* 98–124; Conrad, "Tenant Farmers," 1412–14; Daniel, "Cotton Culture," 34–35; idem, *Breaking the Land,* 3–22; Dethloff, "Agriculture," 11–16; Fite, *Cotton Fields No More,* 1–90; Giles, "Agricultural Revolution," 177–211; Kirby, "Plantations," 26–31; Willis, *Forgotten Time.* See also Ayers, *Promise of the New South,* 187–213; McCandliss, *Base Book of Mississippi Agriculture.*

78. On land values after 1850, see Pressly and Scofield, *Farm Real Estate Values in the United States by Counties.*

79. Woodward, *Origins of the New South,* 180. Cf. Twain, *Life on the Mississippi,* 288–91.

80. See, e.g., the classic monographs *Human Factors in Cotton Culture* and *Human Geography of the South* by Vance. Cf. Earle, "Myth of the Southern Soil Miner," 201–10.

81. Kirby, "Plantations," 27.

82. Wade Hampton III to my Dear Anna [Mrs. Thomas L. Preston], April 27, 1875, in Cauthen, *Family Letters of the Three Wade Hamptons,* 151.

83. Buchanan, *Holt Collier,* 141.

84. Morgan, *Yazoo,* 29.

85. While in charge of the plantations, the well-lettered Montgomery corresponded continuously with his former master and left a fascinating trail of letters, to be consulted in the Davis Papers.

Davis sold the plantations to Montgomery for $300,000 "in gold coin of the United States or its equivalent, payable on or before the first day of January one thousand eight hundred and seventy six, with interest at the rate of 5% five per cent in gold or 7% seven per cent in paper currency lawful money of the United States per annum. The annual amount of said interest to be paid in advance each year, commencing the first day of January 1867, and to continue in the like manner until the full amount of purchase money is paid. And for the said amount of purchase money I[,] Joseph E. Davis[,] do reserve a mortgage." [A draft of a] Bill of Sale for Brierfield and Hurricane plantations, November 15, 1866, box 1, folder A7/2, Davis Papers.

86. Ben[jamin Montgomery] to Very Kind Sir [Joseph E. Davis], April 15, 1867, box 1, folder A3/10, Davis Papers.

87. [Montgomery] to [Davis], November 14, 1867, box 1, folder A3/32; [Montgomery] to [Davis], December 11, 1867, box 1, folder A3/33, Davis Papers.

88. [Montgomery] to [Davis], December 19, 1867, box 1, folder A3/34; [Montgomery] to [Davis], December 30, 1867, box 1, folder A3/35, Davis Papers. In many southern counties, sheriffs customarily arrested vagrants and placed them in work camps.

89. [Montgomery] to [Davis], July 4, 1867, box 1, folder A3/20, Davis Papers.

90. On this fascinating saga, which culminated in the establishment of an all-black town of Mound Bayou in Bolivar County, see the contemporary descriptions by Banks, *Negro Town and Colony;* Lee, "Mound Bayou." Cf. "Biographical Sketches of Some Prominent Negroes"; Sillers, "Isaiah T. Montgomery;" Townsend, "Mound Bayou." Hermann's *Pursuit of a Dream* provides the most comprehensive account.

91. Thomas Gale to Anna [Gale], November 4, 1869, in Gale, "Thomas Gale's New South Letters," 233.

92. Humphreys, "Autobiography," 246.

93. Consult the account of the 1858 Levee Board and the 1867 Liquidating Levee Board in chapter 5.

94. W[illiam] I. Hindman to Mr[. Andrew] Brown, August 12, 1866, box 6, folder 1, Brown-Learned Lumber Company Papers.

95. W[illia]m I. Hindman to Mr[. Andrew] Brown, December 30, 1866, box 6, folder 1, Brown-Learned Lumber Company Papers.

96. *Statistics of the Population of the United States at the Tenth Census,* 397–98. See also table 1; Kelley, "Levee Building and the Settlement of the Yazoo Basin," 303; Claude P. Smith, "Official Efforts by the State of Mississippi to Encourage Immigration." For a contemporary view on immigration into the Delta, see e.g. Gossom, "Our County Prospects."

97. Berry Smith, an ex-slave from Scott County, in Rawick, *American Slave,* 132.

98. Isaac Stier, an ex-slave from Lauderdale County, in Rawick, *American Slave,* 147.

99. James Lucas, an ex-slave from Adams County, in Rawick, *American Slave,*

97–98. Lucas belonged for a while to Jefferson Davis and also worked at the Brierfield Plantation in Warren County.

100. Darling, "Landownership," 168–70. Willis's "On the New South Frontier" and *Forgotten Time* document the rise and fall of black land ownership in the Delta.

101. A vast amount of research has been devoted to the subject of tenancy in southern agriculture. The following account of its general characteristics is based on Aiken, "Decline of Sharecropping in the Lower Mississippi River Valley," 151–65; Daniel, *Shadow of Slavery*; idem, "Metamorphosis of Slavery," 88–99; Vance, *Human Factors in Cotton Culture*, 11–79. On tenancy in Mississippi and the Delta, see also Brandfon, "End of Immigration to the Cotton Fields," 591–611. There are also significant surveys of the subject by the federal government. In addition to Du Bois's "Negro Farmer," 511–40, especially useful have been the U.S. Department of Agriculture bulletins by Boeger and Goldenweiser, *Study of the Tenant Systems of Farming in the Yazoo-Mississippi Delta*; Langsford and Thibodeaux, *Plantation Organization and Operation in the Yazoo-Mississippi Delta Area*. For a planter's view of tenancy, consult Percy, *Lanterns on the Levee*, 270–84.

102. "Cotton Investigation," 134.

103. For discussions of the different methods of renting land in the Delta, see Boeger and Goldenweiser, *Study of the Tenant Systems of Farming in the Yazoo-Mississippi Delta*, 6–7; Langsford and Thibodeaux, *Plantation Organization and Operation in the Yazoo-Mississippi Delta Area*, 4–6.

104. Harrison, "Among the Southerners."

105. Oshinsky, *"Worse Than Slavery,"* 34–53.

106. On the Chinese minority in Mississippi and the Delta, see Loewen, *Mississippi Chinese*, 1–31; Quan, *Lotus among the Magnolias*, 1–21. On the Sunnyside plantation, see Barry, *Rising Tide*, 111–21; Brandfon, *Cotton Kingdom of the New South*, 144–48; Whayne, *Shadows over Sunnyside*. Barry's *Rising Tide* contains a skillfully written account of Percy's paternalistic ideology and his attempts to secure a cheap and sufficient labor force for the Delta planters.

107. Evans, *Big Road Blues*, 167–264, describes the evolution of a local blues tradition in the Drew area, underlining the continuous interaction between the two localities.

108. Berry Smith, an ex-slave from Scott County, in Rawick, *American Slave*, 132.

109. Boeger and Goldenweiser, *Study of the Tenant Systems of Farming in the*

*Yazoo-Mississippi Delta,* 3, 8; Langsford and Thibodeaux, *Plantation Organization and Operation in the Yazoo-Mississippi Delta Area,* 13. The U.S. Census Bureau classified the land operated by each tenant as a separate farm. Therefore, a plantation with 24 tenant families plus some acreage cultivated by the plantation owner himself is indicated in census reports as 25 separate farms.

110. Boeger and Goldenweiser, *Study of the Tenant Systems of Farming in the Yazoo-Mississippi Delta,* 18.

111. Langsford and Thibodeaux, *Plantation Organization and Operation in the Yazoo-Mississippi Delta Area,* 3, 13.

112. On both black and white migration from the South, see Kirby, *Rural Worlds Lost,* 309–33. For total figures by decade, see ibid., 320. Cf. Marks, *Farewell—We're Good and Gone,* 1.

113. For the birth places of prominent Mississippi blues musicians, consult "Mississippi Blues Musicians."

114. Interestingly, the first scholarly description of black folk music in the Delta was provided by archaeologist Charles Peabody of Harvard University, who published "Notes on Negro Music" as an additional result of his excavations of Mississippian mounds in Coahoma County. See also Palmer, *Deep Blues,* 23–25.

In addition to Evans, *Big Road Blues,* 16–105; Palmer, *Deep Blues,* 23–47; good introductions to the Delta blues and its evolution include Lomax, *Land Where the Blues Began;* Oliver, *Screening the Blues,* 1–25; idem, *Blues Fell This Morning,* 1–11. For an early recorded blues performance on the effects of drought, consult Patton, "Dry Well Blues."

115. Hesse-Wartegg, *Travels on the Lower Mississippi,* 94.

116. "Immigration." Cf. Loring and Atkinson, *Cotton Culture and the South.*

117. See, e.g., "Letter from North Leflore"; J. B. Wilson, *Handbook of Yazoo County.* Railroads were seen as essential for the development of local infrastructure. W. F. Tucker, president of the Vicksburg and Nashville Railroad Co., was eager to point out to the residents of the eastern Delta that "[y]ou need the Vicksburg & Nashville Railroad, and the Railroad needs your assistance—without it, the road can never be built." See "Vicksburg & Nashville Railroad."

118. Gossom, "Our County Prospects." However, cf. Pressly and Scofield, *Farm Real Estate Values in the United States by Counties,* 54–55. Two years later, the price of forested bottomland acreage along the Yazoo in Warren County was reported at $10 per acre. See "Statistics of Forestry," 279.

119. Harrison, *Levee Districts and Levee Building in Mississippi,* 51–52. The lands purchased by the Delta and Pine Land Company had originally come to the

market in 1881 and were obtained by an individual named Eugene C. Gordon. After several transactions, the acreage became the company's property in 1886. On the development of the company into an enormous agribusiness operation, see the dissertation by Dong, "From Postbellum Plantation to Modern Agribusiness." See also (Mrs.) Early C. Ewing, "Delta and Pine Land Company"; idem, "Additional Facts Regarding the History of the Delta and Pine Land Company"; Early Ewing, Jr., "Delta and Pine Land Company."

120. Harrison, *Levee Districts and Levee Building in Mississippi*, 154–55; Langsford and Thibodeaux, *Plantation Organization and Operation in the Yazoo-Mississippi Delta Area*, 13; [Tompkins], "Mississippi Alluvial Lands," 482. The mileage data used by Langsford and Thibodeaux was obtained from the railroad companies and pertain to the entire Delta, including all or parts of the nineteen counties. On the Delta railroads, see also [William F. Gray], "Railroads"; "History of Railroads in Bolivar County." General accounts include Stover, *Railroads of the South*; idem, *History of the Illinois Central Railroad*.

121. Mills, "New Life for the River of Death," 291; Weeks, *Clarksdale and Coahoma County*, 59–66. On the river commerce on the Mississippi and its tributaries, see also Harry P. Owens, *Steamboats and the Cotton Economy*; Quick and Quick, *Mississippi Steamboatin'*. Useful government documents include Brock, *Report on the Internal Commerce*; Switzler, *Report on the Internal Commerce* (1886 and 1888).

122. Skene, *Choice Pickings in the Yazoo Valley*; *Yazoo-Mississippi Delta*. See also *Hardwood Lumber and Farming Industries of Mississippi*, 37.

123. Harrison, *Alluvial Empire*, 215.

124. Blanton to W. Z. Hutchinson, January 11, 1909, file 84, Blanton-Smith Family Letters.

125. Andrews, *Railroads and Farming*, 24–25.

126. Ibid., 8–9. See also Gates, "Federal Land Policy in the South"; idem, "Overview of American Land Policy;" idem, "Federal Land Policies in the Southern Public Land States."

127. In addition to *Call of the Alluvial Empire*, consult promotional pamphlets such as Maule, *Yazoo-Mississippi Delta*; Merry, *About The South* and *Yazoo-Mississippi Valley*; *Mississippi Is Calling You*; *Of the Greatness of Mississippi*; Skene, *600,000 Acres Railroad Lands For Sale* and *Choice Pickings in the Yazoo Valley*; "Yazoo-Mississippi Delta"; *Yazoo-Mississippi Delta*. See also "Mississippi."

128. See table 1; [Tompkins], "Mississippi Alluvial Lands," 485, 499. On the growth of Greenville in Washington County, see Twain, *Life on the Mississippi*,

288. On the growth of Leflore County, see Halsell, "Migration Into and Settlement of Leflore County in the Later Periods."

129. *Negro Population*, 783–84 (table 2). See also Langsford and Thibodeaux, *Plantation Organization and Operation in the Yazoo-Mississippi Delta Area*, 9.

130. [Tompkins], "Mississippi Alluvial Lands," 485, 499.

131. Harrison, *Alluvial Empire*, 216; Pressly and Scofield, *Farm Real Estate Values in the United States by Counties*, 54–55; Otto, *Final Frontiers*, 51; Dong, "From Postbellum Plantation to Modern Agribusiness," 44–72; Early Ewing, Jr., "Delta and Pine Land Company," 1–2.

132. Oshinsky, *"Worse Than Slavery,"* 109–10, 154–55, 223–24.

133. Harrison, *Alluvial Empire*, 216; Langsford and Thibodeaux, *Plantation Organization and Operation in the Yazoo-Mississippi Delta Area*, 12–15; Pressly and Scofield, *Farm Real Estate Values in the United States by Counties*, 54–55.

134. Harrison, *Alluvial Empire*, 306.

135. Hamilton, *Trials of the Earth*. For Hamilton's Delta experiences, see 51–216, 237–48.

136. Ibid., 139, 152–53, 191–92.

137. Ibid., 220, 243.

138. Palmer, *Deep Blues*, 49–54.

139. Langsford and Thibodeaux, *Plantation Organization and Operation in the Yazoo-Mississippi Delta Area*, 3, 13. See also Kirby, "Transformation of Southern Plantations," 257–61.

140. Hesse-Wartegg, *Travels on the Lower Mississippi*, 113.

141. Agelasto et al., "Cotton Situation," 349–57.

142. Arnold, "Bo-Weavil Blues." Cf. Patton, "Mississippi Bo Weavil Blues." See also Oliver, *Blues Fell This Morning*, 16–18.

143. See, e.g., Higgs, "Boll Weevil," 335–50.

144. Otto, *Final Frontiers*, 42–43, 62.

145. Agelasto et al., "Cotton Situation," 327, 329, 372.

146. Helms, "Boll Weevil," 32. For example, cotton production in Georgia declined by over 50 percent between 1919 and 1921, while the harvest in Mississippi fell only by some 15 percent. See Agelasto et al., "Cotton Situation," 333 (figure 9).

147. Agelasto et al., "Cotton Situation," 334–35 (figure 10), 348 (figure 21). See also Woodman, *King Cotton and His Retainers*, 315–59.

148. Aldridge, "Fertilizer Industry," 431; Dong, "From Postbellum Plantation to Modern Agribusiness," 90–91.

149. Agelasto et al., "Cotton Situation," 334–35 (figure 10).

150. Ibid., 323–25, 326 (figure 3), 370–72.

151. Ibid., 327–30; Stover, *Forest Resources of the Delta Section of Mississippi*, 1. The most valued type of cotton, Fancy Sea Island, had fiber over two inches long, while the fiber of the so-called Asiatic varieties (*G. herbaceum* and others), grown mostly for domestic use in southern Asia, could be as short as three-fourths of an inch.

152. Michael Williams, "Clearing the United States Forests," 14–15.

153. *Historical Statistics of the United States*, 460 (table K 55). The area of land in farms in Florida increased from 2,374,000 to 4,364,000 acres.

154. Hesse-Wartegg, *Travels on the Lower Mississippi*, 76.

155. Cf. table 2. In 1880, farmland in the Delta was valued at an average of $14 per acre.

156. See table 2. See also Boeger and Goldenweiser, *Study of the Tenant Systems of Farming in the Yazoo-Mississippi Delta*, 3; Langsford and Thibodeaux, *Plantation Organization and Operation in the Yazoo-Mississippi Delta Area*, 12–15. In the Delta cotton kingdom, the other crops of any importance were corn and hay, grown primarily to feed the livestock used in cotton cultivation. Of these, hay gained importance only after the turn of the century, as more land was cleared and range grazing became more restricted.

## Chapter 5. Taming the Rivers

1. Wittfogel, "Hydraulic Civilizations," 152–53. See also Worster, "History as Natural History," 4–6.

2. Worster, *Rivers of Empire*; Wittfogel, "Hydraulic Civilizations," 153.

3. The following examination of the history of flood control and water management in the Yazoo-Mississippi Delta is largely based on the definitive research on the subject by Robert W. Harrison. Especially useful were Harrison's magisterial monographs, *Levee Districts and Levee Building in Mississippi* and *Alluvial Empire*.

4. Harrison, "Early State Flood-Control Legislation in the Mississippi Alluvial Valley," 105–6; idem, *Levee Districts and Levee Building in Mississippi*, 2; Harrison and Mooney, *Flood Control and Water Management in the Yazoo-Mississippi Delta*, 2.

5. Jenkins, "Mississippi River and the Efforts to Confine It in Its Channel," 304–5; Thorne and Curry, *Cultural Resources Survey*, 138; Daniel, *Deep'n as It Come*, 3; Harrison, *Flood Control in the Mississippi Alluvial Valley*. Mills, *Of Men and Rivers*, 105, considers the floods of 1844, 1903, 1913, 1922, and 1927 "superfloods."

6. Thorne and Curry, *Cultural Resources Survey*, 10–11.

7. On the general history of flood control in the Southeast, see Cowdrey, *This Land, This South*, 68–69, 95–98, 143–45; Blake, "Flood Control and Drainage," 335–37. Harrison, "Early State Flood-Control Legislation in the Mississippi Alluvial Valley," 107–12, describes levee construction during the French colonial period.

8. Nevertheless, the state of Mississippi had authorized the erection of a levee in Warren County already in 1819. See Kelley, "Levee Building and the Settlement of the Yazoo Basin," 288.

9. Basil Hall, *Travels in North America in the Years 1827 and 1828*, 350–51.

10. Ibid., 353.

11. Audubon, *Delineations of American Scenery and Character*, 31.

12. Harrison, "Levee Building in Mississippi Before the Civil War," 64–69.

13. Glenn, "Mississippi," 42–43. The need for a working levee system as a prerequisite for economic growth had become evident by the early 1840s. See Wills, "Southern Traveler's Diary," 32.

14. Harrison, *Flood Control in the Mississippi Alluvial Valley;* idem, *Alluvial Empire*, 58–59; Mills, *Of Men and Rivers*, 8–9.

15. Harrison, "Levee Building in Mississippi Before the Civil War," 69; idem, *Levee Districts and Levee Building in Mississippi*, 10–11; idem, *Alluvial Empire*, 67–74; Harrison and Mooney, *Flood Control and Water Management in the Yazoo-Mississippi Delta*, 3. On the Swamp Land Grants in Arkansas and Louisiana, see Harrison, *Alluvial Empire*, 74–85.

16. See, for example, Basil Hall, *Travels in North America in the Years 1827 and 1828*, 363–64.

17. Harrison, *Alluvial Empire*, 143–48; Mills, *Of Men and Rivers*, 12–26. The first official survey of the Ohio and the Mississippi, from Louisville to Balize in the mouth of the Mississippi, was initiated in 1820. Probably the most influential study of all times for river engineering in the United States proved to be Humphreys and Abbot, *Report Upon the Physics and Hydraulics of the Mississippi River*. See also Reuss, "Andrew A. Humphreys and the Development of Hydraulic Engineering," 1–10, 32–33. On the Yazoo, see Mills, "New Life for the River of Death," 288–89.

18. Harrison, *Levee Districts and Levee Building in Mississippi* 7, 10–14; idem, *Alluvial Empire*, 92. Harrison estimates that altogether some three million dollars had been spent for levees in the Yazoo-Mississippi Delta by 1858.

19. Hampton to My Dear Mary [Fisher Hampton], April 16, 1858, in Cauthen, *Family Letters of the Three Wade Hamptons*, 60–61.

20. Harrison, "Levee Building in Mississippi Before the Civil War," 81. Cf. "Levees and Overflows of the Mississippi." See also *By-Laws and Ordinances of the Board of Levee Commissioners for the State of Mississippi.*

21. A typed manuscript signed by Charles Swift, 8, James L. Alcorn and Family Papers. See also Alcorn, "Letters," 196–97. Alcorn's success as a politician-planter continued after the Civil War. He eventually owned some twelve thousand acres in the Delta, producing twelve to eighteen hundred bales of cotton annually. On Alcorn's career, see Pereyra, "James Lusk Alcorn and a Unified Levee System."

22. *Report of the President and Engineer in Chief to the Legislature.* See also Harrison, *Levee Districts and Levee Building in Mississippi,* 15–25; idem, *Flood Control in the Mississippi Alluvial Valley;* idem, *Alluvial Empire,* 93–96; Harrison and Mooney, *Flood Control and Water Management in the Yazoo-Mississippi Delta,* 4. By various legal devices, nonresident landholders continued to evade full taxation and caused Alcorn to report that "[i]t is at war with the plainest dictates of justice that non-residents should be permitted to hold vast bodies of land exempt from taxation, in a state of wilderness, adding nothing to the wealth and prosperity of the State." In "Report of the President of the Board of Levee Commissioners," *Mississippi Senate Journal,* 1862 (November 25, 1861), appendix, 118, as quoted in Harrison, "Levee Building in Mississippi Before the Civil War," 93.

23. Harrison, *Levee Districts and Levee Building in Mississippi,* 7–8.

24. Sillers, "Flood Control in Bolivar County," 9–10. The only implements used in the construction of these early levees were spades and wheelbarrows run on planks laid on the ground.

25. Harrison, *Levee Districts and Levee Building in Mississippi,* 25–27; Harrison and Mooney, *Flood Control and Water Management in the Yazoo-Mississippi Delta,* 4.

The levee at the Yazoo Pass was among the largest flood-control structures at the time. The length of the twenty-eight-foot-high structure exceeded one thousand feet with a base width of three hundred feet. See Mills, *Of Men and Rivers,* 31–33; Harrison, *Levee Districts and Levee Building in Mississippi,* 7.

26. Harrison, *Levee Districts and Levee Building in Mississippi,* 40–52; idem, *Alluvial Empire,* 96–99.

Because of multiple forfeitures of Delta lands to the Liquidating Levee Board; the Board of Levee Commissioners for Bolivar, Washington, and Issaquena Counties; and the Levee Board of the State of Mississippi, District No. 1, questions concerning tax titles in the Delta remained extremely complicated. In 1876, the state legislature transferred the land titles of the two latter organizations to the state and abolished the Liquidating Levee Board. The following year, the landmark case of *Green v. Gibbs* allowed for the sale of the former Liquidating Levee Board lands. Under this and subsequent legislation, the Delta lands in question passed to nongovernmental ownership. For a more painstaking portrayal of this complex process, see Harrison, *Levee Districts and Levee Building in Mississippi,* 43–52.

27. Harrison and Mooney, *Flood Control and Water Management in the Yazoo-Mississippi Delta,* 5.

28. On the Board of Levee Commissioners for Bolivar, Washington, and Issaquena Counties, see Harrison, *Levee Districts and Levee Building in Mississippi,* 27–36; Hurt, *Mississippi,* 23–24. On the activities of the Board of Mississippi Levee Commissioners before 1884, see idem, *Levee Districts and Levee Building in Mississippi,* 52–69; Harrison, *Alluvial Empire,* 99–104. See also Sillers, "Levees of the Mississippi Levee District in Bolivar County," 81–97.

The Board of Mississippi Levee Commissioners is also known as the Lower Yazoo Levee Board or the Greenville Levee Board. By the mid–twentieth century, it also included Sharkey County (created in 1876) and most of Humphreys County (created in 1918). For the historical boundaries of these two counties, see Long, *Mississippi,* 88, 171.

29. McCoy, "When the Levee Breaks."

30. Harrison, *Levee Districts and Levee Building in Mississippi,* 69–70; idem, *Alluvial Empire,* 148–54. See also McBride and McLaurin, "Origin of the Mississippi River Commission," 389–411. The Vicksburg District of the U.S. Army Corps of Engineers was founded in 1873. Its actions in the Delta initially centered on snagging and removal of war wrecks from the channels of the Yazoo and its tributaries. See Mills, *Of Men and Rivers,* 51–52, 69–70. For general histories of the New Orleans and Vicksburg Districts of the U.S. Army Corps of Engineers, consult Cowdrey, *Delta Engineers,* and Mills, *Of Men and Rivers.*

31. Twain, *Life on the Mississippi,* 224.

32. [R. H. Bowman] to Robert Lowry, March 23, 1882, 2E536, Natchez Trace Steamboat Collection. Cf. Twain, *Life on the Mississippi,* 252.

33. Harrison, "Formative Years of the Yazoo-Mississippi Delta Levee District," 236–44. See also Dabney, "Brief History of the Operations of the Yazoo-Mississippi Delta Board-Organization," 295–302; Harrison, *Levee Districts and Levee Building in Mississippi,* 75–76; idem, *Alluvial Empire,* 104–6; Harrison and Mooney, *Flood Control and Water Management in the Yazoo-Mississippi Delta,* 5–6.

The counties in the upper Delta had in 1871 organized the Levee Board of the State of Mississippi, District No. 1. The district soon went bankrupt, leaving behind unpaid bonds worth hundreds of thousands of dollars. See Harrison, *Levee Districts and Levee Building in Mississippi*, 37–39.

The Board of Yazoo-Mississippi Delta Levee Commissioners is alternatively known as the Upper Yazoo Levee Board or the Clarksdale Levee Board.

On the nineteenth-century plans for improvement of the tributaries, see *Yazoo, Tallahatchie, and Sunflower River Improvement Convention*.

34. Harrison, *Alluvial Empire*, 211, 223–25. Idem, *Alluvial Empire*, 211–51, traces the history of drainage development in the Yazoo-Mississippi Delta. See also Robinson, "Reclamation and Irrigation," 355–56. On the development of drainage in Bolivar County, see Boone, "Bolivar County Drainage," 36–38.

35. On levee construction in the Delta during the 1880s and 1890s, see Harrison, *Levee Districts and Levee Building in Mississippi*, 83–116, 120–39; idem, *Alluvial Empire*, 106–17.

36. Thomas Dabney in *Reports of the Chief Engineer, Yazoo-Mississippi Delta Levee District, 1884–1900*, 293–94, as quoted in Harrison and Mooney, *Flood Control and Water Management in the Yazoo-Mississippi Delta*, 6.

37. On the 1897 flood, see Harrison, *Levee Districts and Levee Building in Mississippi*, 140–53; idem, *Alluvial Empire*, 115–16.

38. "Initial Convention." See also Harrison, *Levee Districts and Levee Building in Mississippi*, 116–20.

39. Harrison, *Levee Districts and Levee Building in Mississippi*, 196–98, 202–5. During the nineteenth century, enormous mattresses of 250 to 300 feet in width with a standard length of 1,000 feet were woven from willow stalks and used for revetment purposes. On the techniques and development of revetment work, see Mills, *Of Men and Rivers*, 83–87.

40. Harrison, *Levee Districts and Levee Building in Mississippi*, 188.

41. On the expanding levee construction in the Delta between 1897 and 1916, consult Harrison, *Levee Districts and Levee Building in Mississippi*, 154–69, 188–96, 198–202, 217–24; idem, *Alluvial Empire*, 119–32.

42. H.W.B. [Henry Waring Ball], "Memoranda of the Carter Family in Virginia," bound manuscript written before 1934, 141–42, box 1, folder 15, Ball Papers.

43. On the floods of 1903 and 1912, see Harrison, *Levee Districts and Levee Building in Mississippi*, 169–88, 205–13; idem, *Alluvial Empire*, 120–22, 129–30.

44. Narrative of Berry Smith, an ex-slave from Scott County, in Rawick, *American Slave*, 133.

45. Harrison, *Levee Districts and Levee Building in Mississippi*, 225–27; idem, *Flood Control in the Mississippi Alluvial Valley;* idem, *Alluvial Empire,* 132–34, 158–59; Mills, *Of Men and Rivers,* 116.

46. Harrison, *Levee Districts and Levee Building in Mississippi,* 227–29. The 1917 legislation was augmented by the Flood Control Act of 1923, which provided sixty million dollars over a ten-year period. See Harrison, *Alluvial Empire,* 158.

47. West, "Brief History of Levee Building in the Mississippi Levee District," 306. See also Harrison and Mooney, *Flood Control and Water Management in the Yazoo-Mississippi Delta,* 7. On levee construction between 1917 and 1927, see Harrison, *Levee Districts and Levee Building in Mississippi,* 229–33.

48. [Tompkins], "Mississippi Alluvial Lands," 501; Mills, *Of Men and Rivers,* 89–92; idem, "New Life for the River of Death," 292–93.

49. Harrison, *Alluvial Empire,* 211.

50. Harrison, *Flood Control in the Mississippi Alluvial Valley;* idem, *Alluvial Empire,* 211–12; Harrison and Mooney, *Flood Control and Water Management in the Yazoo-Mississippi Delta,* 16–18. Cf. Langsford and Thibodeaux, *Plantation Organization and Operation in the Yazoo-Mississippi Delta Area,* 13 (table 3). See also *Drainage of Agricultural Lands,* 196–203.

On the plan for the Tallahatchie Drainage District, see Harrison, *Alluvial Empire,* 217–20. For descriptions of the drainage districts in existence in 1961, see Harrison, *Alluvial Empire,* 229–51.

51. Harrison, *Alluvial Empire,* 215–17, 226–27.

52. Daniel, *Deep'n as It Come,* 8; Harrison, *Levee Districts and Levee Building in Mississippi,* 233–35. Comprehensive monographs of the 1927 flood include Daniel, *Deep'n as It Come;* and Barry, *Rising Tide.*

53. Daniel, *Deep'n as It Come,* 7; Harrison, *Levee Districts and Levee Building in Mississippi,* 234; Mills, *Of Men and Rivers,* 123–26.

54. Sam[uel Dunn Finlay II] to Dear Mudder [Ada McPeak Finlay], Sunday, [April 1927], file 94, Blanton-Smith Family Letters.

55. Oliver, *Screening the Blues,* 22–23; Oliver, *Blues Fell This Morning,* 215–25. See also Morrison, "Downhome Tragedy," 267–68.

56. Patton, "High Water Everywhere—Parts I and II." For further examples of the flood blues genre, consult e.g. Delaney, "Tallahatchie River Blues"; Jefferson, "Rising High Water Blues." On Patton, see Herzhaft, *Encyclopedia of the Blues,* 277–78; Palmer, *Deep Blues,* 48–89.

57. Daniel, *Deep'n as It Come,* 8, 107–8, 121–22, 153–56, 183–85; Harrison, *Alluvial Empire,* 159–61; Otto, *Final Frontiers,* 70; Dong, "From Postbel-

lum Plantation to Modern Agribusiness," 110–13. On the restraining of the black population, see also Barry, *Rising Tide*, 303–35. Cf. Percy, *Lanterns on the Levee*, 257–58.

58. Daniel, *Deep'n as It Come*, 68–73; Harrison, *Levee Districts and Levee Building in Mississippi*, 234–35; idem, *Alluvial Empire*, 137–38; Mills, *Of Men and Rivers*, 132–36. See also J. S. Allen, "Levees 1926–46." The scientific aspect of the grandiose project included the creation of an expert organ, the Waterways Experiment Station, in Clinton, Mississippi, just south of Vicksburg and the Delta. Among the most spectacular activities of the laboratory was the construction of a scale model of the entire Mississippi Basin covering some two hundred acres.

In the mid-1970s, the Army Corps of Engineers described the project's hypothetical flood as one 11 percent greater than the flood of 1927 at the mouth of the Arkansas and 29 percent greater at the latitude of Red River Landing, amounting to over three million cubic feet per second at that location. See *Flood Control in the Lower Mississippi Valley*, 11.

59. Smallwood, "Water Use," 368. For an overview of the TVA, see Opie, *Nature's Nation*, 338–40.

60. Harrison, *Levee Districts and Levee Building in Mississippi*, 235.

61. After the turn of the century, willow trees in the amounts needed for revetment purposes were increasingly hard to come by, and by 1917 a working concrete mattress had been developed. See Mills, *Of Men and Rivers*, 84, 86.

62. Ker, *Travels through the Western Interior of the United States*, 44.

63. Harrison, *Levee Districts and Levee Building in Mississippi*, 235; Harrison, *Flood Control in the Mississippi Alluvial Valley*, figure 3A. The effects of the cutoff program had been simulated on the river model at the Waterways Experiment Station.

64. Harrison, *Alluvial Empire*, 299; Langsford and Thibodeaux, *Plantation Organization and Operation in the Yazoo-Mississippi Delta Area*, 7.

65. For more information, consult U.S. Congress documents *Survey of Little Tallahatchie River Watershed in Mississippi; Yazoo River (Lower Tributaries), Miss.; Survey of Yazoo River Watershed in Mississippi*.

66. Harrison, *Alluvial Empire*, 223, 298–99; Harrison and Mooney, *Flood Control and Water Management in the Yazoo-Mississippi Delta*, 10; Mills, *Of Men and Rivers*, 160–63.

67. Harrison, *Flood Control in the Mississippi Alluvial Valley*.

68. MacDonald, Frayer, and Clauser, *Documentation, Chronology, and Future Projections of Bottomland Hardwood Habitat Losses in the Lower Mississippi Alluvial Plain*, 22, 74.

69. On the human modification of the Atchafalaya, see McPhee, *Control of Nature*, 3–92; Reuss, "Army Corps of Engineers and Flood-Control Politics on the Lower Mississippi," 133–40. The definitive monograph on the subject is Reuss, *Designing the Bayous*.

70. Cry, "Surface Waters of the Lower Mississippi River Region," 68, 73; Harrison, *Alluvial Empire*, 168–69; Harrison and Mooney, *Flood Control and Water Management in the Yazoo-Mississippi Delta*, 10–15; Mills, *Of Men and Rivers*, 153–60; idem, "New Life for the River of Death," 294–99; Reuss, "Army Corps of Engineers and Flood-Control Politics on the Lower Mississippi," 140–48.

71. Faulkner, *Big Woods*, 170.

## Chapter 6. Bounties of the Bottomland

1. On timber use in the South Atlantic region during the colonial period, see Silver, *New Face on the Countryside*, 115–38.

The naval stores industry developed early in the South. The exploitation of pines, especially the longleaf pine (*Pinus palustris*), for their resinous juices was of prime importance to the British navy during the era of wooden vessels, and the production of tar, turpentine, and pitch was supported by royal bounties from 1705 to the Revolution. During the nineteenth century, the center of industry moved from the Carolinas southward to Georgia and Florida as the coastal pine forests became exhausted. See Perry, "Naval Stores Industry of the Old South."

2. Cox et al., *This Well-Wooded Land*, 266 (figure 2); *Historical Statistics of the United States*, 542 (table L 113). A board foot measures 1 foot × 1 foot × 1 inch, and twelve board feet equal one cubic foot. Throughout the study, "billion" is the American billion (thousand million).

The following account of the general development of forest industries in the southeastern United States is based on Thomas D. Clark, "Impact of Timber Industry on the South;" idem, *Greening of the South*, 1–25; Cowdrey, *This Land, This South*, 52–55, 89–95, 111–14; Cox et al., *This Well-Wooded Land*, 94–101, 122, 164–65, 188; Maxwell, "The Impact of Forestry on the Gulf South," 31–35; MacCleery, *American Forests*, 8–24; Walker, *Southern Forest*, 1–145; Michael Williams, "Clearing the United States Forests," 12–28; idem, *Americans and Their Forests*, 238–88; idem, "Clearing of the Forests," 159–60.

3. On the development of the log drive and the continental transport system of timber, see Michael Williams, "Clearing the United States Forests," 24–25.

4. *Historical Statistics of the United States*, 542 (tables L 113, 118, and 119).

The states in the South and South Atlantic regions are Alabama, Arkansas, Florida, Georgia, Louisiana, Mississippi, Oklahoma, and Texas; and North Carolina, South Carolina, and Virginia.

5. The history of Mississippi lumber industry has been extensively studied by Nollie W. Hickman. Any account of the subject has to employ his research, beginning with his dissertation, "History of Forest Industries in the Longleaf Pine Belt of Louisiana and Mississippi." Other useful studies by Hickman include a monograph, *Mississippi Harvest,* and several articles: "Mississippi Lumber Industry from 1840 to 1950"; "Logging and Rafting Timber in South Mississippi"; "Lumber Industry in South Mississippi"; "Mississippi Forests." However, his work is first and foremost concerned with the development of Mississippi pine industry in the central and southern parts of the state, and pays only fleeting attention to hardwood lumbering in the Delta. This is largely the case also with James Fickle's recent *Mississippi Forests and Forestry.*

6. Hickman, "Mississippi Lumber Industry from 1840 to 1950," 132.

7. Bromme, *Mississippi,* 9–10.

8. Hickman, "Mississippi Lumber Industry from 1840 to 1950," 132. In the 1850s, southern lumber production more than doubled.

9. Michael Williams, "Clearing the United States Forests," 18; Clawson, "Forests in the Long Sweep of American History," 1172.

10. Silver, *New Face on the Countryside,* 133–35. See also Schob, "Woodhawks and Cordwood," 124–33. The elimination of forest canopy from the drainage basin of a stream is likely to increase evaporation, runoff, and erosion, which can cause considerable fluctuations in the water level.

11. Basil Hall, *Travels in North America in the Years 1827 and 1828,* 348.

12. *Compendium of the Enumeration of the Inhabitants and Statistics of the United States,* 228. See also Michael Williams, "Products of the Forest," 6–11.

13. Hickman, "Mississippi Lumber Industry from 1840 to 1950," 132–34.

14. Government agencies usually classified cypress with pines and other softwoods. For example, the USDA's *Timber Depletion* discusses cypress in connection with upland pines of the South.

15. Cowdrey, *This Land, This South,* 92–93, 97–98; Herndon, "Forest Products of Colonial Georgia," 135; Silver, *New Face on the Countryside,* 119; Mattoon, *Southern Cypress,* 10–11; Moore, "Cypress Lumber Industry of the Lower Mississippi Valley," 25–47. For early reports on the value of cypress, see, e.g., Baird, *View of the Valley of Mississippi,* 31–32; Ker, *Travels through the Western Interior of the United States,* 33–34.

16. Cortese, "Memphis," 137; Bowman, "Early History and Archaeology of Yazoo County," 430. On Andrew Brown, see the definitive biography by Moore, *Andrew Brown and Cypress Lumbering in the Old Southwest.* The enterprising Scotsman himself published on the subject of cypress lumbering in the lower Mississippi Valley. See Dickeson and Brown, *Report on the Cypress Timber of Mississippi and Louisiana.*

17. Hesse-Wartegg, *Travels on the Lower Mississippi,* 88.

18. W[illia]m I. Hindman to Mr[. Andrew] Brown, December 30, 1866, box 6, folder 1, and February 11, 1867, box 6, folder 2, Brown-Learned Lumber Company Papers.

19. Geo[.] S[.] Irving to R[ufus] F. Learned, September 4, 1882, box 19, folder 3, Brown-Learned Lumber Company Papers.

20. Dr. Charles Mohr's 1880 report upon the forests of the Yazoo-Mississippi Delta, in Sargent, *Report on the Forests of North America,* 536.

21. Mattoon, *Southern Cypress,* 11–13; Bowman, "Early History and Archaeology of Yazoo County," 430. Girdling, or, cutting a ring through the bark of a living tree, will kill the tree and allow it to dry up while still standing.

For example, in August 1866, a Delta logging crew foreman wrote his employer that "if we cut timber . . . we better commens a bou[t] middel of September so as to be in good time an[d] condition [for the float in the spring.]" W[illiam] I. Hindman to Mr[. Andrew] Brown, August 12, 1866, box 6, folder 1, Brown-Learned Lumber Company Papers.

22. Charles Helme to Dear Brother, January 28, [1847], box 1, James W. Helme Papers.

23. Cotton Plantation Account and Record Book, box 1, vol. 1, Panther Burn Plantation Records.

24. Michael Williams, *Americans and Their Forests,* 238; *Timber Depletion,* 13–14.

25. Hickman, "Mississippi Lumber Industry from 1840 to 1950," 133.

26. Hough, *Report upon Forestry,* 482.

27. Michael Williams, *Americans and Their Forests,* 239–40. Cf. Fickle, *The New South and the "New Competition,"* 1–8.

28. Cowdrey, *This Land, This South,* 111–14; Michael Williams, *Americans and Their Forests,* 241–44. Some 5.5 million acres of federal land were sold in the five southeastern public land states (Arkansas, Louisiana, Mississippi, Alabama, and Florida) after the Civil War.

29. Geo. C. Starke to Dear Madam [Martha Rebecca Blanton], January 6, 1881, Cooper transcripts of Blanton-Smith Letters. Cf. Tamlin Avent to Dear

Madam [Mary Greenway Smith], March 9, 1869, file 98, Blanton-Smith Family Letters.

30. Hickman, "Mississippi Lumber Industry from 1840 to 1950," 134; Michael Williams, *Americans and Their Forests,* 239–44, 263–69. In contrast to the purchases of federal lands in the Delta, homestead entries in the pine region were usually made in order to acquire only the timber from the land and seldom resulted in permanent agricultural acreage.

31. Hickman, "Mississippi Lumber Industry from 1840 to 1950," 133–34.

32. Ibid., 134–35.

33. *Manufactures,* 630 (table 1).

34. Ibid.; Hickman, "Mississippi Lumber Industry from 1840 to 1950," 136.

35. Hickman, "Mississippi Lumber Industry from 1840 to 1950," 136; Silver, "Paul Bunyan Comes to Mississippi," 104–12; Michael Williams, *Americans and Their Forests,* 268–69, 280–82.

36. Hickman, "Mississippi Lumber Industry from 1840 to 1950," 136; J. M. McBroom to Walter Sillers, Sr., [1916], Miscellaneous Papers, folder 106, Sillers Family Papers.

37. Hickman, "Mississippi Lumber Industry from 1840 to 1950," 136; Michael Williams, *Americans and Their Forests,* 251–52.

38. Hickman, "Mississippi Lumber Industry from 1840 to 1950," 136.

39. Harper, *Preliminary Report on the Geology and Agriculture of the State of Mississippi;* Hilgard, *Report on the Geology and Agriculture of the State of Mississippi;* Hurt, *Mississippi,* 75–89; Wall, *State of Mississippi.* See also Wailes, *Report on the Agriculture and Geology of Mississippi.*

40. Lowe, "Note on the Flora of Mississippi"; idem, *Plants of Mississippi.* See also his *Mississippi.*

41. Sargent, *Report on the Forests of North America,* 530.

42. Dr. Charles Mohr's 1880 report from the Tensas [Tensaw] River region of Alabama, in Sargent, *Report on the Forests of North America,* 526.

43. Ibid.

44. Dr. Charles Mohr's 1880 report upon the forests of the Yazoo-Mississippi Delta, in Sargent, *Report on the Forests of North America,* 535.

45. Ibid., 535–36.

46. Ibid., 536.

47. Ibid., 534–35.

48. Ibid., 536.

49. Dr. Charles Mohr's 1880 report from the Tensas [Tensaw] River region of Alabama, in Sargent, *Report on the Forests of North America,* 525. Efficient

cutting of cypress stands required highly skilled labor, and experienced northern woodsmen (including Michigan Finns) could be lured to southeastern lumbering sites by good pay. See Trowell and Izlar, "Jackson's Folly," 195.

50. See e.g. Henry E. Wilson to R[ufus] F. Learned, September 7, 1880, box 17, folder 4, Brown-Learned Lumber Company Papers.

51. Harrison, *Levee Districts and Levee Building in Mississippi*, 68.

52. A substantial amount of the correspondence and financial records of the Andrew Brown and Son/Rufus F. Learned Lumber Company has been preserved at the Archives and Special Collections of the University of Mississippi Library in Oxford. Unfortunately, much of the material is severely damaged.

53. Hindman to Brown, February 20, 1866, box 6, folder 1, Brown-Learned Lumber Company Papers.

54. Hindman to Brown, November 12, 1866, box 6, folder 1; February 11, 1867, box 6, folder 2; Brown-Learned Lumber Company Papers.

55. Hindman to Mr[. John A.] Kline, December 8, 1866, box 6, folder 1, Brown-Learned Lumber Company Papers.

56. Hindman to Brown, December 10, 1866, box 6, folder 1, Brown-Learned Lumber Company Papers.

57. Hindman to Brown, December 30, 1866, box 6, folder 1; February 11, 1867, box 6, folder 2; and February 26, 1867, box 6, folder 2, Brown-Learned Lumber Company Papers.

58. Hindman to Brown, March 10, 1867; March 11, 1867; and May 4, 1867, all in box 6, folder 2, Brown-Learned Lumber Company Papers.

59. Hindman to Brown, December 16, 1867, box 6, folder 3; Andrew Brown Journal, January 1861–December 1870, 402, Brown-Learned Lumber Company Papers.

60. On the history of the R. F. Learned Lumber Company during the late nineteenth century, see the dissertation by Crawford, "History of the R. F. Learned Lumber Company."

61. Wilson to Learned, April 15, 1879, box 16, folder 2, Brown-Learned Lumber Company Papers.

62. Wilson to Learned, November 1, 1879, box 16, folder 7, Brown-Learned Lumber Company Papers.

63. Wilson to Learned, January 28, [1881?], box 6, folder 5, Brown-Learned Lumber Company Papers.

64. Wilson to Learned, January 31, 1882, box 19, folder 1, Brown-Learned Lumber Company Papers.

65. Wilson to Learned, February [3?], 1882, box 19, folder 1, Brown-Learned Lumber Company Papers.

66. Wilson to Learned, April 26, 1882, box 19, folder 2, Brown-Learned Lumber Company Papers.

67. Wilson to Learned, May 2, 1882, box 19, folder 2, Brown-Learned Lumber Company Papers.

68. Wilson to Learned, August 30, 1882, box 19, folder 3, Brown-Learned Lumber Company Papers.

69. W[illiam] I. Hindman to Mr[. Andrew] Brown, October 29, 1866, box 6, folder 1, Brown-Learned Lumber Company Papers.

70. Henry Wilson to R[ufus] F. Learned, April 15, 1879, box 16, folder 2; Henry Wilson to R[ufus] F. Learned, November 1, 1879, box 16, folder 7; and Henry Wilson to R[ufus] F. Learned, December 14, 1879, box 16, folder 8, Brown-Learned Lumber Company Papers.

71. Wilson to Leonard [Learned], March 19, 1879, box 16, folder 2, Brown-Learned Lumber Company Papers.

72. Wilson to Learned, September 21, 1879, box 16, folder 5; and November 1, 1879, box 16, folder 7, Brown-Learned Lumber Company Papers.

73. Wilson to Learned, August 30, 1882, box 19, folder 3, Brown-Learned Lumber Company Papers.

74. Mattoon, *Southern Cypress,* 11–13; Michael Williams, *Americans and Their Forests,* 248–50. See also Mancil, "Pullboat Logging."

75. Mattoon, *Southern Cypress,* 13–19.

76. Ibid., 51–53.

77. Michael Williams, "Clearing the United States Forests," 16, 24–25.

78. Dr. Charles Mohr's 1880 report upon the forests of the Yazoo-Mississippi Delta, in Sargent, *Report on the Forests of North America,* 535.

79. "Statistics of Forestry," 279–81.

80. Dr. Charles Mohr's 1880 report upon the forests of the Yazoo-Mississippi Delta, in Sargent, *Report on the Forests of North America,* 535.

81. *Industrial and Manufacturing Schedules of Mississippi.*

82. Sargent, *Report on the Forests of North America,* 531.

83. Egleston, *Report on Forestry,* 220–21.

84. Dr. Charles Mohr's 1880 report upon the forests of the Yazoo-Mississippi Delta, in Sargent, *Report on the Forests of North America,* 535.

85. Ibid., 536.

86. Clement to Bobet Brothers, May 25, 1873, folder 13; and [August] 28, 1873, folder 18, Bobet Brothers Papers.

87. Dyche to Bobet Bros[.], August 29, 1873, folder 18; Paul to Mr BoBet and Brothers, May 1, 1873, folder 12; Dawson to Mr B, April 11, 1873, folder 10, Bobet Brothers Papers.

88. Cortese, "Memphis," 137–38. There had, however, been a significant wood manufacturing industry in the city already before the Civil War. See "Southwestern Cities," 607.

89. Egleston, "Report of Chief of Division of Forestry," 192.

90. Gilmore & Co. to Abbey, February 22, 1882, box 1, folder 13; Thomas Kane & Co. to Abbey, December 24, 1881, Abbey Family Papers. See also letters from Edwin W. Adams & Co., October 25, 1881, box 1, folder 12; and D. E. Cheney, October 26, 1881, box 1, folder 12. Abbey's correspondence with the *Southern Lumberman* is located in box 1, folder 15. Thomas & Harris to Abbey, January 6, 1887, box 1, folder 21, offered Long Beach Prime Select Oysters from Harrison County at four dollars per thousand.

91. Hurt, *Mississippi*, 80–83, 85.

92. Ker, *Travels through the Western Interior of the United States*, 43.

93. Schultz, *Travels on an Inland Voyage*, 124.

94. Nancy Bieller to E. W. Brazleton, July 20, 1837, 2E551, Nancy Bieller Letter.

95. Sargent, *Report on the Forests of North America*, 537–38.

96. Faulkner, "The Bear," *Go Down, Moses*, 321. This short story is, of course, a classic fictional account of the human takeover of the Delta. The young Isaac McCaslin conveys the hunter's feelings of loss on the retreat of the wilderness on 318–21. On the relationship between the McCaslin character and the natural world, see also Buell, "Faulkner and the Claims of the Natural World," 5–12; Simpson, "Ike McCaslin and the Second Fall of Man," 202–9; Utley, "Pride and Humility," 167–87.

97. [Tompkins], "Mississippi Alluvial Lands," 499.

98. Irving to Learned, September 4, 1882, box 19, folder 3, Brown-Learned Lumber Company Papers.

99. Ibid. With the expansion of hardwood lumbering into new areas, the loggers had to face new occupational hazards. Log jams became common along the upper tributaries, and breaking a log jam by explosives or by removing individual logs was an extremely dangerous task. See, e.g., LeRoy Barry Allen Manuscript, 30.

100. Irving to Learned, September 4, 1882, box 19, folder 3, Brown-Learned Lumber Company Papers.

101. "Greenwood Has—Greenwood Wants."

102. [Tompkins], "Mississippi Alluvial Lands," 515.

103. LeRoy Barry Allen Manuscript, 23–30.

104. Hickman, "Mississippi Lumber Industry from 1840 to 1950," 136; Silver, "Paul Bunyan Comes to Mississippi," 96–98.

105. [Tompkins], "Mississippi Alluvial Lands," 485; Heavrin, *Boxes, Baskets and Boards,* 53, 63–65, 118–19.

106. Hamilton, *Trials of the Earth,* 54, 60.

107. Ibid., 62–66.

108. Ibid., 69–80, 83–85, 89–91.

109. Ibid., 97–98, 100–101.

110. *Timber Depletion,* 14, 60–61; *Hardwood Lumber and Farming Industries of Mississippi,* 3–4, 53. See also Dunston, *Preliminary Examination of the Forest Conditions of Mississippi,* 31–33; *Lumber Industry,* 73–77, 195–96. Cf. See also "State Shares Pride of Delta Timber," 13.

Reminiscences by southern hardwood operators, such as the incredibly subtitled Hanlon's *Delta Harvest: An Authentic Story of a Hardwood Harvest Interwoven with Intriguing Romance,* can make amusing reading and offer interesting insights to the business, but provide little information on the extent of existing stands and the amount cut. Cf. Kellogg, *Kellogg Story.*

111. *Timber Depletion,* 26.

112. Anonymous editor of a Memphis newspaper in 1899, as quoted in Cortese, "Memphis," 138.

113. Article in *The Commercial Appeal* (March 1905), as quoted in Cortese, "Memphis," 138.

114. Cortese, "Memphis," 138. See also Pendleton, "Short History of the National Hardwood Lumber Association," 127–30.

115. *Lumber Industry,* 196.

116. Trager to Allen, November 8, [1908]; and [December, 1908?], box 1, folder 25, Allen and Family Papers.

117. *Timber Depletion,* 14.

118. Ibid., 26–27. Between 1900 and 1920, the proportion of the hardwood lumber cut in the lower Mississippi Valley increased from 14 to 25 percent of the total U.S. hardwood cut.

119. Ibid., 27.

120. Eldredge, *Preliminary Report on the Forest Survey of the Bottom-Land Hardwood Unit in Mississippi,* 2. Cf. Ashe, "Mississippi Valley as a Permanent Hardwood Region," 61–68; Lentz, "Report of the Mississippi Valley Hardwood Investigations," 7–15; *Second Biennial Report of the Mississippi State Forestry Commission.*

121. Shipley, "Story of the Chicago Mill and Lumber Company," 61–66.

122. Ibid., 63–66.

123. Calhoun to Paepcke Leicht Lumber Company (hereafter PLLC), November

1, 1915 (copy); Soper to PLLC, November 2, 1915 (copy), both in box 1, folder 2, Chicago Mill and Lumber Company Papers.

124. Soper to McClelland, November 26, 1915; [Berry] to McMahon, December 7, 1915 (copy); McMahon to PLLC, January 1, 1916; all in box 1, folder 2, Chicago Mill and Lumber Company Papers.

125. Paepcke to Berry, May 14, 1919; [Berry] to Paepcke, May 16, 1919 (copy); both in box 1, folder 2, Chicago Mill and Lumber Company Papers.

126. Paepcke to Berry, May 23, 1919; [Berry] to Paepcke, May 26, 1919 (copy); both in box 1, folder 2, Chicago Mill and Lumber Company Papers.

127. Calhoun Timber Estimate, Panther Burn Plantation, May 1919; Calhoun to Berry, June 3, 1919; both in box 1, folder 2; map relocated to box 22, folder 2, Chicago Mill and Lumber Company Papers.

128. Deed of Sale, October 4, 1917 (copy); Deed of Sale, March 27, 1918 (copy); Hall to PLLC, April 24, 1919 (copy); all in box 1, folder 5; Map made by J. N. Hall, April 24, 1919, box 22, folder 5, Chicago Mill and Lumber Company Papers.

129. Timber contract 159 between PLLC and Smith, August 27, 1919; Berry to Accounting Department, August 29 and 30, 1919 (copies); Berry to Accounting Department, September 24, 1919 (copy); L. E. Smith account; bankruptcy notice issued by H. C. McCabee, November 29, 1920; all in box 5, folder 72, Chicago Mill and Lumber Company Papers.

130. [Berry] to Paepcke, June 4, 1919 (copy), box 1, folder 2, Chicago Mill and Lumber Company Papers.

131. Satterfield to Berry, June 7, 1919; [Berry] to PLLC, June 9, 1919 (copy); [Berry] to PLLC, June 27, 1919 (copy); all in box 1, folder 2, Chicago Mill and Lumber Company Papers.

132. Report of accident, June 20, 1919; Testimony by Cha[rle]s W. Ford, June 19, 1919; both in box 3, folder 17, Chicago Mill and Lumber Company Papers.

133. [Berry] to Satterfield, June 21, 1919 (copy); [Berry] to Insurance Department, June 21, 1919 (copy); both in box 3, folder 17, Chicago Mill and Lumber Company Papers.

134. Satterfield to PLLC, June 30, 1919, box 1, folder 2, Chicago Mill and Lumber Company Papers.

135. [Berry] to Satterfield, July 2, 1919 (copy), box 1, folder 2, Chicago Mill and Lumber Company Papers.

136. [Berry] to PLLC, July 15, 1919 (copy); [Berry] to PLLC, July 17, 1919 (copy); both in box 1, folder 2, Chicago Mill and Lumber Company Papers.

137. [Berry] to PLLC, July 30, 1919 (copy), box 1, folder 2, Chicago Mill and Lumber Company Papers.

138. [Soper] to PLLC, July 21, 1915 (copy); "Estimate of Timber on Land East of Hampton, Miss.;" [Berry] to PLLC, September 8, 1921 (copy); C. C. and J. T. Wellford to PLLC, February 8, 1922; all in box 1, folder 4, Chicago Mill and Lumber Company Papers.

139. Two letters by [Berry] to PLLC, August 4, 1916 (copies), both in box 1, folder 4, Chicago Mill and Lumber Company Papers.

140. [Berry] to PLLC, September 26, 1916 (copy); [Berry] to Gary and [Berry] to PLLC, October 12, 1916 (copies); A. O. Ratcliff to Berry, November 4, 1916; all in box 1, folder 4, Chicago Mill and Lumber Company Papers.

141. Wellford & Sons to PLLC, September 27, 1919; [Berry] to Wellford & Sons, September 29, 1919 (copy); [Berry] to PLLC, September 8, 1921 (copy), all in box 1, folder 4, Chicago Mill and Lumber Company Papers.

142. C. C. and J. T. Wellford to PLLC, February 8, 1922; PLLC to C. C. and J. T. Wellford, February 13, 1922 (copy); both in box 1, folder 4, Chicago Mill and Lumber Company Papers.

143. Bill of sale, January 10, 1913 (copy), box 2, folder 10, Chicago Mill and Lumber Company Papers.

144. [Berry] to PLLC, June 7, 1920 (copy); "Sheldon-Tract" plat prepared by G. W. Calhoun, 1921; both in box 2, folder 10, Chicago Mill and Lumber Company Papers.

145. [Berry] to PLLC, June 7, 1920 (copy); [Berry] to Satterfield, March 24, 1921 (copy); Satterfield to Sheldon, May 27, 1922 (copy); all in box 2, folder 10, Chicago Mill and Lumber Company Papers.

146. Satterfield to Sheldon, May 27, 1922 (copy), box 2, folder 10, Chicago Mill and Lumber Company Papers.

147. Telegram from PLLC, Chicago, to PLLC, Greenville, June 3; [Berry] to Accounting Department, July 2, 1918 (copy); [Berry] to PLLC, July 29, 1918 (copy); all in box 3, folder 14, Chicago Mill and Lumber Company Papers.

148. [Berry] to PLLC, July 29, 1918 (copy); [Berry] to PLLC, August 2, 1918 (copy), both in box 3, folder 14, Chicago Mill and Lumber Company Papers.

149. [Berry] to PLLC, August 7, 1918 (copy); Gray by McPherson, General Manager, to PLLC, August 16, 1918; [Berry] to PLLC, August 19, 1918 (copy); all in box 3, folder 14, Chicago Mill and Lumber Company Papers.

150. [Berry] to PLLC, November 1, 1920 (copy), box 3, folder 14, Chicago Mill and Lumber Company Papers.

151. Logging contract between PLLC and Johnson, October 22, 1909, box 5, folder 40, Chicago Mill and Lumber Company Papers, 1903–1945. Boxes 4 and 5 of the collection contain contracted logging operations of the Greenville office.

152. Shipley, "Story of the Chicago Mill and Lumber Company," 66–69.

153. *Timber Depletion*, 27.

154. Quoted in Hough, *Report on Forestry*, 144–45.

155. Egleston, *Report on Forestry*, 193.

156. See, e.g., a report from Friar's Point, Coahoma County, in Egleston, *Report on Forestry*, 219.

157. Featherstonhaugh, *Excursion Through the Slave States*, vol. 1, 343–345. See also 352–55.

158. Starr, "American Forests," 210–34.

159. Egleston, "Report of Chief of Division of Forestry," 191.

160. Hurt, *Mississippi*, 75.

161. On the history of southern forestry in general, see, e.g., Linnartz, "Forestry," 467–68. On Mississippi, consult Fickle, *Mississippi Forests and Forestry*, 120–56.

162. Hickman, "Mississippi Lumber Industry from 1840 to 1950," 137.

163. Kerr, "From Timber to Famine," 140; Read, "Forestry's Progress in Louisiana," 143–44; Linnartz, "Forestry," 467.

164. The Southern Forest Experiment Station was founded in 1921, with Louisiana's state forester, R. D. Forbes, as its director. The highly conservationist attitude towards forest utilization adopted by the federal government, faced with decreasing production and rapidly rising prices of lumber in the early 1920s, is evident in publications such as Greeley et al., "Timber," 83–180; Reynolds and Pierson, *Lumber Cut of the United States*, 1–23.

165. Prunty, "Pulp and Paper Industry," 1013–14. See also Fickle, *Mississippi Forests and Forestry*, 139–42.

166. Cortese, "Memphis," 138.

167. Kerr, "From Timber to Famine," 142.

168. Ibid. See also Putnam and Bull, *Trees of the Bottomlands of the Mississippi River Delta Region*; Eldredge, *Preliminary Report on the Forest Survey of the Bottom-Land Hardwood Unit in Mississippi*; Eldredge, *Southern Forest Survey*; Stover, *Forest Resources of the Delta Section of Mississippi*; Duerr, *Basic Data on Forest Area and Timber Volumes from the Southern Forest Survey*.

## Chapter 7. A Transformed Landscape

1. Michael Williams, "Clearing the United States Forests," 25.

2. Michael Williams, *Americans and Their Forests*, 238.

3. *Lumber Industry*, 74–75.

4. *Timber Depletion*, 14.

5. Eldredge, *Preliminary Report on the Forest Survey of the Bottom-Land*

*Hardwood Unit in Mississippi*, 1–2; Stover, *Forest Resources of the Delta Section of Mississippi*, 1–2. The strips of alluvial bottomland along the Mississippi south of Vicksburg supported similar vegetation but were not included in the unit, presumably for their small size. Eldredge reported the size of the survey area as 4,420,400 acres, while Stover omitted the 9,262-acre Pittman Island in Issaquena County from the unit and used the figure of 4,411,100 acres for tabulation purposes.

Sample plots in the Delta were taken at intervals of 660 feet on parallel compass lines run ten miles apart. The timber stand in the sample plots was counted and divided into diameter classes and species groups. In addition, growth rates of individual tree species were obtained from randomly selected sample trees.

6. Eldredge, *Preliminary Report on the Forest Survey of the Bottom-Land Hardwood Unit in Mississippi*, 2–3; Stover, *Forest Resources of the Delta Section of Mississippi*, 1–2.

7. Stover, *Forest Resources of the Delta Section of Mississippi*, 6.

8. Faulkner, "Delta Autumn," *Go Down, Moses*, 343.

9. Lehrbas, "Fine Stand of Hardwoods Found on Jackson Point," 17.

10. Eldredge, *Preliminary Report on the Forest Survey of the Bottom-Land Hardwood Unit in Mississippi*, 4–5.

11. Eldredge, *Preliminary Report on the Forest Survey of the Bottom-Land Hardwood Unit in Mississippi*, 4–7; Stover, *Forest Resources of the Delta Section of Mississippi*, 6–9.

12. On faunal extinctions in the South, see Echternacht and Harris, "Fauna and Wildlife of the Southeastern United States," 112–14. On the two bird species mentioned, see also Greenway, *Extinct and Vanishing Birds of the World*, 322–27, 357–60; Saikku, "Extinction of the Carolina Parakeet"; idem, "'Home in the Big Forest.'" Disappeared ungulates of the Delta include the wapiti (elk) and the bison. In addition to the bobcat and the cougar, the Delta bottomlands may recently have supported felines such as the ocelot (*Felis pardalis*) and even the jaguar (*Felis onca*). The ocelot as a species was described from a type specimen obtained from Arkansas, across the Mississippi from the Delta.

13. On island biogeography, see MacArthur and Wilson, *Theory of Island Biogeography*; Burdick et al., "Faunal Changes and Bottomland Hardwood Forest Loss in the Tensas Watershed," 283, 287.

14. Burdick et al., "Faunal Changes and Bottomland Hardwood Forest Loss in the Tensas Watershed," 282; Cairns et al., "Impacts Associated with Southeastern Bottomland Hardwood Forest Ecosystems," 303–26; Jahn, "Values of Riparian Habitats to Natural Ecosystems," 157–58; Odum, "Ecological Importance of

the Riparian Zone," 2–4; Wharton et al., "Fauna of Bottomland Hardwoods in Southeastern United States," 127.

15. Eckleberry, "Search for the Rare Ivorybill," 198–99. On the history of the Ivory-billed Woodpecker in the Singer Tract, see also Tanner, *Ivory-billed Woodpecker,* 31–39; idem, "Present Status of the Ivory-billed Woodpecker," 57; Cokinos, *Hope Is the Thing with Feathers,* 61–117.

16. Eckleberry, "Search for the Rare Ivorybill," 199–200; Cokinos, *Hope Is the Thing with Feathers,* 98–102.

17. Cokinos, *Hope Is the Thing with Feathers,* 104–5; Eckleberry, "Search for the Rare Ivorybill," 207.

18. M. G. Vaiden to James Bond in 1963, as quoted in Jackson, "History of Ivory-billed Woodpeckers in Mississippi," 6.

19. Jackson, "History of Ivory-billed Woodpeckers in Mississippi," 9. See also http://www.birds.cornell.edu/publications/ birdscope/ Summer2002/ivory_bill_ab sent.html (accessed February 16, 2004).

20. "Extinction" in the following discussion is defined as the disappearance since the European conquest of a species that was established in the pre-Columbian Yazoo-Mississippi Delta. The following discussion on the process of extinction is based on Dawson et al., "Report of the Scientific Advisory Panel on the Spotted Owl," 212–20; Gilpin and Soulé, "Minimum Viable Populations," 19–34; Järvinen and Miettinen, *Sammuuko suuri suku?,* 87–94; Simberloff, "Proximate Causes of Extinction," 262–66; Edward O. Wilson, *Diversity of Life,* 203–30.

21. King, "Ecological Basis of Extinction in Birds," 905; Ziswiler, *Extinct and Vanishing Animals,* 55–60.

22. Sometimes the ultimate cause of a species' extinction is the proximate cause as well: in 1894 the whole population of the flightless Stephen Island Wren (*Xenicus lyelli*) was extirpated by a cat belonging to the keeper of the island's lighthouse. See Greenway, *Extinct and Vanishing Birds of the World,* 367–68.

23. See, e.g., Jacobs Manuscript, 109.

24. Audubon, *Delineations of American Scenery and Character,* 110.

25. Hamilton, *Trials of the Earth,* 138.

26. See, e.g., William Mason Worthington to Dear brother [Albert D. Worthington], November 10, 1857, box 1, folder 4, Worthington Family Papers. On the cultural history of southern and Mississippi hunting, see e.g. Bruce, "Hunting"; Prewitt, "Best of All Breathing." Cf. Seay, "Southern Outdoors."

27. Humphreys, "Autobiography," 241.

28. For example, see William Mason Worthington to Dear brother [Albert D.

Worthington], November 10, 1857. See also Amanda Worthington to Dear Son [Albert D. Worthington], October 27, [1857]; and November 3, 1857. All three in box 1, folder 4, Worthington Family Papers.

29. Chambers Diaries, no. 3 (January 1, 1858–December 31, 1858), September 22, 1858.

30. Worthington to Dear brother [Albert D. Worthington], November 10, 1857, box 1, folder 4, Worthington Family Papers.

31. Jacobs Manuscript, 107–8.

32. De Puy van Buren, *Jottings of a Year's Sojourn in the South*, 87.

33. Wade Hampton II to My Dear Mary [Fisher Hampton], April 22, 1855; November 17, 1855, in Cauthen, *Family Letters of the Three Wade Hamptons*, 39–40.

34. Wade Hampton II to My Dear Mary [Fisher Hampton], November 7, 1857; Wade Hampton III to Mary Fisher Hampton, May 22, 1857; November 2, 1857; and November 8, 1857; in Cauthen, *Family Letters of the Three Wade Hamptons*, 48, 52–53.

35. Wade Hampton III to My Dear Fisher [Mary Fisher Hampton], May 14, 1866; Wade Hampton III to my Dear Anna [Mrs. Thomas L. Preston], May 12, 1875, in Cauthen, *Family Letters of the Three Wade Hamptons*, 120, 151.

36. For a record of a wolf in Washington County, see William Mason Worthington to Dear brother [Albert D. Worthington], November 10, 1857, box 1, folder 4, Worthington Family Papers.

37. Jacobs Manuscript, 109–10.

38. Glenbar Plantation ledger entry, October 1, 1893, as quoted in Buchanan, *Holt Collier*, 149.

39. Buchanan, *Holt Collier*, is the only full-length biography. For a detailed description of the 1902 Delta hunt, see 151–83. Incidentally, Collier's old master, Colonel Howell Hinds, had been killed in 1868 by Dr. Orville M. Blanton, the Delta planter and old hunting mate of Wade Hampton III. See 125–31.

40. For the history of hunting on the Carrier Lumber and Manufacturing Company lands in the northern Delta, see Silver, "Paul Bunyan Comes to Mississippi," 112–19.

41. Audubon, *Ornithological Biography*, 343.

42. Featherstonhaugh, *Excursion Through the Slave States*, vol. 2, 11–12.

43. Schultz, *Travels on an Inland Voyage*, 118. The hunting party was below the St. Francis, on the Arkansas side of the Mississippi.

44. Buchanan, *Holt Collier*, 36.

45. De Puy van Buren, *Jottings of a Year's Sojourn in the South*, 197–98.

46. See also Askins, *Restoring North American Birds,* 88–96; Harris, "Faunal Significance of Fragmentation of Southeastern Bottomland Forests," 126–32; Mader, Aust, and Lea, "Changes in Functional Values of a Forested Wetland Following Timber Harvesting Practices," 149–53.

47. Marschner, *Land Use and Its Patterns in the United States,* 83. For an excellent overview of the process, see Aiken, *Cotton Plantation South Since the Civil War,* 63–96.

48. Kirby, *Rural Worlds Lost,* 63; Dong, "From Postbellum Plantation to Modern Agribusiness," 164–65; Oshinsky, *"Worse Than Slavery,"* 224. On the New Deal and the Delta planters, see also Cobb, *Most Southern Place on Earth,* 184–208.

49. Aiken, *Cotton Plantation South Since the Civil War,* 97–132; Kirby, *Rural Worlds Lost,* 72, 338–40; Kirby, "Plantations," 27–28.

Population of the ten core counties in 2000 was 245,952, or, almost 8,000 less than in 1910. Cf. *Population, Vol. 1,* 583–84 (table 3) with http://www.census .gov/census2000/states/ms.html (accessed February 16, 2004). See also table 1.

50. Sternitzke, "Impact of Changing Land Use on Delta Hardwood Forests," 25–27; Daniel, "Cotton Culture," 34–35; Shipley, "Story of the Chicago Mill and Lumber Company," 73–74.

51. MacDonald, Frayer, and Clauser, *Documentation, Chronology, and Future Projections of Bottomland Hardwood Habitat Losses in the Lower Mississippi Alluvial Plain,* 23.

52. Siniard, "Soybean," 1154–55; Rasmussen, "Soybeans," 47; Dong, "From Postbellum Plantation to Modern Agribusiness," 265–66. Soybean yields remained low until special varieties for narrow ranges of latitude were developed. Planting of the crop for beans (i.e., oil) gained momentum after 1915. The annual return for soybeans has been manifold higher than that the landowner receives from natural or planted stands of hardwoods, or even from cultivation of many crop varieties. See R. Eugene Turner, Stephen W. Forsythe, and Nancy J. Craig, "Bottomland Hardwood Forest Land Resources of the Southeastern United States," 24.

53. Bell and Wilson, "Kudzu," 383–84.

54. Shipley, "Story of the Chicago Mill and Lumber Company," 70–71; Hickman, "Mississippi Forests," 224–26; Stover, *Forest Resources of the Delta Section of Mississippi,* 6–7, 18.

55. Fickle, "Introduction," 9.

56. Harrison, *Alluvial Empire,* 307–9, 313–15; Fickle, *Mississippi Forests and*

*Forestry,* 161–63. See also *Mississippi Forests,* 6–8; Sternitzke and Putnam, *Forests of the Mississippi Delta,* 1–2.

57. For the USFWS position, consult "Planning-Aid Report on the Yazoo Backwater Area Project," available at http://yazoobackwater.fws.gov/ (accessed February 16, 2004).

58. Thorne and Curry, *Cultural Resources Survey,* 17, 33.

59. Harrison and Mooney, *Flood Control and Water Management in the Yazoo-Mississippi Delta,* 21; Cross and Wales, *Atlas of Mississippi,* 86–88.

60. Cross and Wales, *Atlas of Mississippi,* 106–8, 120–22; Harrison, *Alluvial Empire,* 325–26; Harrison and Mooney, *Flood Control and Water Management in the Yazoo-Mississippi Delta,* 21. See also Robinson, "Reclamation and Irrigation," 356.

61. Hickman, "Mississippi Forests," 227–29.

62. Mattoon, *Southern Cypress,* 23; Sharitz and Mitsch, "Southern Floodplain Forests," 328; William H. Martin and Stephen C. Boyce, "Introduction," 11, 13. In the states of Virginia, North Carolina, South Carolina, Georgia, and Florida, only about 5 percent of the remaining bottomland hardwood forests are in public ownership. See Langdon et al., "Extent, Condition, Management, and Research Needs of Bottomland Hardwood-Cypress Forests in the Southeastern United States," 72 (table 3.1).

"Loss of Missouri's Lowland Hardwood Ecosystem" by Korte and Frederickson convincingly illustrates the demise of the bottomland hardwood forest in southeastern Missouri. They estimate that these forests covered some 2.4 million acres of the region in the late eighteenth century. By 1870, agricultural development had reduced the forests to 2.1 million acres. Lumbering then added to the agricultural clearing, and by 1920 only 1.3 million acres of bottomland forest remained. Since then, loss of forested habitat has accelerated as a result of government assisted drainage, agricultural mechanization, and increased soybean production. Consequently only 98,000 acres of bottomland hardwood forest prevailed by 1975, and much of this land was fragmented into blocks of less than 1,000 acres.

63. Shands and Healy, *Lands Nobody Wanted,* 6 (table 1).

64. Fickle, *Mississippi Forests and Forestry,* 151–56; Cross and Wales, *Atlas of Mississippi,* 116–18; *Endangered Species of Mississippi.* On the Delta NF, see also http://www.southernregion.fs.fed.us/mississippi/delta/ (accessed February 16, 2004).

65. William H. Martin and Stephen C. Boyce, "Introduction," 13. See also

Ernst and Brown, "Conserving Endangered Species on Southern Forested Wetlands," 135–45.

66. The following discussion on the NWRs is based on the individual refuge fact sheets, available at http://southeast.fws.gov/refuges/index.html#mississippi/ (accessed February 16, 2004).

67. On the lands managed by the Mississippi Department of Wildlife, Fisheries, and Parks, see http://www.mdwfp.com/ (accessed February 16, 2004). On the Anderson-Tully Company lands, see Heavrin, *Boxes, Baskets, and Boards*, 119.

On the bird species mentioned, consult "Least Tern" and "Wood Stork" in *Endangered Species of Mississippi*. Among other Delta species in this book considered threatened or endangered by the U.S. Fish and Wildlife Service are the pallid sturgeon (*Scaphirhynchus albus*) and the pondberry (*Lindera melissifolia*).

68. The following account of ecological revolutions is based on Merchant, "Theoretical Structure of Ecological Revolutions," 265–74. For a fuller treatise on her theory, see her *Ecological Revolutions*.

69. Harrison and Mooney, *Flood Control and Water Management in the Yazoo-Mississippi Delta*, 15.

70. Belt, "1973 Flood and Man's Constriction of the Mississippi River," 684; Barry, *Rising Tide*, 423–26.

71. Faulkner, *Big Woods*, 166.

72. Faulkner, *Big Woods*, 201.

73. Faulkner, "Delta Autumn," *Go Down, Moses*, 349, 353–54.

# Bibliography

## Primary Sources

UNPUBLISHED

Abbey, Richard and Family. Papers, 1843–1891. Special Collections, Archives and Library Division, Mississippi Department of Archives and History.

Alcorn, James L. and Family. Papers, 1839–1906. Special Collections, Archives and Library Division, Mississippi Department of Archives and History.

Allen, James and Family. Papers, 1840–1960. Special Collections, Archives and Library Division, Mississippi Department of Archives and History.

Allen, LeRoy Barry. Photocopied manuscript, "Autobiography of LeRoy Barry Allen," 1966. Box 6, Mississippi Authors Small Manuscripts, Archives and Special Collections, John Davis Williams Library, University of Mississippi.

Ball, Henry Waring. Papers, 1884–1911; 1913–34; n.d. Special Collections, Archives and Library Division, Mississippi Department of Archives and History.

Bieller, Nancy. Letter, 1837. Natchez Trace Collection, Eugene C. Barker Texas History Center, Center for American History, University of Texas at Austin.

Blanton-Smith Family. Letters, 1844–1927. Archives and Special Collections, John Davis Williams Library, University of Mississippi.

Bobet Brothers. Papers, 1865–1873. Special Collections, Manuscripts Department, Howard-Tilton Memorial Library, Tulane University.

Brown, Andrew and Son/Rufus F. Learned Lumber Company. Papers, 1833–1958. Archives and Special Collections, John Davis Williams Library, University of Mississippi.

Chambers, Rowland. Diaries, 1849–1863. Louisiana and Lower Mississippi Valley Collections, Hill Memorial Library, Louisiana State University.

Chicago Mill and Lumber Company. Papers, 1903–1945. Special Collections, Archives and Library Division, Mississippi Department of Archives and History.

Cooper, Georgie Blanton Finlay. Blanton-Smith Letters, 1844–1927. Photocopies of manuscript letters with typed transcriptions. [La Habra, Calif.: Georgie Blanton Finlay Cooper, 1986.] Special Collections, Archives and Library Division, Mississippi Department of Archives and History.

Davis, Joseph E. Papers, 1865–1870. Archives and Special Collections, John Davis Williams Library, University of Mississippi.

Helme, James W. Papers, 1844–1938. Michigan Historical Collections, Bentley Historical Library, University of Michigan.

Jacobs, Annie E. Manuscript, "The Master of Doro Plantation—An Epic of the Old South," n.d. Special Collections, Archives and Library Division, Mississippi Department of Archives and History.

Kiger Family. Papers, 1841–1885. Natchez Trace Collection, Eugene C. Barker Texas History Center, Center for American History, University of Texas at Austin.

Natchez Trace Steamboat Collection. Correspondence, 1836–1907. Natchez Trace Collection, Eugene C. Barker Texas History Center, Center for American History, University of Texas at Austin.

Panther Burn Plantation. Records, 1859–1862; 1864–1865; 1879; 1881–1883. Special Collections, Archives and Library Division, Mississippi Department of Archives and History.

Polk, Alexander Hamilton. Papers. Special Collections, Manuscripts Department, Howard-Tilton Memorial Library, Tulane University.

Sillers Family. Papers. The Walter Sillers, Sr., Collection. University Archives, Roberts Memorial Library, Delta State University.

Taylor, Calvin and Family. Papers, 1822–1913. Louisiana and Lower Mississippi Valley Collections, Hill Memorial Library, Louisiana State University.

Wilkinson, Micajah. Papers, 1853–1935. The Merrit M. Shilg Memorial Collection, Louisiana and Lower Mississippi Valley Collections, Hill Memorial Library, Louisiana State University.

Worthington Family. Papers, 1820–1878, n.d. Typewritten copies of letters and diaries made in 1938 by Eunice J. Stockwell. Special Collections, Archives and Library Division, Mississippi Department of Archives and History.

PUBLISHED

Affleck, Thomas. "The Duties of an Overseer." In *Cotton Plantation Record and Account Book, No. 3. Suitable for a Force of 40 Hands, or Under.* 8th ed. Vicksburg, Miss.: R. H. Crump and Sons, 1849.

Alcorn, James Lusk. "Letters of James Lusk Alcorn." Edited by Percy L. Rainwater. *Journal of Southern History* 3 (May 1937): 196–209.

Ashe, W. W. "The Mississippi Valley as a Permanent Hardwood Region: The Use of Its Submarginal Lands." *Southern Lumberman* 145 (December 15, 1932): 61–68.

Audubon, John James. *Audubon's America: The Narratives and Experiences of John James Audubon.* Edited by Donald Culross Peattie. Boston: Houghton Mifflin, 1940.

———. *Delineations of American Scenery and Character.* New York: G. A. Baker, 1926.

———. *Ornithological Biography, Or an Account of the Habits of the Birds of the United States.* Vol. 1. Philadelphia: E. L. Carey and A. Hart, 1832.

Baily, Francis. *Journal of a Tour in Unsettled Parts of North America in 1796 & 1797.* 1856. Reprint, edited by Jack D. L. Holmes, Travels on the Western Waters, Carbondale: Southern Illinois University Press, 1969.

Baird, Robert. *View of the Valley of Mississippi: or the Emigrant's and Traveller's Guide to the West; Containing a General Description of that Entire Country; and Also, Notices of the Soil, Productions, Rivers, and Other Channels of Intercourse and Trade: and Likewise of the Cities and Towns, Progress of Education, &c., of Each State and Territory.* Philadelphia: H. S. Tanner, 1832.

Banks, Charles. *Negro Town and Colony, Mound Bayou, Bolivar County, Mississippi: Opportunities Open to Negro Farmers and Settlers.* Mound Bayou, Miss.: Demonstrator Print, n.d.

Boone, W. W. "Bolivar County Drainage." In *Imperial Bolivar,* edited by William F. Gray. Cleveland, Miss.: Bolivar Commercial, 1923.

Bowman, Robert. "Early History and Archaeology of Yazoo County." *Publications of the Mississippi Historical Society* 8 (1904): 427–41.

Bromme, Traugott. *Mississippi: A Geographic-Statistic-Topographic Sketch for Immigrants and Friends of Geography and Ethnology.* 1837. Translated by Charles F. Heartman. Reprint, Hattiesburg, Miss.: Book Farm, 1942.

*The Call of the Alluvial Empire: Containing Authentic Information about the Alluvial Region of the Lower Mississippi Valley, Particularly the States of Arkansas, Tennessee, Mississippi, and Louisiana.* Memphis, Tenn.: Southern Alluvial Land Association, 1919.

Catesby, Marcus. *Catesby's Birds of Colonial America.* Edited by Alan Feduccia. Chapel Hill: University of North Carolina Press, 1985.

Cauthen, Charles E., ed. *Family Letters of the Three Wade Hamptons, 1782–1901.* South Caroliniana, no. 4. Columbia: University of South Carolina Press, 1953.

Cohn, David L. *Where I Was Born and Raised.* Boston: Houghton Mifflin, 1948.

Collot, Victor. *A Journey in North America, Containing a Survey of the Countries Watered by the Mississipi, Ohio, Missouri, and Other Affluing Rivers; With Exact Observations on the Course and Soundings of These Rivers; And on the Towns, Villages, Hamlets and Farms of That Part of the New-World; Followed by Philosophical, Political, Military and Commercial Remarks and by a Projected Line of Frontiers and General Limits.* Vol. 2. Paris: Arthur Bertrand, 1826.

Cuming, F. *Sketches of a Tour to the Western Country Through the States of Ohio and Kentucky: A Voyage Down the Ohio and Mississippi Rivers, and a Trip through the Mississippi Territory, and Part of West Florida, Commenced at Philadelphia in the Winter of 1807, and Concluded in 1809.* Reprinted in *Early Western Travels, 1748–1846: A Series of Annotated Reprints of Some of the Best and Rarest Contemporary Volumes of Travel, Descriptions of the Aborigines and Social and Economic Conditions of the Middle and Far West, during the Period of Early American Settlement,* edited by Reuben Gold Thwaites. Vol. 4. Cleveland, Ohio: Arthur H. Clark, 1904.

Dabney, A. L. "A Brief History of the Operations of the Yazoo-Mississippi Delta Board-Organization." In *Riparian Lands of the Mississippi River, Past—Present—Prospective,* edited by Frank H. Tompkins. [Chicago: A. L. Swift, 1901.]

De Bow, J. D. B. "Agricultural Survey of Mississippi." *De Bow's Review* 23 (December 1857): 644–50.

De Puy van Buren, A. *Jottings of a Year's Sojourn in the South; Or First Impres-*

sions of the Country and Its People; With a Glimpse at School-Teaching in That Southern Land, And Reminiscenes of Distinguished Men. Battle Creek, Mich.: Review and Herald, 1859.

Dickeson, Montroville W., and Andrew Brown. *Report on the Cypress Timber of Mississippi and Louisiana*. Philadelphia, Pa.: J. H. Jones, 1848.

Dorson, Richard M. "Negro Tales from Bolivar County, Mississippi." *Southern Folklore Quarterly* 19 (June 1955): 104–16.

Eckleberry, Don. "Search for the Rare Ivorybill." In *Discovery: Great Moments in the Lives of Outstanding Naturalists,* edited by John K. Terres. Philadelphia and New York: J. B. Lippincott, 1961.

Faulkner, William. *Go Down, Moses*. New York: Random House, 1942.

———. *Requiem for a Nun*. New York: Random House, 1951.

Featherstonhaugh, George William. *Excursion Through the Slave States, From Washington on the Potomac to the Frontier of Mexico; With Sketches of Popular Manners and Geological Notices*. 2 vols. London: John Murray, 1844.

Flint, Timothy. *Recollections of the Last Ten Years*. 1822. Reprint, edited by C. Hartley Grattan, New York: Alfred A. Knopf, 1932.

Gale, Thomas. "Thomas Gale's New South Letters." Edited by William Warren Rogers. *Journal of Mississippi History* 28 (August 1966): 228–36.

Glenn, D. C. "Mississippi." *DeBow's Review* 7 (July 1849): 38–44.

Gossom, W. F. "Our County Prospects." *Greenwood Times,* July 5, 1873.

"Greenwood Has—Greenwood Wants." *Greenwood New Era,* February 16, 1894.

Hall, Basil. *Travels in North America in the Years 1827 and 1828*. 3rd ed. Vol. 3. Edinburgh and London: Robert Cadell, Simpkin and Marshall, 1830.

Hall, James. "A Brief History of the Mississippi Territory, To Which Is Prefixed a Summary View of the Country between the Settlements on Cumberland-River, and the Territory." 1801. Reprinted in *Publications of the Mississippi Historical Society* 9 (1906): 539–76.

Hamilton, Mary. *Trials of the Earth: The Autobiography of Mary Hamilton*. Edited by Helen Dick Davis. Jackson: University Press of Mississippi, 1992.

Hanlon, Howard A. *Delta Harvest: An Authentic Story of a Hardwood Harvest Interwoven with Intriguing Romance*. Watkins Glen, N.Y.: Watkins Review, 1966.

*The Hardwood Lumber and Farming Industries of Mississippi as Shown in Moving Pictures*. San Francisco: Panama Pacific Exposition and Lamb-Fish Lumber Co., 1915.

Harrison, T. S. "Among the Southerners." Printed broadside, May 9, [ca. 1888]. Louisiana and Lower Mississippi Valley Collections, Hill Memorial Library, Louisiana State University.

Hesse-Wartegg, Ernst von. *Travels on the Lower Mississippi, 1879–1880: A Memoir.* Edited and translated by Frederic Trautmann. Columbia: University of Missouri Press, 1990. Originally published as *Mississippi-Fahrten: Reisebilder aus dem amerikanischen Süden, 1879–1880* (1881).

Hodgson, Adam. *Remarks during a Journey through North America, in the Years 1819, 1820, and 1821.* New York: Samuel Whiting, 1823.

Humphreys, Benjamin Grubb. "The Autobiography of Benjamin Grubb Humphreys, August 26, 1808–December 20, 1882." Edited by Percy L. Rainwater. *Mississippi Valley Historical Review* 21 (June 1934–March 1935): 231–55.

Hutchins, Thomas. *An Historical Narrative and Topographical Description of Louisiana and West-Florida.* 1784. Reprint, Floridiana Facsimile and Reprint Series. Gainesville: University of Florida Press, 1968.

"Immigration." *Greenwood Times,* August 9, 1873.

"Initial Convention." *Bolivar County Review* (Rosedale, Miss.), October 2, 1890.

Jefferson, Thomas. *The Writings of Thomas Jefferson.* Vol. 17. Edited by Albert Ellery Burgh. Washington, D.C.: Thomas Jefferson Memorial Association, 1907.

Ker, Henry. *Travels through the Western Interior of the United States, from the Year 1808 up to the Year 1816, With a Particular Description of a Great Part of Mexico, or New-Spain.* Elizabethtown, N.J.: privately printed, 1816.

Lee, G. A. "Mound Bayou, the Negro City of Mississippi." *Voice of the Negro* (January 1906): 36–41.

Lehrbas, Mark. "Fine Stand of Hardwoods Found on Jackson Point, Mississippi." *Southern Lumberman* 145 (October 15, 1932): 17.

Lentz, G. H. "Report of the Mississippi Valley Hardwood Investigations." *Proceedings of the Eleventh Southern Forestry Congress* (1929): 107–15.

"Letter from North Leflore." *Greenwood Oriental,* July 27, 1877.

"The Levees and Overflows of the Mississippi." *DeBow's Review* 25 (October 1858): 436–42.

Lewis, Henry. *The Valley of Mississippi Illustrated.* Edited by Bertha L. Heilbron. Translated by A. Hermina Poatgieter. St. Paul: Minnesota Historical Society, 1967. Originally published as *Das Illustrirte Mississippithal* (1854).

Loring, F. W., and C. F. Atkinson. *Cotton Culture and the South, Considered with Reference to Emigration.* Boston: A. Williams, 1869.

Lyell, Sir Charles. *A Second Visit to the United States of North America.* 2nd ed. Vol. 2. London: John Murray, 1850.

Maule, C. A. *Yazoo-Mississippi Delta: The Homeseekers Opportunity.* N.p.: Illinois Central Railroad, Passenger Department, n.d.

Merry, J. F. *About the South on Lines of the Illinois Central and Yazoo and Mississippi Valley Railroads.* 4th ed. Manchester, Iowa: Illinois Central Railroad, Passenger Department, 1906.

————. *The Yazoo-Mississippi Valley: A Pamphlet Full of Information for Home Seekers and Investors.* 4th ed. N.p.: Illinois Central Railroad, Passenger Department, 1910.

"Mississippi." *Manufacturers Record* 63 (March 27, 1913, pt. 2): 111–12.

*Mississippi Is Calling You.* N.p.: Illinois Central Railroad, Passenger Department, n.d.

Morgan, Albert T. *Yazoo; Or, On the Picket Line of Freedom in the South.* 1889. Reprint, New York: Russell and Russell, 1968.

"Northern Portion of Warren County, Mississippi; Showing Land Fronting on Lower Yazoo River and Tributaries, Including Deer Creek." Annotated map (B2.325/V48), Natchez Trace Map Collection, Eugene C. Barker Texas History Center, Center for American History, University of Texas at Austin.

Nutt, Rush[worth]. "Nutt's Trip to the Chickasaw Country." Edited by Jesse D. Jennings. *Journal of Mississippi History* 9 (January 1947): 34–61.

Nuttall, Thomas. *A Journal of Travels into the Arkansas Territory During the Year 1819.* 1821. Reprint, edited by Savoie Lottinville, Norman: University of Oklahoma Press, 1980.

*Of the Greatness of Mississippi.* Pamphlet reproduced from the August and September 1927 issues of *Illinois Central Magazine.* N.p., n.d.

Olmstead, Frederick Law. *Journeys and Explorations in the Cotton Kingdom: A Traveller's Observations on Cotton and Slavery in the American Slave States.* Vol. 2. London: Sampson Low, 1861.

Paulding, James Kirke. "The Mississippi." 1843. Reprinted as "James Kirke Paulding on the Mississippi, 1842." Edited by Mentor Lee Williams. *Journal of Mississippi History* 10 (October 1948): 317–44.

Penicaut, M. "Annals of Louisiana, From the Establishment of the First Colony under M. D'Iberville, To the Departure of the Author to France, In 1722. Including an Account of the Manners, Customs, and Religion of the Numerous Indian Tribes of That Country." Translated by B. F. French. In *Historical Collections of Louisiana and Florida: Including Translations of Original Manu-*

scripts Relating to Their Discovery and Settlement, with Numerous Historical and Biographical Notes, edited by B. F. French. New York: J. Sabin and Sons, 1869.

Percy, William Alexander. *Lanterns on the Levee: Recollections of a Planter's Son.* New York: Alfred A. Knopf, 1941.

Pittman, Captain Philip. *The Present State of the European Settlements on the Mississippi, With a Geographical Description of That River Illustrated by Plans and Draughts.* 1770. Reprint, edited by Frank Heywood Hodder, Cleveland, Ohio: Arthur H. Clark, 1906.

Price, Rev. Mr. "The Mississippi Swamp." *DeBow's Review* 7 (July 1849): 53–56.

Rawick, George P., ed. "Mississippi Narratives, Prepared by the Federal Writers' Project of the Works Progress Administration for the State of Mississippi." In *Oklahoma and Mississippi Narratives,* vol. 7 in *The American Slave: A Composite Autobiography, Series 1,* Contributions in Afro-American and African Studies, no. 11. Westport, Conn.: Greenwood, 1972.

Ruffin, Thomas. *The Papers of Thomas Ruffin.* Vol. 2. Edited by J. G. De Roulhac Hamilton. Publications of the North Carolina Historical Commission. Raleigh, N.C.: Edwards and Broughton, 1918.

Schultz, Christian. *Travels on an Inland Voyage through the States of New-York, Pennsylvania, Virginia, Ohio, Kentucky, and Tennessee, and through the Territories of Indiana, Louisiana, Mississippi, and New Orleans; Performed in the Years 1807 and 1808; Including a Tour of Nearly Six Thousand Miles.* Vol. 2. New York: Isaac Riley, 1810.

Sessions, James Oliver Hazard Perry. "Diary of James Oliver Hazard Perry Sessions of Rokeby Plantation, on the Yazoo, January 1, 1862–June 1872." Edited by Claude E. Fike. *Journal of Mississippi History* 39 (August 1977): 239–54.

Skene, E. P. *Choice Pickings in the Yazoo Valley: Railroad Lands For Sale Owned by the Yazoo & Mississippi Valley Railroad Co. in the Famous Yazoo Valley of Mississippi.* Chicago: Illinois Central Railroad, 1899.

———. *600,000 Acres Railroad Lands For Sale Owned by the Yazoo & Mississippi Valley Railroad Co. in the Famous Yazoo Valley of Mississippi.* Chicago: Illinois Central Railroad, 1896.

"Southwestern Cities." *DeBow's Review* 17 (December 1854): 606–8.

"State Shares Pride of Delta Timber." *Mississippi Forests and Parks* 12 (April 1946): 13.

[Tompkins, Frank H.] "Mississippi Alluvial Lands." In *Riparian Lands of the*

*Mississippi River, Past—Present—Prospective,* edited by Frank H. Tompkins. [Chicago: A. L. Swift, 1901.]

Twain, Mark. *Life on the Mississippi.* 1883. Reprint, New York: Book-of-the-Month Club, 1992.

"Vicksburg & Nashville Railroad." *Greenwood Times,* July 5, 1873.

West, C. H. "Brief History of Levee Building in the Mississippi Levee District." In *Riparian Lands of the Mississippi River, Past—Present—Prospective,* edited by Frank H. Tompkins. [Chicago: A. L. Swift, 1901.]

Whitaker, Arthur P., ed. "The South Carolina Yazoo Company." *Mississippi Valley Historical Review* 16 (December 1929): 383–94.

Williams, Tennessee. *Cat on a Hot Tin Roof.* New York: James Laughlin, 1955.

Wills, William H. "A Southern Traveler's Diary, 1840." *Publications of the Southern History Association* 8 (January 1904): 23–39.

Wilson, Alexander, and Charles Lucian Bonaparte. *American Ornithology; Or the Natural History of the Birds of the United States.* Vol. 1. Edinburgh: Constable and Co., 1831.

Wilson, J. B. *Handbook of Yazoo County, Miss., for the Encouragement of Immigration and Capital to the County.* [Yazoo City, Miss.:] n.p., 1884.

"The Yazoo-Mississippi Delta." In *Mississippi To-Day,* edited and compiled by Munro Nichols. Gulfport, Miss.: Gulf States Publishing Co., n.d.

*The Yazoo-Mississippi Delta: The Garden Spot of America.* Greenwood, Miss.: n.p., n.d.

*Yazoo, Tallahatchie, and Sunflower River Improvement Convention, Vicksburg, Miss., Wednesday, Jan. 13, 1892.* Vicksburg, Miss.: Vicksburg Printing and Publishing Co., 1891.

## SOUND RECORDINGS

Arnold, Kokomo. "Bo-Weavil Blues." Decca 7191. Reissued in *Kokomo Arnold/ Peetie Wheatstraw.* Blues Classics 4.

Delaney, Mattie. "Tallahatchie River Blues." Vocalion 1480. Reissued in *Mississippi Blues, Vol. 1 (1928–1937).* Document DOCD 5157.

Jefferson, Blind Lemon. "Rising High Water Blues." Paramount 12474. Reissued in *Blind Lemon Jefferson: Complete Recorded Works in Chronological Order, Vol. 2 (1927).* Document DOCD 5018.

McCoy, "Kansas" Joe. "When the Levee Breaks." Columbia 14439. Reissued in *Memphis Minnie and Kansas Joe: 1929–1934 Recordings in Chronological Order, Vol. 1 (18 June, 1929–24 May, 1930).* Document DOCD 5028.

Patton, Charley. "Dry Well Blues." Paramount 13070. Reissued in *Charley Patton: Complete Recorded Works in Chronological Order, Vol. 3 (December 1929 to 1 February, 1930)*. Document DOCD 5011.

―――. "High Water Everywhere—Parts I and II." Paramount 12909. Reissued in *Charley Patton: Complete Recorded Works in Chronological Order, Vol. 2 (Late November/Early December, 1929)*. Document DOCD 5010.

―――. "Mississippi Bo Weavil Blues." Paramount 12805. Reissued in *Charley Patton: Complete Recorded Works in Chronological Order, Vol. 1 (14 June, 1929 to Late November/Early December, 1929)*. Document DOCD 5009.

## Government Documents

Agelasto, A. M., C. B. Doyle, G. S. Meloy, and O. C. Stine. "The Cotton Situation." In *Yearbook of Agriculture 1921*. U.S. Department of Agriculture. Washington, D.C.: Government Printing Office, 1922.

*Agriculture of the United States in 1860; Compiled from the Original Returns of the Eighth Census*. Eighth Census of the United States, 1860. U.S. Department of the Interior, Census Office. Washington, D.C.: Government Printing Office, 1864.

*Agriculture. Vol. 5, Pt. 1: Farms, Live Stock, and Animal Products*. Twelfth Census of the United States, 1900. U.S. Department of Commerce and Labor, Bureau of the Census. Washington, D.C.: Government Printing Office, 1902.

*Agriculture, 1909 and 1910. Vol. 5. General Report and Analysis*. Thirteenth Census of the United States, 1910. U.S. Department of Commerce, Bureau of the Census. Washington, D.C.: Government Printing Office, 1913.

*Agriculture. Vol. 6, Pt. 2. Reports for States, with Statistics for Counties and a Summary for the United States and the North, South, and West. The Southern States*. Fourteenth Census of the United States, 1920. U.S. Department of Commerce, Bureau of the Census. Washington, D.C.: Government Printing Office, 1922.

*Agriculture. Vol. 2, Pt. 2—The Southern States. Reports by States, with Statistics for Counties and a Summary for the United States*. Fifteenth Census of the United States, 1930. U.S. Department of Commerce, Bureau of the Census. Washington, D.C.: Government Printing Office, 1932.

Andrews, Frank. *Railroads and Farming: Some Influences Affecting the Progress of Agriculture*. U.S. Department of Agriculture. Bulletin, no. 100. Washington, D.C.: Government Printing Office, 1912.

Bailey, Robert G. *Description of the Ecoregions of the United States.* Ogden, Utah: U.S. Department of Agriculture, Forest Service, 1978.

Best, Louis B., Dean F. Stauffer, and Anthony R. Geier. "Evaluating the Effects of Habitat Alteration on Birds and Small Mammals Occupying Riparian Communities." In *Strategies for Protection and Management of Floodplain Wetlands and Other Riparian Ecosystems: Proceedings of the Symposium, December 11–13, 1978, Callaway Gardens, Georgia,* edited by R. R. Johnson and J. F. McCormick. General Technical Report WO-12. Washington, D.C.: U.S. Department of Agriculture, Forest Service, 1979.

Boeger, Ernest A., and Emanuel A. Goldenweiser. *A Study of the Tenant Systems of Farming in the Yazoo-Mississippi Delta.* U.S. Department of Agriculture. Bulletin, no. 337. Washington, D.C.: Government Printing Office, 1916.

Brock, S. G. *Report on the Internal Commerce of the United States for the Year 1891. Pt. 2 of Commerce and Navigation: The Commerce of the Great Lakes, the Mississippi River and Its Tributaries.* U.S. Treasury Department, Bureau of Statistics. Washington, D.C.: Government Printing Office, 1892.

*By-Laws and Ordinances of the Board of Levee Commissioners for the State of Mississippi: To Which Is Prefixed The Levee Law Passed at the Called Session of Legislature, November 1858.* The Board of Levee Commissioners for the State of Mississippi. Vicksburg, Miss.: Southern Sun Book and Job Print, 1858.

*Compendium of the Enumeration of the Inhabitants and Statistics of the United States.* Sixth Census of the United States, 1840. Washington, D.C.: U.S. Department of State, 1841.

Connaway, John and Sam McGahey. *Archaeological Survey and Salvage in the Yazoo-Mississippi Delta and in Hinds County.* Mississippi Archaeological Survey, Preliminary Report. Jackson: Mississippi Archaeological Survey, 1970.

"Cotton Investigation." In *Report of the Commissioner of Agriculture, 1876.* U.S. Department of Agriculture. Washington, D.C.: Government Printing Office, 1877.

Cowdrey, Albert E. *The Delta Engineers: A History of the United States Army Corps of Engineers in the New Orleans District.* [New Orleans: U.S. Army Engineer District, New Orleans,] 1971.

"Descriptive Notes of Choctaw District. Ranges 5 & 6 West." In Field Notes, Surveys, and Plats, RG 28, Secretary of State (Formerly Land Office), Choctaw District Township Descriptions [1830s], Series 1109. Official Archives, Archives and Library Division, Mississippi Department of Archives and History.

Dickson, James G. "Forest Bird Communities of the Bottomland Hardwoods." In *Proceedings of the Workshop Management of Southern Forests for Nongame*

*Birds, January 24–26, 1978, Atlanta, Georgia.* General Technical Report SE-14. Asheville, N.C.: U.S. Department of Agriculture, Forest Service, Southeastern Forest Experiment Station, 1978.

*Drainage of Agricultural Lands. Reports by States with Statistics for Counties, a Summary for the United States, and a Synopsis of Drainage Laws.* Fifteenth Census of the United States, 1930. U.S. Department of Commerce, Bureau of the Census. Washington, D.C.: Government Printing Office, 1932.

Du Bois, W. E. Burghardt. "The Negro Farmer." In *Special Reports. Supplementary Analysis and Derivative Tables.* Twelfth Census of the United States, 1900. U.S. Department of Commerce and Labor, Bureau of the Census. Washington, D.C.: Government Printing Office, 1906.

Duerr, William A. *Basic Data on Forest Area and Timber Volumes from the Southern Forest Survey, 1932–36.* Forest Survey Release, no. 54. New Orleans, La.: U.S. Department of Agriculture, Forest Service, Southern Forest Experiment Station, 1946.

Dunston, C. E. *Preliminary Examination of the Forest Conditions of Mississippi.* Bulletin, no. 7. Jackson, Miss.: State Geological Survey, 1910.

Egleston, Nathaniel H. "Report of Chief of Division of Forestry." In *Report of the Commissioner of Agriculture, 1885.* U.S. Department of Agriculture. Washington, D.C.: Government Printing Office, 1885.

———. *Report on Forestry.* Vol. 4. [U.S. Congress Document.] Washington, D.C.: Government Printing Office, 1884.

Eldredge, I. F. *Preliminary Report on the Forest Survey of the Bottom-Land Hardwood Unit in Mississippi.* Forest Survey Release, no. 6. New Orleans, La.: U.S. Department of Agriculture, Forest Service, Southern Forest Experiment Station, 1934.

———. *The Southern Forest Survey (The Forest Survey in the South).* Occasional Papers, no. 31 (revised). New Orleans, La.: U.S. Department of Agriculture, Forest Service, Southern Forest Experiment Station, 1934.

*Endangered Species of Mississippi.* Jackson: Mississippi Department of Wildlife, Fisheries, and Parks, Museum of Natural Science, 1995.

Ernst, John P., and Valerie Brown. "Conserving Endangered Species on Southern Forested Wetlands." In *Forested Wetlands of the Southern United States: Proceedings of the Symposium, Orlando, Florida, July 12–14, 1988,* edited by Donal D. Hook and Russ Lea. General Technical Report SE-50. Asheville, N.C.: U.S. Department of Agriculture, Forest Service, Southeastern Forest Experiment Station, 1989.

*Flood Control in the Lower Mississippi Valley.* Vicksburg, Miss.: Public Affairs

Office, Mississippi River Commission, and U.S. Army Engineer Division, Lower Mississippi Valley, 1976.

Greeley, W. B., Earle H. Clapp, Herbert A. Smith, Raphael Zon, W. N. Sparhawk, Ward Shephard, and J. Kittredge. "Timber: Mine or Crop?" In *Yearbook of Agriculture 1922*. U.S. Department of Agriculture. Washington, D.C.: Government Printing Office, 1923.

Hamel, Paul B., H. E. LeGrand, Jr., M. R. Lennartz, and S. A. Gauthereaux, Jr. *Bird-Habitat Relationships on Southeastern Forest Lands*. General Technical Report SE-22. Asheville, N.C.: U.S. Department of Agriculture, Forest Service, Southeastern Forest Experiment Station, 1982.

Harper, Lewis. *Preliminary Report on the Geology and Agriculture of the State of Mississippi*. Jackson, Miss.: E. Barksdale, State Printer, 1857.

Harris, Larry D. "The Faunal Significance of Fragmentation of Southeastern Bottomland Forests." In *Forested Wetlands of the Southern United States: Proceedings of the Symposium, Orlando, Florida, July 12–14, 1988*, edited by Donal D. Hook and Russ Lea. General Technical Report SE-50. Asheville, N.C.: U.S. Department of Agriculture, Forest Service, Southeastern Forest Experiment Station, 1989.

Hilgard, Eugene W. *Report on Cotton Production in the United States; Also Embracing Agricultural and Physico-Geographical Descriptions of the Several Cotton States and of California. Pt. 1. Mississippi Valley and Southwestern States*. Tenth Census of the United States, 1880. U.S. Department of the Interior, Census Office. Washington, D.C.: Government Printing Office, 1884.

———. *Report on the Geology and Agriculture of the State of Mississippi*. Jackson, Miss.: E. Barksdale, State Printer, 1860.

*Historical Statistics of the United States, Colonial Times to 1970. Pt. 1*. U.S. Department of Commerce, Bureau of the Census. Washington, D.C.: Government Printing Office, 1975.

Hough, Franklin B. *Report on Forestry, Submitted to Congress by the Commissioner of Agriculture*. [Vol. 3.] 47th Cong., 1st sess. Miscellaneous Document, no. 38. Washington, D.C.: Government Printing Office, 1882.

———. *Report upon Forestry*. [Vol. 1.]. Prepared under the Direction of the Commissioner of Agriculture. Washington, D.C.: Government Printing Office, 1878.

Humphreys, Capt. A. A., and Lieut. H. L. Abbot. *Report Upon the Physics and Hydraulics of the Mississippi River; Upon the Protection of the Alluvial Region Against Overflow; and Upon the Deepening of the Mouths: Based Upon Surveys and Investigations Made under the Acts of Congress Directing the To-*

*pographical and Hydrographical Survey of the Delta of the Mississippi River, with Such Investigations as Might Lead to Determine the Most Practicable Plan for Securing It from Inundation, and the Best Mode of Deepening the Channels at the Mouths of the River.* Professional Paper, no. 4. Philadelphia: United States Army, Corps of Topographical Engineers, 1861.

Hurt, A. B. *Mississippi: Its Climate, Soil, Productions, and Agricultural Capabilities.* U.S. Department of Agriculture. Miscellaneous Special Report, no. 3. Washington, D.C.: Government Printing Office, 1884.

*Industrial and Manufacturing Schedules of Mississippi, 1880: Adams-Yazoo.* Tenth Census of the United States, 1880. Official Archives, Archives and Library Division, Mississippi Department of Archives and History.

Jahn, Laurence R. "Values of Riparian Habitats to Natural Ecosystems." In *Strategies for Protection and Management of Floodplain Wetlands and Other Riparian Ecosystems: Proceedings of the Symposium, December 11–13, 1978, Callaway Gardens, Georgia,* edited by R. R. Johnson and J. F. McCormick. General Technical Report WO-12. Washington, D.C.: U.S. Department of Agriculture, Forest Service, 1979.

Langsford, E. L., and B. H. Thibodeaux. *Plantation Organization and Operation in the Yazoo-Mississippi Delta Area.* U.S. Department of Agriculture. Technical Bulletin, no. 682. Washington, D.C.: Government Printing Office, 1939.

Little, Elbert L., Jr. *Checklist of United States Trees: Native and Naturalized.* Tree and Range Plant Name Committee, U.S. Department of Agriculture, Forest Service. Agriculture Handbook, no. 541. Washington, D.C.: Government Printing Office, 1979.

Lowe, Ephraim Noble. *Mississippi: Its Geology, Geography, Soils, and Mineral Resources.* Bulletin, no. 12. Jackson, Miss.: State Geological Survey, 1915.

———. "Note on the Flora of Mississippi." In *Forest Conditions of Mississippi: Being a Reprint with Additions of Bulletins Nos. 5 and 7.* Bulletin, no. 11. Jackson, Miss.: State Geological Survey, 1913.

———. *Plants of Mississippi: A List of Flowering Plants and Ferns.* Bulletin, no. 17. Jackson, Miss.: State Geological Survey, 1921.

*The Lumber Industry. Pt. 1: Standing Timber.* U.S. Department of Commerce and Labor, Bureau of Corporations. Washington, D.C.: Government Printing Office, 1913.

MacCleery, Douglas W. *American Forests: A History of Resiliency and Recovery.* FS-540. Durham, N.C.: U.S. Department of Agriculture, Forest Service, in cooperation with Forest History Society, 1992.

MacDonald, Purificacion O., Warren E. Frayer, and Jerome K. Clauser. *Documentation, Chronology, and Future Projections of Bottomland Hardwood*

*Habitat Losses in the Lower Mississippi Alluvial Plain, Vol. 1: Basic Report.* Washington, D.C.: U.S. Department of the Interior, Fish and Wildlife Service, Ecological Services, 1979.

Mader, Stephen F, W. Michael Aust, and Russ Lea. "Changes in Functional Values of a Forested Wetland Following Timber Harvesting Practices." In *Forested Wetlands of the Southern United States: Proceedings of the Symposium, Orlando, Florida, July 12–14, 1988,* edited by Donal D. Hook and Russ Lea. General Technical Report SE-50. Asheville, N.C.: U.S. Department of Agriculture, Forest Service, Southeastern Forest Experiment Station, 1989.

*Manufactures, 1909. Vol. 9. Reports by States, with Statistics for Principal Cities.* Thirteenth Census of the United States, 1910. U.S. Department of Commerce, Bureau of the Census. Washington, D.C.: Government Printing Office, 1913.

Marschner, Francis J. *Land Use and Its Patterns in the United States.* U.S. Department of Agriculture. Agriculture Handbook, no. 153. Washington, D.C.: Government Printing Office, 1959.

Mattoon, Wilbur R. *The Southern Cypress.* U.S. Department of Agriculture. Bulletin, no. 272. Washington, D.C.: Government Printing Office, 1915.

McCandliss, D. A. *Base Book of Mississippi Agriculture, 1866–1953.* Jackson, Miss.: U.S. Department of Agriculture and Mississippi Department of Agriculture and Commerce, 1955.

Mills, Gary B. *Of Men and Rivers: The Story of the Vicksburg District.* Vicksburg, Miss.: U.S. Army Engineer District, Vicksburg, 1978.

*Mississippi Forests.* Forest Survey Release, no. 81. New Orleans, La.: U.S. Department of Agriculture, Forest Service, Southern Forest Experiment Station, 1958.

*Negro Population, 1790–1915.* U.S. Department of Commerce, Bureau of the Census. Washington, D.C.: Government Printing Office, 1918.

Odum, Eugene P. "Ecological Importance of the Riparian Zone." In *Strategies for Protection and Management of Floodplain Wetlands and Other Riparian Ecosystems: Proceedings of the Symposium, December 11–13, 1978, Callaway Gardens, Georgia,* edited by R. R. Johnson and J. F. McCormick. General Technical Report WO-12. Washington, D.C.: U.S. Department of Agriculture, Forest Service, 1979.

*Population, Vol. 1. Number and Distribution of Inhabitants. Total Population for States, Counties, and Townships or Other Minor Civil Divisions; for Urban and Rural Areas; and for Cities and Other Incorporated Places.* Fifteenth Census of the United States, 1930. U.S. Department of Commerce, Bureau of the Census. Washington, D.C.: Government Printing Office, 1931.

Putnam, John A. *Management of Bottomland Hardwoods.* Occasional Papers, no. 116. New Orleans, La.: U.S. Department of Agriculture, Forest Service, Southern Forest Experiment Station, 1951.

Putnam, John A., and Henry Bull. *The Trees of the Bottomlands of the Mississippi River Delta Region.* Occasional Papers, no. 27. New Orleans, La.: U.S. Department of Agriculture, Forest Service, Southern Forest Experiment Station, 1932.

Putnam, John A., George M. Furnival, and J. S. McKnight. *Management and Inventory of Southern Hardwoods.* U.S. Department of Agriculture, Forest Service. Agriculture Handbook, no. 181. Washington, D.C.: Government Printing Office, 1960.

*Report of the President and Engineer in Chief to the Legislature; With the Treasurer's Report of the Levee Commissioners, April, 1860.* The Board of Levee Commissioners for the State of Mississippi. Vicksburg, Miss.: Southern Sun Book and Job Print, 1860.

*Report on the Productions of Agriculture as Returned at the Tenth Census (June 1, 1880).* Vol. 3 of the Tenth Census of the United States, 1880. U.S. Department of the Interior, Census Office. Washington, D.C.: Government Printing Office, 1883.

"Report of the Statistician." In *Report of the Commissioner of Agriculture, 1885.* U.S. Department of Agriculture. Washington, D.C.: Government Printing Office, 1885.

*Report on the Statistics of Agriculture in the United States at the Eleventh Census: 1890.* Eleventh Census of the United States, 1890. U.S. Department of the Interior, Census Office. Washington, D.C.: Government Printing Office, 1895.

Reuss, Martin. *Designing the Bayous: The Control of Water in the Atchafalya Basin, 1800–1995.* Alexandria, Va.: U.S. Army Corps of Engineers, Office of History, 1998.

Reynolds, Robert V., and Albert H. Pierson. *Lumber Cut of the United States, 1870–1920: Declining Production and High Prices as Related to Forest Exhaustion.* U.S. Department of Agriculture. Bulletin, no. 1119. Washington, D.C.: Government Printing Office, 1923.

Sargent, Charles S. *Report on the Forests of North America (Exclusive of Mexico).* Vol. 9 of the Tenth Census of the United States, 1880. U.S. Department of the Interior, Census Office. Washington, D.C.: Government Printing Office, 1884.

*Second Biennial Report of the Mississippi State Forestry Commission.* Jackson, Miss.: Mississippi State Forestry Commission, 1929.

*The Seventh Census of the United States: 1850. Embracing a Statistical View of Each of the States and Territories, Arranged by Counties, Towns, Etc.* Seventh

Census of the United States, 1850. U.S. Department of State. Washington, D.C.: Robert Armstrong, 1853.

Shantz, H. L., and R. Zon. "Natural Vegetation." In *Atlas of American Agriculture*, pt. 1, sec. E. Washington, D.C.: U.S. Department of Agriculture, 1924.

Shugart, H. H., T. M. Smith, J. T. Kitchings, and R. L. Kroodsma. "The Relationship of Nongame Birds to Southern Forest Types and Successional Stages." In *Proceedings of the Workshop Management of Southern Forests for Nongame Birds, January 24–26, 1978, Atlanta, Georgia.* General Technical Report SE-14. Asheville, N.C.: U.S. Department of Agriculture, Forest Service, Southeastern Forest Experiment Station, 1978.

Starr, Frederick, Jr. "American Forests: Their Destruction and Preservation." In *Report of the Commissioner of Agriculture for the Year 1865.* U.S. Department of Agriculture. Washington, D.C.: Government Printing Office, 1866.

"Statistics of Forestry." In *Report of the Commissioner of Agriculture for the Year 1875.* U.S. Department of Agriculture. Washington, D.C.: Government Printing Office, 1876.

*Statistics of the Population of the United States at the Tenth Census (June 1, 1880).* Vol. 1 of the Tenth Census of the United States, 1880. U.S. Department of the Interior, Census Office. Washington, D.C.: Government Printing Office, 1883.

*The Statistics of the Wealth and Industry of the United States; Embracing the Tables of Wealth, Taxation, and Public Indebtedness; Of Agriculture; Manufactures; Mining; and the Fisheries. With Which are Reproduced, From the Volume on Population, the Major Tables of Occupations. Compiled from the Original Returns of the Ninth Census (June 1, 1870).* Vol. 3 of the Ninth Census of the United States, 1870. U.S. Department of the Interior, Census Office. Washington, D.C.: Government Printing Office, 1872.

Sternitzke, Herbert S., and John A. Putnam. *Forests of the Mississippi Delta.* Forest Survey Release, no. 78. New Orleans, La.: U.S. Department of Agriculture, Forest Service, Southern Forest Experiment Station, 1956.

Stover, W. S. *Forest Resources of the Delta Section of Mississippi: A Progress Report by the Southern Forest Survey.* Forest Survey Release, no. 53. New Orleans, La.: U.S. Department of Agriculture, Forest Service, Southern Forest Experiment Station, 1942.

*Survey of Little Tallahatchie River Watershed in Mississippi.* U.S. House of Representatives document, no. 892, 77th Cong., 2nd sess., 1942.

*Survey of Yazoo River Watershed in Mississippi.* U.S. House of Representatives document, no. 564, 78th Cong., 2nd sess., 1944.

Swanton, John R. *The Indians of the Southeastern United States.* Smithsonian Institution, Bureau of American Ethnology Bulletin, no. 137. Washington, D.C.: Government Printing Office, 1946.

Switzler, Wm. F. *Report on the Internal Commerce of the United States. Pt. 2 of Commerce and Navigation: The Commercial, Industrial Transportation, and Other Interests of the Southern States.* U.S. Treasury Department, Bureau of Statistics. Washington, D.C.: Government Printing Office, 1886.

———. *Report on the Internal Commerce of the United States. Pt. 2 of Commerce and Navigation: Special Report on the Commerce of the Mississippi, Ohio, and Other Rivers, and of the Bridges Which Cross Them.* U.S. Treasury Department, Bureau of Statistics. Washington, D.C.: Government Printing Office, 1888.

*Timber Depletion, Lumber Prices, Lumber Exports, and Concentration of Ownership: Report on Senate Resolution 311.* U.S. Department of Agriculture, Forest Service. Washington, D.C.: Government Printing Office, 1920.

Wailes, Benjamin L. C. *Report on the Agriculture and Geology of Mississippi, Embracing a Sketch of the Social and Natural History of the State.* Jackson, Miss.: E. Barksdale, State Printer, 1854.

Wall, E. G. *The State of Mississippi: Resources, Condition, and Wants.* Mississippi Board of Immigration and Agriculture. Jackson, Miss.: Clarion, 1879.

Wharton, Charles H., Wiley M. Kitchens, Edward C. Pendleton, and Timothy W. Sipe. *The Ecology of Bottomland Hardwood Swamps of the Southeast: A Community Profile.* OBS-81/37. Washington, D.C.: U.S. Department of the Interior, Fish and Wildlife Service, Biological Services, National Coastal Ecosystems Team, 1982.

White, Richard, and William Cronon. "Ecological Change and Indian-White Relations." In *History of Indian-White Relations,* edited by Wilcom E. Washburn. Handbook of North American Indians, vol. 4. Washington, D.C.: Smithsonian Institution, 1984.

*Yazoo River (Lower Tributaries), Miss.* U.S. House of Representatives document, no. 516, 78th Cong., 2nd sess., 1944.

## Secondary Sources

Aiken, Charles S. "Cotton Gins." In *Encyclopedia of Southern Culture,* edited by Charles Reagan Wilson and William Ferris. Chapel Hill: University of North Carolina Press, 1989.

————. *The Cotton Plantation South since the Civil War*. Creating the North American Landscape Series. Baltimore: Johns Hopkins University Press, 1998.

————. "The Decline of Sharecropping in the Lower Mississippi River Valley." In *Man and Environment in the Lower Mississippi Valley*, edited by Sam B. Hilliard. Geoscience and Man, vol. 19. Baton Rouge: Louisiana State University, School of Geoscience, 1978.

Aldridge, James W. "Fertilizer Industry." In *The Encyclopedia of Southern History*, edited by David C. Roller and Robert W. Twyman. Baton Rouge: Louisiana State University Press, 1979.

Allen, J. S. "Levees 1926–46." In *History of Bolivar County, Mississippi*, edited by Wirt A. Williams. Jackson, Miss.: Daughters of the American Revolution, Mississippi Delta Chapter, 1948.

Askins, Robert A. *Restoring North American Birds: Lessons from Landscape Ecology*. New Haven, Conn.: Yale University Press, 2000.

Åström, Sven-Erik. *Natur och byte: Ekologiska synpunkter på Finlands ekonomiska historia*. Ekenäs, Finland: Söderström, 1978.

Ayers, Edward L. *The Promise of the New South: Life after Reconstruction*. New York: Oxford University Press, 1992.

Bailes, Kendall E. "Critical Issues in Environmental History." In *Environmental History: Critical Issues in Comparative Perspective*, edited by Kendall E. Bailes. New York: University Presses of America, 1985.

Bailey, Alfred M. "Ivory-billed Woodpecker's Beak in an Indian Grave in Colorado." *Condor* 41 (July 1939): 164.

Ballas, Donald J. "Historical Geography and American Indian Development." In *A Cultural Geography of North American Indians*, edited by Thomas E. Ross and Tyrel G. Moore. Boulder, Colo.: Westview, 1987.

Barber, Russell J. "Appendix D: Faunal Remains; D.2 Analysis of Molluscs." In Stephen Williams and Jeffrey P. Brain, *Excavations at the Lake George Site, Yazoo County, Mississippi, 1958–1960*. Papers of the Peabody Museum of Archaeology and Ethnology, vol. 74. Cambridge, Mass.: Harvard University Press, 1983.

Barnes, Burton V. "Deciduous Forests of North America." In *Ecosystems of the World 7: Temperate Deciduous Forests*, edited by E. Röhrig and B. Ulrich. Amsterdam: Elsevier Science Publishers, 1991.

Barry, John M. *Rising Tide: The Great Mississippi Flood of 1927 and How It Changed America*. New York: Simon and Schuster, 1997.

Begon, Michael, John L. Harper, and Colin R. Townsend. *Ecology: Individuals,*

*Populations and Communities.* 2nd ed. Boston: Blackwell Scientific Publications, 1990.

Bell, C. Ritchie, and Charles Reagan Wilson. "Kudzu." In *Encyclopedia of Southern Culture,* edited by Charles Reagan Wilson and William Ferris. Chapel Hill: University of North Carolina Press, 1989.

Belmont, John S. "Appendix D: Faunal Remains; D.1 Analysis of the Bone and Shell." In Stephen Williams and Jeffrey P. Brain, *Excavations at the Lake George Site, Yazoo County, Mississippi, 1958–1960.* Papers of the Peabody Museum of Archaeology and Ethnology, vol. 74. Cambridge, Mass.: Harvard University Press, 1983.

Belt, C. B., Jr. "The 1973 Flood and Man's Constriction of the Mississippi River." *Science* 189 (August 29, 1975): 681–84.

"Biographical Sketches of Some Prominent Negroes." In *History of Bolivar County, Mississippi,* edited by Wirt A. Williams. Jackson, Miss.: Daughters of the American Revolution, Mississippi Delta Chapter, 1948.

Björn, Ismo. *Kaikki irti metsästä: Metsän käyttö ja muutos taigan reunalla itäisimmässä Suomessa erätaloudesta vuoteen 2000.* Bibliotheca Historica, no. 49. Helsinki: Suomen Historiallinen Seura, 1999.

Blake, Nelson M. "Flood Control and Drainage." In *Encyclopedia of Southern Culture,* edited by Charles Reagan Wilson and William Ferris. Chapel Hill: University of North Carolina Press, 1989.

Bond, James. *Birds of the West Indies.* Baltimore, Md.: Academy of Natural Sciences of Philadelphia, 1936.

Boyden, Stephen. "Human Ecology and Biohistory: Conceptual Approaches to Understanding Human Situations in the Biosphere." In *Human Ecology: Fragments of Anti-fragmentary Views of the World,* edited by Dieter Steiner and Markus Nauser. London: Routledge, 1993.

Brain, Jeffrey P. "La Salle at the Natchez: An Archaeological and Historical Perspective." In *La Salle and His Legacy: Frenchmen and Indians in the Lower Mississippi Valley,* edited by Patricia K. Galloway. Jackson: University Press of Mississippi, 1982.

———. "Late Prehistoric Settlement Patterning in the Yazoo Basin and Natchez Bluffs Regions of the Lower Mississippi Valley." In *Mississippian Settlement Patterns,* edited by Bruce D. Smith. Studies in Archaeology. New York: Academic Press, 1978.

———. "Tunica Archaeology." In *Traces of Prehistory: Papers in Honor of William G. Haag,* edited by Frederick Hadleigh West and Robert W. Neuman.

Geoscience and Man, vol. 22. Baton Rouge: Louisiana State University, School of Geoscience, 1981.

―――. *Tunica Archaeology*, with contributions by T. M. Hamilton and Arthur Spiess. Papers of the Peabody Museum of Archaeology and Ethnology, vol. 78. Cambridge, Mass.: Harvard University Press, 1988.

―――. *The Tunica-Biloxi*. Indians of North America. New York: Chelsea House, 1990.

―――. "Tunica Triumph." In *Historical Archaeology of the Eastern United States: Papers from the R. J. Russell Symposium*, edited by Robert W. Neuman. Geoscience and Man, vol. 23. Baton Rouge: Louisiana State University, School of Geoscience, 1983.

Brandfon, Robert L. *Cotton Kingdom of the New South: A History of the Yazoo Mississippi Delta from Reconstruction to the Twentieth Century*. Cambridge, Mass.: Harvard University Press, 1967.

―――. "The End of Immigration to the Cotton Fields." *Mississippi Valley Historical Review* 50 (March 1964): 591–611.

―――. "Planters of the New South: An Economic History of the Yazoo-Mississippi Delta." Ph.D. diss., Harvard University, 1962.

Braudel, Fernand. *Écrits sur l'histoire*. Paris: Flammarion, 1969.

―――. *The Mediterranean and the Mediterranean World in the Age of Philip II*. Vol. 1. Translated by Siân Reynolds. London: Fontana/Collins, 1981.

Braun, E. Lucy. *Deciduous Forests of Eastern North America*. Philadelphia: Blakiston, 1950.

Broadfoot, W. M., and H. L. Williston. "Flooding Effects on Southern Forests." *Journal of Forestry* 71 (September 1973): 584–87.

Brockman, C. Frank. *Trees of North America: A Guide to Field Identification*. Rev. ed. New York: Golden, 1986.

Brown, Ian W. "An Archaeological Study of Culture Contact and Change in the Natchez Bluffs Region." In *La Salle and His Legacy: Frenchmen and Indians in the Lower Mississippi Valley*, edited by Patricia K. Galloway. Jackson: University Press of Mississippi, 1982.

Bruce, Dickson D., Jr. "Hunting: Dimensions of Antebellum Southern Culture." *Mississippi Quarterly* 30 (spring 1977): 259–81.

Buchanan, Minor Ferris. *Holt Collier: His Life, His Roosevelt Hunts, and the Origin of the Teddy Bear*. Jackson: Centennial Press of Mississippi, 2002.

Buell, Lawrence. "Faulkner and the Claims of the Natural World." In *Faulkner and the Natural World*, edited by Donald M. Kartiganer and Ann J. Abadie.

Papers from the 23rd Annual Faulkner and Yoknapatawpha Conference, 1996. Jackson: University Press of Mississippi, 1999.

Burdick, David M., Douglas Cushman, Robert Hamilton, and James G. Gosselink. "Faunal Changes and Bottomland Hardwood Forest Loss in the Tensas Watershed, Louisiana." *Conservation Biology* 3 (September 1989): 282–92.

Butzer, Karl W. "The Indian Legacy in the American Landscape." In *The Making of the American Landscape,* edited by Michael P. Conzen. Boston: Unwin Hyman, 1990.

Byrd, Kathleen M., and Robert W. Neuman. "Archaeological Data Relative to Prehistoric Subsistence in the Lower Mississippi River Alluvial Valley." In *Man and Environment in the Lower Mississippi Valley,* edited by Sam B. Hilliard. Geoscience and Man, vol. 19. Baton Rouge: Louisiana State University, School of Geoscience, 1978.

Cairns, John, Jr., Mark M. Brinson, Robert L. Johnson, W. Blake Parker, R. Eugene Turner, and Parley V. Winger. "Impacts Associated with Southeastern Bottomland Hardwood Forest Ecosystems." In *Wetlands of Bottomland Hardwood Forests,* edited by J. R. Clark and J. Benforado. Developments in Agricultural and Managed-Forest Ecology, no. 11. New York: Elsevier Scientific Publishing, 1981.

Campbell, T. N. "Choctaw Subsistence: Ethnographic Notes from the Lincecum Manuscript." Reprint from *The Florida Archaeologist* 12 (1/1959). In *Ethnology of the Southeastern Indians: A Source Book,* edited by Charles M. Hudson. New York: Garland, 1985.

Caplenor, Donald. "Forest Composition on Loessal and Non-loessal Soils in West-Central Mississippi." *Ecology* 49 (early spring 1968): 322–31.

Caporaso, James A. "Dependence, Dependency, and Power in the Global System: A Structural and Behavioral Analysis." *International Organization* 32 (winter 1978): 13–43.

Carter, Hodding. *The Lower Mississippi.* New York: Farrar and Rinehart, 1942.

Cash, Wilbur J. *The Mind of the South.* New York: Alfred A. Knopf, 1941.

Christensen, Norman L. "Landscape History and Ecological Change." *Journal of Forest History* 33 (July 1989): 116–24.

———. "Vegetation of the Southeastern Coastal Plain." In *North American Terrestrial Vegetation,* edited by Michael G. Barbour and William Dwight Billings. Cambridge: Cambridge University Press, 1988.

Clark, J. R., and J. Benforado. Introduction to *Wetlands of Bottomland Hardwood Forests,* edited by J. R. Clark and J. Benforado. Developments in Agricultural and Managed-Forest Ecology, no. 11. New York: Elsevier Scientific Publishing, 1981.

Clark, Thomas D. "Agriculture." In *Encyclopedia of Southern Culture*, edited by Charles Reagan Wilson and William Ferris. Chapel Hill: University of North Carolina Press, 1989.

————. *The Greening of the South: The Recovery of Land and Forest.* Lexington: University Press of Kentucky, 1984.

————. "The Impact of Timber Industry on the South." *Mississippi Quarterly* 25 (spring 1972): 141–64.

Clawson, Marion. "Forests in the Long Sweep of American History." *Science* 204 (June 1979): 1168–74.

Clifton, Ann D. "A Demographic Study of Bolivar County in 1860." *Journal of the Bolivar County Historical Society* 1 (March 1977): 10–20.

Cobb, James C. *The Most Southern Place on Earth: The Mississippi Delta and the Roots of Regional Identity.* New York: Oxford University Press, 1992.

Cokinos, Christopher. *Hope Is the Thing with Feathers: A Personal Chronicle of Vanished Birds.* New York: Tarcher/Putnam, 2000.

Conrad, David E. "Tenant Farmers." In *Encyclopedia of Southern Culture*, edited by Charles Reagan Wilson and William Ferris. Chapel Hill: University of North Carolina Press, 1989.

Cortese, Jim. "Memphis: The Hardwood Capital." *Southern Lumberman* 193 (December 15, 1956): 137–39.

Cotterill, R. S. *The Southern Indians: The Story of the Civilized Tribes before Removal.* Civilization of the American Indian Series, vol. 38. Norman: University of Oklahoma Press, 1954.

Cowdrey, Albert E. *This Land, This South.* New Perspectives on the South. Lexington: University Press of Kentucky, 1983.

Cox, Thomas R., Robert S. Maxwell, Philip Drennon Thomas, and Joseph J. Malone. *This Well-Wooded Land: Americans and Their Forests from Colonial Times to the Present.* Lincoln: University of Nebraska Press, 1985.

Crawford, Charles Wann. "A History of the R. F. Learned Lumber Company, 1865–1900." Ph.D. diss., University of Mississippi, 1968.

Criss, Gail, and Charlotte Hill. "A Demographic Study of Bolivar County in 1850." *Journal of the Bolivar County Historical Society* 1 (March 1977): 3–9.

Cronon, William. *Changes in the Land: Indians, Colonists, and the Ecology of New England.* New York: Hill and Wang, 1983.

————. "Modes of Prophecy and Production: Placing Nature in History." *The Journal of American History* 76 (March 1990): 1122–31.

————. *Nature's Metropolis: Chicago and the Great West.* New York: W. W. Norton, 1991.

————. "The Trouble with Wilderness; or, Getting Back to the Wrong Nature." *Environmental History* 1 (January 1996): 7–28.

————. "The Uses of Environmental History." *Environmental History Review* 17 (fall 1993): 1–22.

Crosby, Alfred W. "Biotic Change in Nineteenth-Century New Zealand." *Environmental Review* 10 (fall 1986): 189–98.

————. *The Columbian Exchange: Biological and Cultural Consequences of 1492.* Westport, Conn.: Greenwood, 1972.

————. *Ecological Imperialism: The Biological Expansion of Europe, 900–1900.* Studies in Environment and History. Cambridge: Cambridge University Press, 1986.

————. "Ecological Imperialism: The Overseas Migration of Western Europeans as a Biological Phenomenon." In *The Ends of the Earth: Perspectives on Modern Environmental History,* edited by Donald Worster. Studies in Environment and History. Cambridge: Cambridge University Press, 1988.

————. "An Enthusiastic Second." *The Journal of American History* 76 (March 1990): 1107–10.

————. "Papua New Guinea, Its Demographic History and Infectious Diseases." In *European Impact and Pacific Influence,* edited by Hermann J. Hiery and John M. MacKenzie. London: The German Historical Institute, 1997.

————. "The Past and Present of Environmental History." *The American Historical Review* 100 (October 1995): 1177–89.

Cross, Ralph D., and Robert W. Wales, eds. *Atlas of Mississippi.* Jackson: University Press of Mississippi, 1974.

Crow, A. Bigler. "Fire Ecology and Fire Management in the Forests of the Lower Mississippi River Valley." In *Man and Environment in the Lower Mississippi Valley,* edited by Sam B. Hilliard. Geoscience and Man, vol. 19. Baton Rouge: Louisiana State University, School of Geoscience, 1978.

Cry, George W. "Surface Waters of the Lower Mississippi River Region." In *Man and Environment in the Lower Mississippi Valley,* edited by Sam B. Hilliard. Geoscience and Man, vol. 19. Baton Rouge: Louisiana State University, School of Geoscience, 1978.

Daniel, Pete. *Breaking the Land: The Transformation of Cotton, Tobacco, and Rice Cultures since 1880.* Urbana: University of Illinois Press, 1985.

————. "Cotton Culture." In *Encyclopedia of Southern Culture,* edited by Charles Reagan Wilson and William Ferris. Chapel Hill: University of North Carolina Press, 1989.

————. *Deep'n as It Come: The 1927 Mississippi River Flood.* New York: Oxford University Press, 1977. Reprint, Fayetteville: University of Arkansas Press, 1996.

———. "The Metamorphosis of Slavery, 1865–1900." *Journal of American History* 66 (June 1979): 88–99.

———. *The Shadow of Slavery: Peonage in the South, 1901–1969.* Urbana: University of Illinois Press, 1972.

Darling, Marsha Jean. "Landownership, Black." In *Encyclopedia of Southern Culture,* edited by Charles Reagan Wilson and William Ferris. Chapel Hill: University of North Carolina Press, 1989.

Davis, Donald E. *Where There Are Mountains: An Environmental History of the Southern Appalachians.* Athens: University of Georgia Press, 2000.

Davis, Margaret Bryan. "Quaternary History and the Stability of Forest Communities." In *Forest Succession: Concepts and Application,* edited by Darrel C. West, Herman H. Shugart, and Daniel B. Botkin. New York: Springer-Verlag, 1981.

Dawson, William R., J. David Ligon, Joseph R. Murphy, J. P. Myers, Daniel Simberloff, and Jared Verner. "Report of the Scientific Advisory Panel on the Spotted Owl." *Condor* 89 (February 1987): 205–29.

Day, Gordon M. "The Indian as an Ecological Factor in the Northeastern Forest." *Ecology* 34 (April 1953): 329–46.

Delcourt, Hazel R. "Presettlement Vegetation of the North of Red River Land District, Louisiana." *Castanea* 41 (June 1976): 122–39.

Delcourt, Hazel R., and Paul A. Delcourt. "The Blufflands: Pleistocene Pathway into the Tunica Hills." *American Midland Naturalist* 94 (October 1975): 385–400.

Delcourt, Paul A., and Hazel R. Delcourt. *Long-Term Forest Dynamics of the Temperate Zone: A Case Study of Late-Quaternary Forests in Eastern North America.* Ecological Studies, vol. 63. New York: Springer-Verlag, 1987.

Delcourt, Paul A., Hazel R. Delcourt, Ronald C. Brister, and Laurence E. Lackey. "Quaternary Vegetation History of the Mississippi Embayment." *Quaternary Research* 13 (January 1980): 111–32.

Delcourt, Paul A., Hazel R. Delcourt, Dan F. Morse, and Phyllis S. Morse. "History, Evolution, and Organization of Vegetation and Human Culture." In *Biodiversity of the Southeastern United States: Lowland Terrestrial Communities,* edited by William H. Martin, Stephen G. Boyce, and Arthur C. Echternacht. New York: John Wiley and Sons, 1993.

Denevan, William M. "Native American Populations in 1492: Recent Research and a Revised Hemispheric Estimate." In *The Native Population of the Amer-*

*icas in 1492,* edited by William M. Denevan. 2nd ed. Madison: University of Wisconsin Press, 1992.

Dethloff, Henry C. "Agriculture." In *The Encyclopedia of Southern History,* edited by David C. Roller and Robert W. Twyman. Baton Rouge: Louisiana State University Press, 1979.

Dickson, James G. "Bird Communities in Oak-Gum-Cypress Forests." In *Bird Conservation 3,* edited by Jerome A. Jackson. Madison: University of Wisconsin Press, 1988.

――――. "Seasonal Populations of Insectivorous Birds in a Mature Bottomland Hardwood Forest in South Louisiana." In *The Role of Insectivorous Birds in Forest Ecosystems,* edited by James G. Dickson, Richard N. Conner, Robert R. Fleet, James C. Kroll, and Jerome A. Jackson. New York: Academic Press, 1979.

Dobyns, Henry F. *Their Number Become Thinned: Native American Population Dynamics in Eastern North America.* Native American Historic Demography Series. Knoxville: University of Tennessee Press, 1983.

Dong, Zhengkai. "From Postbellum Plantation to Modern Agribusiness: A History of the Delta and Pine Land Company." Ph.D. diss., Purdue University, 1993.

Doolittle, William E. *Cultivated Landscapes of Native North America.* Oxford Geographical and Environmental Studies Series. Oxford: Oxford University Press, 2000.

Earle, Carville. "The Myth of the Southern Soil Miner: Macrohistory, Agricultural Innovation, and Environmental Change." In *The Ends of the Earth: Perspectives on Modern Environmental History,* edited by Donald Worster. Studies in Environment and History. Cambridge: Cambridge University Press, 1988.

Echternacht, Arthur C., and Larry D. Harris. "The Fauna and Wildlife of the Southeastern United States." In *Biodiversity of the Southeastern United States: Lowland Terrestrial Communities,* edited by William H. Martin, Stephen G. Boyce, and Arthur C. Echternacht. New York: John Wiley and Sons, 1993.

Evans, David. *Big Road Blues: Tradition and Creativity in the Folk Blues.* Berkeley: University of California Press, 1982. Reprint, New York: Da Capo, 1987.

Ewing, Early C. (Mrs.). "Additional Facts Regarding the History of the Delta and Pine Land Company." In *History of Bolivar County, Mississippi,* edited by Wirt A. Williams. Jackson, Miss.: Daughters of the American Revolution, Mississippi Delta Chapter, 1948.

――――. "The Delta and Pine Land Company." In *History of Bolivar County,*

*Mississippi,* edited by Wirt A. Williams. Jackson, Miss.: Daughters of the American Revolution, Mississippi Delta Chapter, 1948.

Ewing, Early, Jr. "The Delta and Pine Land Company." In *Washington County Historical Society Programs of 1979.* Greenville, Miss.: Washington County Historical Society, 1979.

Fickle, James E. Introduction to *Boxes, Baskets, and Boards: A History of Anderson-Tully Company,* by Charles A. Heavrin. Memphis: Anderson-Tully Co. and Memphis State University Press, 1981.

————. *Mississippi Forests and Forestry.* Jackson: University Press of Mississippi, 2001.

————. *The New South and the "New Competition": Trade Association Development in the Southern Pine Industry.* Urbana: Published for the Forest History Society by the University of Illinois Press, 1980.

Fite, Gilbert C. *Cotton Fields No More: Southern Agriculture, 1865–1980.* Lexington: University Press of Kentucky, 1984.

Flannery, Tim. *The Eternal Frontier: An Ecological History of North America and Its Peoples.* London: William Heinemann, 2001.

Flores, Dan. "Place: An Argument for Bioregional History." *Environmental History Review* 18 (winter 1994): 1–18.

Ford, Richard I. "Patterns of Prehistoric Food Production in North America." In *Prehistoric Food Production in North America,* edited by Richard I. Ford. Anthropological Papers, no. 75. Ann Arbor: University of Michigan Museum of Anthropology, 1985.

Forman, Richard T. T., and Michael Godron. *Landscape Ecology.* New York: John Wiley and Sons, 1986.

Forman, Richard T. T., and Emily W. B. Russell. "Evaluation of Historical Data in Ecology." *Bulletin of the Ecological Society of America* 64 (March 1983): 5–7.

Friedrich, Ernst. "Wesen und Geographische Verbreitung der Raubwirtschaft." *Petermanns Mitteilungen* 50 (1904): 68–79, 92–95.

Galloway, Patricia. "Sources for the La Salle Expedition of 1682." In *La Salle and His Legacy: Frenchmen and Indians in the Lower Mississippi Valley,* edited by Patricia K. Galloway. Jackson: University Press of Mississippi, 1982.

Gates, Paul Wallace. "Federal Land Policies in the Southern Public Land States." In *Southern Agriculture since the Civil War: A Symposium,* edited by George L. Robson, Jr., and Roy V. Scott. Santa Barbara, Calif.: McNally and Loftin, West; and Washington, D.C.: Agricultural History Society, 1979.

————. "Federal Land Policy in the South, 1866–1888." *Journal of Southern History* 6 (August 1940): 303–30.

————. "An Overview of American Land Policy." *Agricultural History* 50 (January 1976): 213–29.

Genovese, Eugene D. *Roll, Jordan, Roll: The World the Slaves Made*. New York: Pantheon Books, 1974.

Gibson, Arrell M. *The Chikasaws*. Civilization of the American Indian Series, vol. 109. Norman: University of Oklahoma Press, 1971.

————. "The Indians of Mississippi." In *A History of Mississippi*, vol. 1, edited by Richard Aubrey McLemore. Hattiesburg: University and College Press of Mississippi, 1973.

Giles, William Lincoln. "Agricultural Revolution, 1890–1970." In *A History of Mississippi*, vol. 2, edited by Richard Aubrey McLemore. Hattiesburg: University and College Press of Mississippi, 1973.

Gilpin, Michael E., and Michael E. Soulé, "Minimum Viable Populations: Processes of Species Extinction." In *Conservation Biology: The Science of Scarcity and Diversity*, edited by Michael E. Soulé. Sunderland, Mass.: Sinauer Associates, 1986.

Glacken, Clarence. *Traces on the Rhodian Shore: Nature and Culture in the Western Thought from the Ancient Times to the End of the Eighteenth Century*. Berkeley: University of California Press, 1967.

Golley, Frank B. *A History of the Ecosystem Concept in Ecology: More Than the Sum of the Parts*. New Haven, Conn.: Yale University Press, 1993.

Gray, Lewis Cecil. *History of Agriculture in the Southern United States to 1860*. 2 vols. 1933. Carnegie Institution of Washington Publication no. 40. Reprint, Gloucester, Mass.: Peter Smith, 1958.

[Gray, William F.] "Railroads: The Yazoo and Mississippi Valley Railroad." In *Imperial Bolivar*, edited by William F. Gray. Cleveland, Miss.: Bolivar Commercial, 1923.

Greenway, James C., Jr. *Extinct and Vanishing Birds of the World*. 2nd rev. ed. New York: Dover, 1967.

Greller, Andrew M. "Deciduous Forest." In *North American Terrestrial Vegetation*, edited by Michael G. Barbour and William Dwight Billings. Cambridge: Cambridge University Press, 1988.

Grim, Ronald E. "Maps of the Township and Range System." In *From Sea Charts to Satellite Images: Interpreting North American History through Maps*, edited by David Buisseret. Chicago: University of Chicago Press, 1990.

Haag, W. G. "A Prehistory of the Lower Mississippi Valley." In *Man and Environment in the Lower Mississippi Valley*, edited by Sam B. Hilliard. Geoscience and Man, vol. 19. Baton Rouge: Louisiana State University, School of Geoscience, 1978.

Haila, Yrjö. "Assessing Ecosystem Health across Spatial Scales." In *Ecosystem Health*, edited by David Rapport, Robert Costanza, Paul R. Epstein, Connie Gaudet, and Richard Levins. Malden, Mass.: Blackwell Science, 1998.

———. "Environmental Problems, Ecological Scales, and Social Deliberation." In *Co-operative Environmental Governance: Public-Private Agreements as a Policy Strategy*, edited by P. Glasbergen. Dordrecht, the Netherlands: Kluwer Academic Publishers, 1998.

———. *Vihreään aikaan: Kirjoituksia ihmisen ekologiasta*. Helsinki: Tutkija-liitto, 1990.

———. "'Wilderness' and the Multiple Layers of Environmental Thought." *Environment and History* 3 (1997): 129–47.

Haila, Yrjö, and Richard Levins. *Humanity and Nature: Ecology, Science, and Society*. London: Pluto, 1992.

Hale, Duane K., and Arrell M. Gibson. *The Chickasaw*. Indians of North America. New York: Chelsea House, 1991.

Halsell, Willie D. "Migration into and Settlement of Leflore County, 1833–1876." *Journal of Mississippi History* 9 (October 1947): 219–37.

———. "Migration into and Settlement of Leflore County in the Later Periods, 1876–1920." *Journal of Mississippi History* 10 (July 1948): 240–60.

Harris, Marvin. *Cannibals and Kings: The Origins of Cultures*. London: Collins, 1978.

Harrison, Robert W. *Alluvial Empire: A Study of State and Local Efforts toward Land Development in the Alluvial Valley of the Lower Mississippi River, Including Flood Control, Land Drainage, Land Clearing, Land Forming*. Vol. 1. Delta Fund in cooperation with U.S. Department of Agriculture, Economic Research Service. Little Rock, Ark.: Pioneer Press, 1961.

———. "Clearing Land in the Mississippi Alluvial Valley." *Arkansas Historical Quarterly* 13 (winter 1954): 352–71.

———. "Early State Flood-Control Legislation in the Mississippi Alluvial Valley." *Journal of Mississippi History* 23 (April 1961): 104–26.

———. *Flood Control in the Mississippi Alluvial Valley*. Stoneville, Miss.: Delta Council, 1952.

———. "The Formative Years of the Yazoo-Mississippi Delta Levee District." *Journal of Mississippi History* 13 (October 1951): 236–48.

———. "Levee Building in Mississippi before the Civil War." *The Journal of Mississippi History* 12 (April 1950): 63–97.

———. *Levee Districts and Levee Building in Mississippi: A Study of State and Local Efforts to Control Mississippi River Floods*. Stoneville, Miss.: Delta Council, the Board of Mississippi Levee Commissioners, Board of Levee Commissioners for the Yazoo-Mississippi Delta, and Mississippi Agricultural Experiment Station, cooperating with U.S. Department of Agriculture, Bureau of Agricultural Economics, 1951.

Harrison, Robert W., and Joseph F. Mooney, Jr. *Flood Control and Water Management in the Yazoo-Mississippi Delta*. Social Research Report Series 93–5. Starkville, Miss.: Social Science Research Center of Mississippi State University, 1993.

Hays, Samuel P. *Conservation and the Gospel of Efficiency: The Progressive Conservation Movement, 1890–1920*. Harvard Historical Monographs, no. 40. Cambridge, Mass.: Harvard University Press, 1959.

Heavrin, Charles A. *Boxes, Baskets, and Boards: A History of Anderson-Tully Company*. Memphis: Anderson-Tully Co. and Memphis State University Press, 1981.

Helms, Douglas. "Boll Weevil." In *Encyclopedia of Southern Culture*, edited by Charles Reagan Wilson and William Ferris. Chapel Hill: University of North Carolina Press, 1989.

Henriksson, Markku. "Land Ownership and Cultural Development." In *Victorian Brand, Indian Brand: The White Shadow on the Native Image*, edited by Naila Clerici. Torino: Il Segnalibro, 1993.

Hermann, Janet Sharp. *The Pursuit of a Dream*. New York: Oxford University Press, 1981.

Herndon, G. Melvin. "Forest Products of Colonial Georgia." *Journal of Forest History* 23 (July 1979): 130–35.

———. "Indian Agriculture in the Southern Colonies." *North Carolina Historical Review* 44 (summer 1967): 283–97.

Herzhaft, Gérard. *Encyclopedia of the Blues*. Translated by Brigitte Debord. Fayetteville: University of Arkansas Press, 1992.

Hickman, Nollie W. "History of Forest Industries in the Longleaf Pine Belt of Louisiana and Mississippi, 1840–1915." Ph.D. diss., University of Texas, 1958.

———. "Logging and Rafting Timber in South Mississippi, 1840–1910." *Journal of Mississippi History* 19 (July 1957): 154–72.

———. "The Lumber Industry in South Mississippi." *Journal of Mississippi History* 20 (October 1958): 211–23.

———. "Mississippi Forests." In *A History of Mississippi*, vol. 2, edited by Richard Aubrey McLemore. Hattiesburg: University and College Press of Mississippi, 1973.

———. *Mississippi Harvest: Lumbering in the Longleaf Pine Belt, 1840–1915*. [Oxford, Miss.]: University of Mississippi, 1962.

————. "Mississippi Lumber Industry from 1840 to 1950." *Southern Lumberman* 193 (December 15, 1956): 132–37.

Higgs, Robert. "The Boll Weevil, the Cotton Economy, and Black Migration, 1910–1930." *Agricultural History* 50 (April 1976): 335–50.

Hilliard, Sam Bowers. *Atlas of Antebellum Southern Agriculture.* Baton Rouge: Louisiana State University Press, 1984.

"History of Railroads in Bolivar County, Mississippi." In *History of Bolivar County, Mississippi,* edited by Wirt A. Williams. Jackson, Miss.: Daughters of the American Revolution, Mississippi Delta Chapter, 1948.

Holloway, Richard G., and Sam Valastro. "Palynological Investigations along the Yazoo River." Appendix 1 in *Cultural Resources Survey of Items 3 and 4, Upper Yazoo River Projects, Mississippi, with a Paleoenvironmental Model of the Lower Yazoo Basin,* by Robert M. Thorne and Hugh K. Curry. Archaeological Papers of the Center for Archaeological Research, no. 3. [Oxford, Miss.]: University of Mississippi, Department of Sociology and Anthropology, Center for Archaeological Research, 1983.

Hudson, Charles M. *The Southeastern Indians.* Knoxville: University of Tennessee Press, 1976.

Hughes, J. Donald. "Whither Environmental History." *ASEH News* 8 (autumn 1997): 1–3.

Humphreys, Margaret. *Yellow Fever and the South.* Baltimore: Johns Hopkins University Press, 1999.

Hurt, R. Douglas. *Indian Agriculture in America.* Lawrence: University Press of Kansas, 1987.

Innis, Harold A. *The Fur Trade in Canada: An Introduction to Canadian Economic History.* Rev. ed., 1956. Reprint, Toronto: University of Toronto Press, 1975.

Jackson, Jerome A. "The History of Ivory-billed Woodpeckers in Mississippi." *Mississippi Kite* 18:1 (July 1988): 3–10.

————. "The Southeastern Pine Forest Ecosystem and Its Birds: Past, Present, and Future." In *Bird Conservation 3,* edited by Jerome A. Jackson. Madison: University of Wisconsin Press, 1988.

Jacobs, Wilbur R. "Indians as Ecologists and Other Environmental Themes in American Frontier History." In *American Indian Environments,* edited by Christopher Vecsey and Robert W. Venables. Syracuse, N.Y.: Syracuse University Press, 1980.

Järvinen, Olli, and Kaarina Miettinen. *Sammuuko suuri suku? Luonnon puolustamisen biologiaa.* Helsinki: Suomen Luonnonsuojelun Tuki, 1987.

Jenkins, Wm. Dunbar. "The Mississippi River and the Efforts to Confine It in Its Channel." *Publications of the Mississippi Historical Society* 6 (1902): 283–306.

Johnson, Jay K. "Prehistoric Mississippi." In *Ethnic Heritage in Mississippi*, edited by Barbara Carpenter. Jackson: Mississippi Humanities Council, 1992.

Jordan, Terry G., and Matti Kaups. *The American Backwoods Frontier: An Ethnic and Ecological Interpretation*. Creating the North American Landscape Series. Baltimore: Johns Hopkins University Press, 1989.

Kelley, Arthell. "Levee Building and the Settlement of the Yazoo Basin." *Southern Quarterly* 1 (July 1963): 285–308.

———. "Some Aspects of the Geography of the Yazoo Basin, Mississippi." Ph.D. diss., University of Nebraska, 1954.

Kellogg, Walter W. *The Kellogg Story: Fifty Years in Southern Hardwoods*. Monroe, La.: Privately printed, 1969.

Kerr, Ed. "From Timber to Famine—and Back Again: The Story of Louisiana's Forest Industries." *Southern Lumberman* 193 (December 15, 1956): 139–43.

Kidwell, Clara Sue. "The Mississippi Choctaws in the Nineteenth Century." In *Ethnic Heritage in Mississippi*, edited by Barbara Carpenter. Jackson: Mississippi Humanities Council, 1992.

King, Warren B. "Ecological Basis of Extinction in Birds." In *Acta XVII Congressus Internationalis Ornithologici*, Vol. 2. Berlin: Verlag der Deutschen Ornithologen-Gesellschaft, 1980.

Kirby, Jack Temple. "Plantations." In *Encyclopedia of Southern Culture*, edited by Charles Reagan Wilson and William Ferris. Chapel Hill: University of North Carolina Press, 1989.

———. *Poquosin: A Study of Rural Landscape and Society*. Studies in Rural Culture. Chapel Hill: University of North Carolina Press, 1995.

———. *Rural Worlds Lost: The American South, 1920–1960*. Baton Rouge: Louisiana State University Press, 1987.

———. "The Transformation of Southern Plantations, ca. 1920–1960." *Agricultural History* 57 (July 1983): 257–76.

Kitchings, J. Thomas, and Barbara T. Walton. "Fauna of the North American Temperate Deciduous Forest." In *Ecosystems of the World 7: Temperate Deciduous Forests*, edited by E. Röhrig and B. Ulrich. Amsterdam: Elsevier Science Publishers, 1991.

Köppen, Wladimir. *Die Klimate der Erde: Grundriss der Klimakunde*. Berlin: Walter de Gruyter, 1923.

Korte, Paul A., and Leigh H. Frederickson. "Loss of Missouri's Lowland Hardwood Ecosystem." In *Transactions of the Forty-second North American Wild-*

*life and Natural Resources Conference,* edited by Kenneth Sabol. Washington, D.C.: Wildlife Management Institute, 1977.

Krech, Shepard, III. *The Ecological Indian: Myth and History.* New York: W. W. Norton, 1999.

Kroeber, Alfred L. *Cultural and Natural Areas of Native North America.* University of California Publications in American Archaeology and Ethnology, vol. 38. 1939. Reprint, Berkeley: University of California Press, 1963.

Küchler, August Wilhelm. *Potential Natural Vegetation of the Coterminous United States (Map and Manual).* American Geographical Society Special Publication, no. 36. New York: American Geographical Society, 1964.

Langdon, O. Gordon, Joe P. McClure, Donal D. Hook, Joe M. Crockett, and Ron Hunt. "Extent, Condition, Management, and Research Needs of Bottomland Hardwood-Cypress Forests in the Southeastern United States." In *Wetlands of Bottomland Hardwood Forests,* edited by J. R. Clark and J. Benforado. Developments in Agricultural and Managed-Forest Ecology, no. 11. New York: Elsevier Scientific Publishing, 1981.

Larson, Lewis H. *Aboriginal Subsistence Technology on the Southeastern Coastal Plain during the Late Prehistoric Period.* Ripley P. Bullen Monographs in Anthropology and History, no. 2. Gainesville: University Presses of Florida, 1980.

Levine, Lawrence. *Black Culture and Black Consciousness: Afro-American Folk Thought from Slavery to Freedom.* New York: Oxford University Press, 1978.

Linnartz, Norwin E. "Forestry." In *The Encyclopedia of Southern History,* edited by David C. Roller and Robert W. Twyman. Baton Rouge: Louisiana State University Press, 1979.

Loewen, James W. *The Mississippi Chinese: Between Black and White.* 2nd ed. Prospect Heights, Ill.: Waveland, 1988.

Lomax, Alan. *The Land Where the Blues Began.* New York: Pantheon Books, 1993.

Long, John E., ed. *Mississippi: Atlas of Historical County Boundaries.* Compiled by Peggy Tuck Sinko. New York: Simon and Schuster, 1993.

Lugo, Ariel E. "A Review of Early Literature on Forested Wetlands in the United States." In *Cypress Swamps,* edited by Katherine Carter Ewel and Howard T. Odum. Gainesville: University Presses of Florida, 1984.

MacArthur, R. H., and E. O. Wilson. *The Theory of Island Biogeography.* Princeton, N.J.: Princeton University Press, 1967.

Mancil, Ervin. "Pullboat Logging." *Journal of Forest History* 24 (July 1980): 135–41.

Marks, Carole. *Farewell—We're Good and Gone: The Great Black Migration.* Blacks in the Diaspora. Bloomington: Indiana University Press, 1989.

Marshall, Richard A. "Archaeological Provinces of Mississippi: A Tentative Definition." In *Anthology of Mississippi Archaeology 1966–1979: A Selection from the Publications of the Mississippi Archaeological Association,* edited by Patricia Galloway. Jackson: Mississippi Archaeological Association and Mississippi Department of Archives and History, 1985.

———. *Indians of Mississippi: An Archaeological Perspective.* [Starkville, Miss.]: Mississippi State University, Cobb Institute of Archaeology, n.d.

———. "The Prehistory of Mississippi." In *A History of Mississippi,* vol. 1, edited by Richard Aubrey McLemore. Hattiesburg: University and College Press of Mississippi, 1973.

Martin, Calvin. "Fire and Forest Structures in the Aboriginal Eastern Forest." *Indian Historian* 6 (summer 1973): 23–26.

Martin, William H., and Stephen C. Boyce. "Introduction: The Southeastern Setting." In *Biodiversity of the Southeastern United States: Lowland Terrestrial Communities,* edited by William H. Martin, Stephen G. Boyce, and Arthur C. Echternacht. New York: John Wiley and Sons, 1993.

Massa, Ilmo. *Pohjoinen luonnonvalloitus: Suunnistus ympäristöhistoriaan Lapissa ja Suomessa.* Helsinki: Gaudeamus, 1994.

———. "Ympäristöhistoria tutkimuskohteena." *Historiallinen Aikakauskirja* 89 (1991): 294–301.

Maxwell, Robert S. "The Impact of Forestry on the Gulf South." *Forest History* 17 (April 1973): 31–35.

McAndrews, John H. "Human Disturbance of North American Forests and Grasslands: The Fossil Pollen Record." In *Vegetation History,* edited by Brian Huntley and Thompson Webb III. Handbook of Vegetation Science, vol. 7. Dordrecht, the Netherlands: Kluwer Academic Publishers, 1988.

McBride, Mary G., and Ann M. McLaurin. "The Origin of the Mississippi River Commission." *Louisiana History* 36 (fall 1995): 389–411.

McEvoy, Arthur F. *The Fisherman's Problem: Ecology and Law in the California Fisheries, 1850–1980.* Studies in Environment and History. Cambridge: Cambridge University Press, 1986.

McNeill, John R. "Observations on the Nature and Culture of Environmental History." *History and Theory* 42 (December 2003): 5–43.

McNeill, William H. *Plagues and Peoples.* New York: Doubleday, 1977.

McPhee, John. *The Control of Nature.* New York, Noonday Press, 1989.

Merchant, Carolyn. *The Death of Nature: Women, Ecology, and the Scientific Revolution.* San Francisco: Harper and Row, 1980.

————. *Ecological Revolutions: Nature, Gender, and Science in New England.* Chapel Hill: North Carolina University Press, 1989.

————. "Gender and Environmental History." *The Journal of American History* 76 (March 1990): 1117–21.

————. "The Theoretical Structure of Ecological Revolutions." *Environmental Review* 11 (winter 1987): 265–74.

Mills, Gary B. "New Life for the River of Death: Development of the Yazoo River Basin, 1873–1977." *Journal of Mississippi History* 41 (November 1979): 287–300.

"Mississippi Blues Musicians." Map. Clarksdale, Miss.: Rooster Blues Records and the Delta Blues Museum, 1995.

Moore, John Hebron. *Agriculture in Ante-Bellum Mississippi.* New York: Bookman Associates, 1958.

————. *Andrew Brown and Cypress Lumbering in the Old Southwest.* Baton Rouge: Louisiana State University Press, 1967.

————. "The Cypress Lumber Industry of the Lower Mississippi Valley during the Colonial Period." *Louisiana History* 24 (winter 1983): 25–48.

————. *The Emergence of the Cotton Kingdom in the Old Southwest: Mississippi, 1770–1860.* Baton Rouge: Louisiana State University Press, 1988.

Morrison, Steven J. "Downhome Tragedy: The Blues and the Mississippi River Flood of 1927." *Southern Folklore* 51:3 (1994): 265–84.

Muller, Robert A., and James E. Willis. "Climatic Variability in the Lower Mississippi River Valley." In *Man and Environment in the Lower Mississippi Valley,* edited by Sam B. Hilliard. Geoscience and Man, vol. 19. Baton Rouge: Louisiana State University, School of Geoscience, 1978.

Muzika, R. M., J. B. Gladden, and J. D. Haddock. "Structural and Functional Aspects of Succession in Southeastern Floodplain Forests Following a Major Disturbance." *American Midland Naturalist* 117 (January 1987): 1–9.

Myllyntaus, Timo. "Environment in Explaining History." In *Encountering the Past in Nature: Essays in Environmental History,* edited by Timo Myllyntaus and Mikko Saikku. Rev. ed. Athens: Ohio University Press, 2001.

Myllyntaus, Timo, and Mikko Saikku. "Environmental History: A New Discipline with Long Traditions." In *Encountering the Past in Nature: Essays in Environmental History,* edited by Timo Myllyntaus and Mikko Saikku. Rev. ed. Athens: Ohio University Press, 2001.

Nash, Roderick Frazier. *The Rights of Nature: A History of Environmental Ethics.* History of American Thought and Culture. Madison: University of Wisconsin Press, 1989.

————. *Wilderness and the American Mind.* 3rd ed. New Haven, Conn.: Yale University Press, 1982.

"New Ways to the New World." *Time International Magazine* (April 17, 2000): 54.

Nice, Margaret Morse, and Leonard Blaine Nice. *The Birds of Oklahoma.* University of Oklahoma Bulletin, no. 20. Norman: University of Oklahoma, 1924.

Odum, Eugene P. *Fundamentals of Ecology.* Philadelphia: W. D. Saunders, 1953.

Odum, Howard T. "Summary: Cypress Swamps and Their Regional Role." In *Cypress Swamps,* edited by Katherine Carter Ewel and Howard T. Odum. Gainesville: University Presses of Florida, 1984.

Oliver, Paul. *Blues Fell This Morning: Meaning in the Blues.* 2nd ed. Cambridge: Cambridge University Press, 1990.

————. *Screening the Blues: Aspects of the Blues Tradition.* London: Cassell, 1968. Reprint, New York: Da Capo, 1989.

Opie, John. "Environmental History: Pitfalls and Opportunities." *Environmental Review* 7 (spring 1983): 8–16.

————. *Nature's Nation: An Environmental History of the United States.* Fort Worth, Tex.: Harcourt Brace, 1998.

Oshinsky, David M. *"Worse Than Slavery:" Parchman Farm and the Ordeal of Jim Crow Justice.* New York: Free Press, 1996.

Otto, John Solomon. *The Final Frontiers, 1880–1930: Settling the Southern Bottomlands.* Contributions in American History, no. 183. Westport, Conn.: Greenwood, 1999.

Owens, Harry P. *Steamboats and the Cotton Economy: River Trade in the Yazoo-Mississippi Delta.* Jackson: University Press of Mississippi, 1990.

Owens, Robert M. "Jeffersonian Benevolence on the Ground: The Indian Land Cession Treaties of William Henry Harrison." *Journal of the Early Republic* 22 (fall 2002): 405–35.

Palmer, Robert. *Deep Blues.* New York: Penguin, 1982.

Patrick, William H., Jr., George Bissmeyer, Donal D. Hook, Victor W. Lambou, Helen M. Leitman, and Charles H. Wharton. "Characteristics of Wetland Ecosystems of Southeastern Bottomland Hardwood Forests." In *Wetlands of Bottomland Hardwood Forests,* edited by J. R. Clark and J. Benforado. Developments in Agricultural and Managed-Forest Ecology, no. 11. New York: Elsevier Scientific Publishing, 1981.

Peabody, Charles. "Notes on Negro Music." *Journal of American Folklore* 16 (1903).

Pendleton, M. B. "A Short History of the National Hardwood Lumber Association." *Southern Lumberman* 193 (December 15, 1956): 127–30.

Pereyra, Lillian. "James Lusk Alcorn and a Unified Levee System." *Journal of Mississippi History* 27 (February 1965): 18–41.

Perry, Percival. "The Naval Stores Industry of the Old South, 1790–1860." *Journal of Southern History* 34 (November 1968): 509–26.

Peterken, G. F. "The Use of Records in Woodland Ecology." *Archives* 14 (autumn 1979): 81–87.

Phillips, Philip. *Archaeological Survey in the Lower Yazoo Basin, Mississippi, 1949–1955.* Papers of the Peabody Museum of Archaeology and Ethnology, vol. 60. Cambridge, Mass.: Harvard University Press, 1970.

Polanyi, Karl. *The Great Transformation: The Political and Economic Origins of Our Time.* New York: Farrar and Rinehart, 1944. Reprint, Boston: Beacon, 1985.

Pollan, Michael. *Second Nature: A Gardener's Education.* New York: Dell, 1992.

Pressly, Thomas J., and William H. Scofield, eds. *Farm Real Estate Values in the United States by Counties, 1850–1959.* Seattle: University of Washington Press, 1965.

Prewitt, Wiley Charles, Jr. "The Best of All Breathing: Hunting and Environmental Change in Mississippi, 1900–1980." Master's thesis, University of Mississippi, 1991.

Prunty, Merle C. "Pulp and Paper Industry." In *The Encyclopedia of Southern History,* edited by David C. Roller and Robert W. Twyman. Baton Rouge: Louisiana State University Press, 1979.

Pyne, Stephen J. *Fire in America: A Cultural History of Wildland and Rural Fire.* Princeton, N.J.: Princeton University Press, 1982.

———. "Firestick History." *The Journal of American History* 76 (March 1990): 1132–41.

Quan, Robert Seto. *Lotus among the Magnolias: The Mississippi Chinese.* Jackson: University Press of Mississippi, 1982.

Quick, Herbert, and Edward Quick. *Mississippi Steamboatin': A History of Steamboating on the Mississippi and Its Tributaries.* New York: Henry Holt and Company, 1926.

Rappaport, Roy. *Pigs for the Ancestors: Ritual in the Ecology of a New Guinea People.* New Haven, Conn.: Yale University Press, 1967.

Rasmussen, Wayne D. "Soybeans." In *Encyclopedia of Southern Culture,* edited by Charles Reagan Wilson and William Ferris. Chapel Hill: University of North Carolina Press, 1989.

Read, Arthur D. "Forestry's Progress in Louisiana." *Southern Lumberman* 193 (December 15, 1956): 143–47.

Reuss, Martin. "Andrew A. Humphreys and the Development of Hydraulic Engineering: Politics and Technology in the Army Corps of Engineers, 1850–1950." *Technology and Culture* 26 (January 1985): 1–33.

———. "The Army Corps of Engineers and Flood-Control Politics on the Lower Mississippi." *Louisiana History* 23 (spring 1982): 131–48.

Robbins, Chandler S., Bertel Bruun, and Herbert S. Zim. *Birds of North America: A Guide to Field Identification.* Illustrated by Arthur Singer. Rev. ed. New York: Golden, 1986.

Robinson, Michael C. "Reclamation and Irrigation." In *Encyclopedia of Southern Culture,* edited by Charles Reagan Wilson and William Ferris. Chapel Hill: University of North Carolina Press, 1989.

Royall, P. Daniel, Paul A. Delcourt, and Hazel R. Delcourt. "Late Quaternary Paleoecology and Paleoenvironments of the Central Mississippi Alluvial Valley." *Bulletin of the Geological Society of America* 103 (February 1991): 157–70.

Russell, Emily W. B. "Indian-Set Fires in the Forests of the Northeastern United States." *Ecology* 64 (February 1983): 78–88.

Sahlins, Marshall. "Poor Man, Rich Man, Big-Man, Chief: Political Types in Melanesia and Polynesia." *Comparative Studies in Society and History* 5 (1963): 285–303.

Saikku, Mikko. "Down by the Riverside: The Disappearing Bottomland Hardwood Forest of Southeastern North America." *Environment and History* 2 (February 1996): 77–95.

———. *The Evolution of a Place: Patterns of Environmental Change in the Yazoo-Mississippi Delta from the Ice Age to the New Deal.* Renvall Institute Publications, no. 12. Helsinki: Renvall Institute for Area and Cultural Studies, 2001.

———. "A extinção nas areas de expansão Europeia—perigo de vida das aves na região do Atlântico Norte." In *História e Meio-Ambiente: O impacto da extensão Europeia.* Colecção Memórias, no. 26. Translated by Teresa Mizon. Funchal, Portugal: Centro de Estudos de História do Atlântico, 1999.

———. "The Extinction of the Carolina Parakeet." *Environmental History Review* 14 (fall 1990): 1–18.

———. "'Home in the Big Forest': Decline of the Ivory-billed Woodpecker and Its Habitat in the United States." In *Encountering the Past in Nature: Essays in*

*Environmental History*, edited by Timo Myllyntaus and Mikko Saikku. Rev. ed. Athens: Ohio University Press, 2001.

————. "The Second Wave in the Southern Woods: European Impact on the Bachman's Warbler and Its Forest Habitat." In *In Search of a Continent: A North American Studies Odyssey*, edited by Mikko Saikku, Maarika Toivonen, and Mikko Toivonen. Helsinki: Renvall Institute for Area and Cultural Studies, 1999.

Scheiber, Harry N., and Harold G. Vatter. *American Economic History*. 9th ed. New York: Harper and Row, 1976.

Schob, David E. "Woodhawks and Cordwood: Steamboat Fuel on the Ohio and Mississippi Rivers, 1820–1860." *Journal of Forest History* 21 (July 1977): 124–33.

Sclater, P. L. "On the General Geographical Distribution of the Members of the Class Aves." *Journal of the Linnaean Society, Zoology* 2 (1858): 130–45.

Seavoy, Ronald E. "Fletcher v. Peck." In *The Encyclopedia of Southern History*, edited by David C. Roller and Robert W. Twyman. Baton Rouge: Louisiana State University Press, 1979.

Seay, James. "The Southern Outdoors: Bass Boats and Bear Hunts." In *The American South: Portrait of a Culture*, edited by Louis D. Rubin, Jr. Baton Rouge: Louisiana State University Press, 1980.

Shands, William E., and Robert G. Healy. *The Lands Nobody Wanted: Policy for National Forests in the Eastern United States*. Conservation Foundation Report. Washington, D.C.: The Conservation Foundation, 1977.

Sharitz, Rebecca R., and William J. Mitsch. "Southern Floodplain Forests." In *Biodiversity of the Southeastern United States: Lowland Terrestrial Communities*, edited by William H. Martin, Stephen G. Boyce, and Arthur C. Echternacht. New York: John Wiley and Sons, 1993.

Shelford, Victor E. *The Ecology of North America*. Urbana: University of Illinois Press, 1963.

Shipley, John R. "The Story of the Chicago Mill and Lumber Company." In *Washington County Historical Society Programs of 1980*. Greenville, Miss.: Washington County Historical Society, 1980.

Sillers, Walter, Sr. "Flood Control in Bolivar County, 1838–1924." *Journal of Mississippi History* 9 (January 1947): 3–20.

————. "Isaiah T. Montgomery." In *History of Bolivar County, Mississippi*, edited by Wirt A. Williams. Jackson, Miss.: Daughters of the American Revolution, Mississippi Delta Chapter, 1948.

————. "Levees of the Mississippi Levee District in Bolivar County." In *History of Bolivar County, Mississippi,* edited by Wirt A. Williams. Jackson, Miss.: Daughters of the American Revolution, Mississippi Delta Chapter, 1948.

Silver, James W. "Paul Bunyan Comes to Mississippi." *Journal of Mississippi History* 19 (April 1957): 93–119.

Silver, Timothy. *A New Face on the Countryside: Indians, Colonists, and Slaves in South Atlantic Forests, 1500–1800.* Studies in Environment and History. Cambridge: Cambridge University Press, 1990.

Simberloff, Daniel. "The Proximate Causes of Extinction." In *Patterns and Processes in the History of Life,* edited by D. M. Raup and D. Jablonski. Berlin: Springer, 1986.

Simmons, I. G. *Changing the Face of the Earth: Culture, Environment, History.* Oxford: Basil Blackwell, 1989.

Simpson, Lewis P. "Ike McCaslin and the Second Fall of Man." In *Bear, Man, and God: Eight Approaches to William Faulkner's "The Bear,"* edited by Francis Lee Utley, Lynn Z. Bloom, and Arthur F. Kinney. 2nd ed. New York: Random House, 1971.

Sinclair, J. D. "Studies of Soil Erosion in Mississippi." *Journal of Forestry* 29 (April 1931): 533–40.

Siniard, L. Arnold. "Soybean." In *The Encyclopedia of Southern History,* edited by David C. Roller and Robert W. Twyman. Baton Rouge: Louisiana State University Press, 1979.

Smallwood, J. B. "Water Use." In *Encyclopedia of Southern Culture,* edited by Charles Reagan Wilson and William Ferris. Chapel Hill: University of North Carolina Press, 1989.

Smith, Bruce D. "The Archaeology of the Southeastern United States, from Dalton to de Soto, 10,500–500 B.P." In *Advances in World Archaeology,* vol. 5, edited by Fred Wendorf and Angela E. Close. Orlando, Fla.: Academic Press, 1986.

———. "Variation in Mississippian Settlement Patterns." In *Mississippian Settlement Patterns,* edited by Bruce D. Smith. Studies in Archaeology. New York: Academic Press, 1978.

Smith, Claude P. "Official Efforts by the State of Mississippi to Encourage Immigration, 1868–1886." *Journal of Mississippi History* 32 (November 1970): 327–40.

Smith, David M. "The Forests of the United States." In *Regional Silviculture of the United States,* edited by John W. Barret. 2nd ed. New York: John Wiley and Sons, 1980.

Smith, David Wm., and Norwin E. Linnartz. "The Southern Hardwood Region." In *Regional Silviculture of the United States,* edited by John W. Barret. 2nd ed. New York: John Wiley and Sons, 1980.

Smith, Frank E. *The Yazoo River*. Reprint, Jackson: University Press of Mississippi, 1988.

Stauffer, Dean F., and Louis B. Best. "Habitat Selection by Birds of Riparian Communities: Evaluating Effects of Habitat Alterations." *Journal of Wildlife Management* 44 (January 1980): 1–15.

Sternitzke, Herbert S. "Impact of Changing Land Use on Delta Hardwood Forests." *Journal of Forestry* 74 (January 1976): 25–27.

Steward, Julian. *Theory of Culture Change: The Methodology of Multilinear Evolution*. Urbana: University of Illinois Press, 1963.

Stewart, Mart. *"What Nature Suffers to Groe": Life, Labor, and Landscape on the Georgia Coast, 1680–1920*. Wormsloe Foundation Publications, no. 19. Athens: University of Georgia Press, 1996.

Stover, John F. *History of the Illinois Central Railroad*. New York: Macmillan, 1975.

———. *The Railroads of the South, 1865–1900: A Study in Finance and Control*. Chapel Hill: University of North Carolina Press, 1955.

Stubbs, John D., Jr. "The Chickasaw Contact with the La Salle Expedition in 1682." In *La Salle and His Legacy: Frenchmen and Indians in the Lower Mississippi Valley*, edited by Patricia K. Galloway. Jackson: University Press of Mississippi, 1982.

Swift, Bryan L., Joseph S. Larson, and Richard M. DeGraaf. "Relationship of Breeding Bird Density and Diversity to Habitat Variables in Forested Wetlands." *Wilson Bulletin* 96 (March 1984): 48–59.

Tanner, James T. *The Ivory-billed Woodpecker*. National Audubon Society Research Report, no. 1. New York: National Audubon Society, 1942.

———. "Present Status of the Ivory-billed Woodpecker." *Wilson Bulletin* 54:1 (March 1942): 57–58.

Thomas, David Hurst. "The World as It Was." In *The Native Americans: An Illustrated History*, edited by Betty Ballantine and Ian Ballantine. Atlanta: Turner, 1993.

Thorne, Robert M. and Hugh K. Curry. *Cultural Resources Survey of Items 3 and 4, Upper Yazoo River Projects, Mississippi, with a Paleoenvironmental Model of the Lower Yazoo Basin*. Appendices by Richard G. Holloway, Sam Valastro, Allen Saltus, and Roger T. Saucier. Archaeological Papers of the Center for Archaeological Research, no. 3. [Oxford, Miss.]: University of Mississippi, Department of Sociology and Anthropology, Center for Archaeological Research, 1983.

*Townsend, Eugenia Dixon. "Mound Bayou." In History of Bolivar County, Mississippi*, edited by Wirt A. Williams. Jackson, Miss.: Daughters of the American Revolution, Mississippi Delta Chapter, 1948.

Trimble, Stanley W. "Nature's Continent." In *The Making of the American Landscape,* edited by Michael P. Conzen. Boston: Unwin Hyman, 1990.

Trowell, C. T., and R. L. Izlar. "Jackson's Folly: The Suwanee Canal Company in the Okefenokee Swamp." *Journal of Forest History* 28 (October 1984): 187–95.

Tuan, Yi-Fu. *Space and Place: The Perspective of Experience.* Minneapolis: University of Minnesota Press, 1977.

Turner, Frederick Jackson, "The Significance of the Frontier in American History." 1894. Reprinted in *Major Problems in the History of the American West,* edited by Clyde A. Milner II. Lexington, Mass.: Heath, 1989.

Turner, R. Eugene, Stephen W. Forsythe, and Nancy J. Craig. "Bottomland Hardwood Forest Land Resources of the Southeastern United States." In *Wetlands of Bottomland Hardwood Forests,* edited by J. R. Clark and J. Benforado. Developments in Agricultural and Managed-Forest Ecology, no. 11. New York: Elsevier Scientific Publishing, 1981.

Urban, Dean L., Robert V. O'Neill, and Herman H. Shugart, Jr. "Landscape Ecology." *BioScience* 37 (February 1987): 119–27.

Usner, Daniel H., Jr. *American Indians in the Lower Mississippi Valley: Social and Economic Histories.* Indians in the Southeast Series. Lincoln: University of Nebraska Press, 1998.

———. "American Indians of the East." In *Encyclopedia of American Social History,* vol. 2, edited by Mary Kupiec Cayton, Elliot J. Gorn, and Peter W. Williams. New York: Charles Scribner's Sons, 1993.

———. *Indians, Settlers, and Slaves in a Frontier Exchange Economy: The Lower Mississippi Valley Before 1783.* Chapel Hill: The Institute of Early American History and Culture and the University of North Carolina Press, 1992.

Utley, Francis Lee. "Pride and Humility: The Cultural Roots of Ike McCaslin." In *Bear, Man, and God: Eight Approaches to William Faulkner's "The Bear,"* edited by Francis Lee Utley, Lynn Z. Bloom, and Arthur F. Kinney. 2nd ed. New York: Random House, 1971.

Utterström, Gustaf. "Climatic Fluctuations and Population Problems in Early Modern History." In *The Ends of the Earth: Perspectives on Modern Environmental History,* edited by Donald Worster. Studies in Environment and History. Cambridge: Cambridge University Press, 1988.

Vance, Rupert Bayless. *Human Factors in Cotton Culture: A Study in the Social Geography of the American South.* Chapel Hill: University of North Carolina Press, 1929.

————. *Human Geography of the South: A Study in Regional Resources and Human Adequacy.* Chapel Hill, University of North Carolina Press, 1932.

Vankat, John L. *The Natural Vegetation of North America: An Introduction.* New York: John Wiley and Sons, 1979.

Vileisis, Ann. *Discovering the Unknown Landscape: A History of America's Wetlands.* Washington, D.C.: Island Press, 1999.

Walker, Laurence C. *The Southern Forest: A Chronicle.* Austin: University of Texas Press, 1991.

Wallerstein, Immanuel. *The Modern World-System: Capitalist Agriculture and the Origin of the European World-Economy in the Sixteenth Century.* New York: Academic Press, 1974.

Webb, Thompson, III. "Eastern North America." In *Vegetation History,* edited by Brian Huntley and Thompson Webb III. Handbook of Vegetation Science, vol. 7. Dordrecht, the Netherlands: Kluwer Academic Publishers, 1988.

————. "The Past 11,000 Years of Vegetational Change in Eastern North America." *BioScience* 31 (July-August 1981): 501–6.

Weeks, Linton. *Clarksdale and Coahoma County: A History.* Clarksdale, Miss.: Carnegie Public Library, 1982.

Wells, Samuel J. "Trail of Treaties: The Choctaws from the American Revolution to Removal." In *Ethnic Heritage in Mississippi,* edited by Barbara Carpenter. Jackson: Mississippi Humanities Council, 1992.

Wharton, Charles H., Victor M. Lambou, John Newsom, Parley V. Winger, L. L. Gaddy, and Rudy Mancke. "The Fauna of Bottomland Hardwoods in Southeastern United States." In *Wetlands of Bottomland Hardwood Forests,* edited by J. R. Clark and J. Benforado. Developments in Agricultural and Managed-Forest Ecology, no. 11. New York: Elsevier Scientific Publishing, 1981.

Whayne, Jeannie M., ed. *Shadows over Sunnyside: An Arkansas Plantation in Transition, 1830–1945.* Fayetteville: University of Arkansas Press, 1993.

White, Lynn, Jr., "The Historic Roots of Our Ecological Crisis," *Science* 155 (March 10, 1967): 1202–7.

White, Richard. "American Environmental History: The Development of a New Historical Field." *Pacific Historical Review* 54 (1985): 297–335.

————. "Environmental History, Ecology, and Meaning." *The Journal of American History* 76 (March 1990): 1111–16.

————. *Land Use, Environment, and Social Change: The Shaping of Island County, Washington.* 1980. Reprint, with a new preface, Seattle: University of Washington Press, 1992.

————. "Native Americans and the Environment." In *Scholars and the Indian*

*Experience: Critical Reviews of Recent Writing in the Social Sciences*, edited by W. R. Swagerty. Bloomington: University of Indiana Press, 1984.

———. *The Roots of Dependency: Subsistence, Environment, and Social Change among the Choctaws, Pawnees, and Navajos*. Lincoln: University of Nebraska Press, 1983.

Williams, Michael. *Americans and Their Forests: A Historical Geography*. Studies in Environment and History. New York, Cambridge University Press, 1989.

———. "The Clearing of the Forests." In *The Making of the American Landscape*, edited by Michael P. Conzen. Boston: Unwin Hyman, 1990.

———. "Clearing the United States Forests: Pivotal Years 1810–1860." *Journal of Historical Geography* 8 (January 1982): 12–28.

———. "Products of the Forest: Mapping the Census of 1840." *Journal of Forest History* 24 (January 1980): 4–23.

Williams, Stephen, and Jeffrey P. Brain. *Excavations at the Lake George Site, Yazoo County, Mississippi, 1958–1960*. Papers of the Peabody Museum of Archaeology and Ethnology, vol. 74. Cambridge, Mass.: Harvard University Press, 1983.

Willis, John C. *Forgotten Time: The Yazoo-Mississippi Delta after the Civil War*. Charlottesville: University Press of Virginia, 2000.

———. "On the New South Frontier: Life in the Yazoo-Mississippi Delta, 1865–1920." Ph.D. diss., University of Virginia, 1991.

Wilson, Edward O. *The Diversity of Life*. Cambridge, Mass.: Harvard University Press, 1992.

Wissler, Clark. *The Relation of Nature to Man in Aboriginal North America*. New York: Oxford University Press, 1926.

Wittfogel, Karl A. "The Hydraulic Civilizations." In *The Man's Role in Changing the Face of the Earth*, edited by William L. Thomas, Jr. Chicago: University of Illinois Press, 1956.

Wolf, Eric R. *Europe and the People without History*. Berkeley: University of California Press, 1982.

Woodman, Harold D. *King Cotton and His Retainers: Financing and Marketing the Cotton Crop of the South, 1800–1925*. Lexington: University of Kentucky Press, 1968. Reprint, Columbia: University of South Carolina Press, 1990.

Woods, Clyde. *Development Arrested: Race, Power, and the Blues in the Mississippi Delta*. The Haymarket Series. London: Verso, 1998.

Woodward, C. Vann. *Origins of the New South: 1877–1913*. Baton Rouge: Louisiana State University Press, 1951.

Worster, Donald. "Doing Environmental History." In *The Ends of the Earth:*

*Perspectives on Modern Environmental History,* edited by Donald Worster. Studies in Environment and History. Cambridge: Cambridge University Press, 1988.

———. *Dust Bowl: The Southern Plains in the 1930s.* New York: Oxford University Press, 1979.

———. "Ecology of Order and Chaos." *Environmental History Review* 14 (spring-summer 1990): 1–18.

———. "History as Natural History: An Essay in Theory and Method." *Pacific Historical Review* 53 (February 1984): 1–19.

———. "Nature and the Disorder of History." *Environmental History Review* 18 (summer 1994): 1–15.

———. *Nature's Economy: A History of Ecological Ideas.* Studies in Environment and History. New ed. Cambridge: Cambridge University Press, 1985.

———. *Rivers of Empire: Water, Aridity, and the Growth of the American West.* New York: Oxford University Press, 1985.

———. "Seeing Beyond Culture." *The Journal of American History* 76 (March 1990): 1142–47.

———. "Transformations of the Earth: Toward an Agroecological Perspective in History." *The Journal of American History* 76 (March 1990): 1087–106.

Wunder, John. "American Law and Order Comes to Mississippi Territory: The Making of Sargent's Code." *Journal of Mississippi History* 38 (1976): 131–55.

Zelinsky, Wilbur. "Landscapes." In *Encyclopedia of American Social History,* vol. 2, edited by Mary Kupiec Cayton, Elliot J. Gorn, and Peter W. Williams. New York: Charles Scribner's Sons, 1993.

Ziswiler, Vinzenz. *Extinct and Vanishing Animals.* Rev. English ed. by Fred and Pille Bunnell. New York: Springer, 1967.

# Index

Unless otherwise indicated, settlements, towns, and counties are located in Mississippi.

Abbey, Richard, 190, 225–26
Adair, James, 70
Adams County, 29
Agricultural Adjustment Administration (AAA), 236
agriculture: colonial, 87–88; land clearing for purposes of, 109, 134–35, 220, 237, 238–39, 240–41, 242; mechanization of cotton, 132, 237; on National Wildlife Refuges, 246; post–World War II, in bottomlands, 238–39; regional shifts in cotton, 132; slash-and-burn, 62–63, 106; swidden, 62–63; utilization of cutovers for, 186, 216–17. *See also* corn; cotton; plantations
Alabama River, 31
alcohol, consumption by levee workers and loggers, 147, 185, 196
Alcorn, James Lusk, 146
Allen, Charles B., 198
Allen, LeRoy Barry, 193
Alliance Trust Co., 211

alligator, American, 47, 235, 245, 267
  (n. 32)
Altamaha River, 31
American Forestry Association, 217
American Hardwoods Manufacturers
  Association, 198
amphibians, 46–47
Anderson-Tully Company, 194, 246
Andrew Brown Lumber Company:
  becomes Learned Lumber Company,
  183; logging operations by, 181–83
Apalachicola River, 31
Arkabutla Reservoir, 164
Arkansas Oak Flooring Company,
  214
Arkansas River, 31, 72, 163
Arnold, Kokomo, 131
ash, 41, 169, 187, 197, 198; green, 42
Ashe, W. W., 217
Atchafalaya River, 31, 163, 174
Audubon, John James, 42, 49–50, 67,
  143, 222, 234

Bailey, Robert E., 27
Baily, Francis, 47
baldcypress. *See* cypress
Baton Rouge, La., 161, 180
batture land, 37, 162, 244
bear, black, 47, 143, 231, 245, 252;
  hunting of, by Native Americans,
  66; population declines of, 83, 233,
  234; sport hunting of, 112, 232–34
beaver, 47
beech, 43
Belzoni, 238
Benton, Thomas Hart, 144
Berry, C. Fred, 200–214
Berry, James W., 206, 211, 212–13,
  214
Berryman, Clifford, 234

Beulah, 213
Bienville, Jean Baptiste Le Moyne,
  Sieur de, 73, 142
Big Black River, 167
Biltmore Forest School, 216
bioregion: concept of, 24–25; Delta as,
  247
birch, river, 42
birds, 48–50, ; hunting of, 232, 234–
  35; on Mississippi National Wildlife
  Refuges, 245; Native Americans
  and, 67–68; as pest control, 114;
  in slave folklore, 101–2. *See also
  individual species*
bison, 47, 64, 83, 230
Black Bayou, 181
Black Creek, 179
Black Creek Wilderness Area, 244
blacks: emigration of, from Delta, 123;
  immigration of, to Delta, 114–16,
  127; as landowners, 116–17, 122–
  23. *See also* blues; sharecroppers;
  slaves; tenancy
Blanton, Orville M., 108, 125, 232,
  309 (n. 39)
Blantonia, 103
blues, 123; on boll weevil, 131; on
  flooding, 149, 157–58
Blytheville, Leachville & Arkansas
  Southern Railroad, 200
Bobet brothers (Edward J. and
  Alphonse), stave purchases by,
  188–89
Bogue Phalia River, 179
Bolivar County, 28, 41, 98; agriculture
  in, 97, 109, 129, 222; attempts
  at flood control in, 148; cotton
  production in, 110, 126; crevasse
  in, 157; cypress in, 179–80, 187;
  hardwoods in, 187; immigrants in,

147; logging in, 194, 212–13, 228;
population of, 106, 127; sawmills
in, 187; wildlife in, 50, 233
boll weevil, 129, 130–32; in blues, 131
Bond, James, 270 (n. 57)
bottomland hardwood forests: animal
communities of, 45–50; classification of, 44–45; climate of, 31–34;
decline of, 214, 221, 240, 242–43;
definition of, 30; development of,
33–34, 37–38; effects of fire on,
44, 215; effects of flooding on,
36–37; effects of war effort on,
212–13, 228; extent of, 30–31;
old-growth, 223, 225–26, 243;
plant communities of, 40–45; soils
of, 37, 38–39; succession in, 41–44;
surveys of, 196, 214, 219, 221–26;
synonyms for, 30; topography of,
37–38, 45. *See also* hardwoods
"Bo-Weavil Blues," 131
Bowman, R. H., 150
boxelder, 41, 43, 95
Brandfon, Robert L., 4
Braudel, Fernand, 7, 15–16, 24
Brown, Andrew, 169, 181–83
buffalo. *See* bison
buffalofish, 100, 102
butternut, 64

Caernarvon, 159
Cahokia, 60
Cairo, Ill., 27, 150, 200
Calhoun, G. W., 201, 202–3, 206
Calhoun, John C., 144
Calhoun County, 193
Cameron, Paul C., 99, 103, 108–9
cane, 41, 42, 44, 115; as indicator of
good soil, 97; Native American uses
of, 68. *See also* canebrakes

canebrakes, 40, 41, 47, 95–96;
clearing of, 102–3; as pastureland,
83, 103. *See also* cane
Cape Fear River, 31
Carrier Lumber Manufacturing
Company, 194
Carroll County, 28; sawmills in,
187
Carter, Alfred Grayson, 152
Cary, Austin, 217
Catesby, Mark, 67
catfish, 66, 102; farming of, 238, 242
Chakchiuma, 72, 73, 78
Chambers, Rowland, 32, 232
Chapman, H. H., 217
Charleston, 194
Chattahoochee River, 31
Chicago Mill and Lumber Company:
closing of Greenville plant by, 239;
enters agribusiness, 238; founding
of, 214; operations of, on Singer
Tract, 227–28
Chickasaw, 72, 73, 74, 75, 79;
mixed-blood, 82–85; treaties of,
with U.S. government, 85
Choctaw, 70, 72, 73, 74, 75, 79, 98;
in fur trade, 80–81; mixed-blood,
82–85, 96; treaties of, with U.S.
government, 85
Choula, 72, 73
Civil War, effects of: on Delta
landscape, 187; on levee system,
147; on southern economy, 110–11,
112, 117, 172–73
Claiborne County, 29
Clark, Charles, 97–98
Clarksdale, 124
Clay, Henry, 144
Clement, W. B., 188–89
Cleveland, 129

climate: effects of, on Native
  Americans, 54–56; flooding
  enhanced by changes in, 141;
  historical changes of, 33–34; of
  South, 29, 31–32
Coahoma County, 28, 43, 106, 124,
  146, 222, 223, 306 (n. 156); cypress
  in, 179–80; lack of flood control in,
  149; land sales in, 98
Cobb, James C., 4
Cohn, David L., 26
Cold River, 95
Coldwater River, 28, 141, 161,
  163–64
Coles Creek, 184
Collier, Holt, 234
Collot, Victor, 95
Columbus and Greenville Railroad,
  124
Concordia Island, 41, 194–95
convict labor, 120, 149
Copiah County, 121
corn: Euro-American cultivation of,
  99, 100, 246; Native American
  cultivation of, 59, 60–61, 62–64,
  68
cotton, 90–91; culture, 91; Delta
  production of, 110, 126, 134,
  135–37; Egyptian, 132, 134; gins,
  91–92; long-staple, 134; pests of,
  130–32; Sea Island, 90, 132, 134;
  short-staple, 134; in Southern
  economy, 92, 133–34; upland, 134;
  U.S. production of, 132–34, 236;
  varieties of, 134
cotton bollworm, 130–31
cotton budworm, 130–31
cottonwood, 41, 42, 68, 187, 225;
  demand for, 200, 213–14; uses for,
  197

cougar, 47, 226, 231, 236, 252; sport
  hunting of, 232–33
crevasses, 142, 152, 156, 157, 159
Cronon, William, 12
crop lien system, 110–11, 117, 118,
  119
Crosby, Alfred W., 18
Crystal Springs, 121
Cuming, F., 105
Curry, Hugh K., 94, 95
cypress, 42, 95, 96, 243, 246;
  in California Brake, 179, 183;
  depletion in Delta of, 178–80, 225;
  in Hollywood Brake, 184; logging
  of, 167, 169–71, 180–81, 185–86;
  Native American use of, 58, 68; on
  Panther Burn Plantation, 201, 203,
  206; surveys of, 178–80; uses for,
  169, 197; value of, 185–86

Dahomy, 213
Darnell-Love Lumber Company, 201,
  204, 206–7
Davis, Jefferson, 112
Davis, Joseph E., 112–13
Davis Bend (Island), 112–14
Dawson, S. S., 189
deer, white-tailed, 47, 244, 245;
  hunting of, by Euro-Americans,
  143, 232, 233; hunting of,
  by Native Americans, 64–66;
  population dynamics of, 78–79, 81,
  83. *See also* fur trade
Deer Creek, 28, 96, 141, 179
de la Vega, Garciliaso, 62
Delcourt, Hazel R., 96
Delta, the: antebellum settlement
  of, 96–99, 100–101, 102–3;
  backwater area of, 161–62, 222–23,
  225, 241; "border" counties of,

28–29, 187; climate history of, 34; "core" counties of, 28, 187; cotton production in, 110, 126, 134, 135–37; decline of animal populations in, 226–27, 228–29, 231–36; definitions of, 2, 26, 27–29; drainage development in, 100, 151, 154–56, 162–63; as ecological complex, 50–51; environmental change in, 2–3, 85–86, 222–26, 247–55; ethnic minorities in, 120–21, 147, 195–96; European and American conquest of, 72, 74–75, 76; groundwater in, 241–42; hardwood reserves in, 187–88; historiography of, 3–4; hydrology of; 35–36, 43–44; irrigation in, 242; land clearing in, 108–10, 135, 136 (table 2), 137, 222; land forfeitures in, 147–48, 239; land sales in, 98, 124, 198; land surveys of, 44, 94–96; land values in, 98, 110, 124, 127, 128, 135, 198; Native American tribes of, 71–72; ownership of timberlands in, 199; population of, 106, 107 (table 1), 126, 135, 237; predominance of sharecropping in, 122; small farmers in, 114–17, 123–27; soils of, 39–40; taxation in, 147–48, 150–51; topography of, 34–35; transportation in, 97, 124, 145; value of, as animal habitat, 45–46; weather of, 33. *See also individual counties*

Delta City, 95

Delta Cooperage Company, 193–94

Delta National Forest, 243–44

Delta Pine and Land Company: decline of tenancy on lands of, 237; development of, 127; effects of 1927 flood on, 159; land purchases by, 124, 174; as recipient of AAA payments, 236; as soybean planting seed supplier, 239; use of fertilizers by, 133

dependency: definition of, 80; Native American, 80–85

de Soto, Hernando, 62, 72, 77, 78, 140

DeSoto County, 28; lack of flood control in, 149

de Spain, Major (literary character), 232. *See also* Faulkner, William

disease, 104–5, 119

Dismal Swamp, 31

Dockery, William "Will," 129

dogwood, 41, 43, 95

Dogwood Ridge, 43

Drew, 121

Dyche, J. K., 189

Eastern Deciduous Forest, 29–30

Eckleberry, Don, 228

ecohistorical formations, 19–20; in Delta, 248, 250–51 (table 5); identification of, 20

ecohistorical periods, 15–19; definition of, 16; in Delta, 247–48, 250–51 (table 5)

ecological complex, 22; Delta as, 50–51; versus ecosystem, 21–22

ecological imperialism, 18, 261–62 (n. 33)

ecological revolutions, theory of, 248–49; as applied to Delta, 249, 252

ecology, scientific, 21

ecoregion: concept of, 26–27; Delta as, 247

ecosystem, concept of, 21. *See also* ecological complex

ecotones, 45

Egleston, Nathaniel H., 216

elm, water, 42

elms, 41, 95; American, 42; uses for, 197

Enid Reservoir, 164

environmental change, 13; study of, 13–15

environmental historiography: approaches to, 8–10; characteristics of, 7, 11–12; definitions of, 5–6; and ecological imperialism, 18; history of, 5; importance of, 6–7, 12, 247; methodology of, 10–11; scaling in, 15, 20–24; sources for, 4–5, 10–11; of the South, 3–4. *See also* ecohistorical formations; ecohistorical periods; ecological revolutions, theory of; *and individual authors*

erosion, 40, 55, 130, 245–46; and flood of 1927, 159; geological role of, in creating bottomlands, 35, 38, 44; and logging, 220; and Native American agriculture, 63; and regulation of river flows, 241

extinctions: biology of, 226–27, 229–31; in Delta: 226, 227–29, 235–36

Faisonia, 150

Faulkner, William, 1, 109–10, 191, 223, 253–54; epigraphs by, 1, 26, 52, 87, 138, 165, 220; Ike McCaslin character of, 164, 255, 302 (n. 96); Major de Spain character of, 232

Fayette, 97

Featherstonhaugh, G. W., 215

fertilizers, 133, 237

Fickle, James E., 4

fish, 46, 102, 246; buffalofish, 100, 102; catfish, 66, 102, 238, 242

Fish, Stuyvesant, 233

fishing: by Native Americans, 55, 64, 66; prohibited in Yazoo National Wildlife Refuge, 246; by settlers in the Delta, 100, 102

flood control: conflicting interests in, 148, 151–52; construction of reservoirs for, 164; demands for, 144; environmental effects in Delta of, 252–54; federal involvement in, 149, 152, 153, 159–61; French attempts at, 141–42; importance of, 138–39; local attempts at, 143–49; technical developments in, 153, 161. *See also* flooding; levees; Mississippi floods

flooding: backwater, 35, 36, 157; effects of, on animals, 46; effects of, on plants, 41–44; headwater, 35, 36; in the Lower Mississippi Valley, 35, 140–41; and Native Americans, 62; types of, 35. *See also* flood control; Mississippi floods

Flores, Dan, 21, 24–25

Foote, Huger Lee, 234

Ford, Charles W., 205

forest conservation, 214–18; legislation on, 214–15, 216, 217, 221; and prevention of fires, 214–15, 217

forestry, scientific, 216–19, 240, 242, 243

forests: Eastern Deciduous Forest, 29–30; River Bottom Forest, 30; Southeastern Bottomland Hardwood Forest, 30; Southern Floodplain Forest, 27, 29, 30. *See also* bottomland hardwood forests; hardwoods

Fort St. Pierre, 74

Friedrich, Ernst, 21

fur trade, 73, 80–81, 143

Gale, Thomas, 102

Gary, W. W., 208–9

George T. Houston & Company, 191–92

Grady, Henry W., 172

Grant, U. S., 147

grapes, 41, 43

grasslands, in the Delta, 70, 74, 83, 95

Gray, Allan, 212–13, 228, 245

Great Southern Lumber Company, 217

Greeley, Horace, 144

Greene County, 244

Greenville, 153, 179, 238, 246; flood of 1927 in, 157, 158; founding of, 97; lumber industry in, 193, 199–200, 214; yellow fever epidemic in, 104

Greenwood, 164, 246; attempts to attract settlers to, 123–24; as center for cotton marketing, 134; founding of, 97; lumber industry in, 193

Grenada County, 28, 245

Grenada Reservoir, 164

Griffin, Y. F., 100–101

Griswold, James F., 228

gums, 41, 95, 187. *See also* sweetgum

hackberry, 41, 42; uses for, 197

Haila, Yrjö, 13, 16, 19–20, 22, 260 (n. 23)

Hall, Basil, 142–43, 168

Hall, J. N., 203

Hamilton, Frank, 101, 128–29, 194–96

Hamilton, Jones S., 120

Hamilton, Mary, 44, 128–29, 194–96

Hammond, James H., 92

Hampton, Wade, I, 98–99

Hampton, Wade, II, 99, 105

Hampton, Wade, III, 99, 104–5, 111–12, 146, 232–33

Hardtner, Henry E., 217

Hardwood Manufacturers Association of the United States, 198

hardwoods: exhaustion of, 194, 197, 199, 212–13, 240; logging of, 186–87, 189–90, 193–98; purchase prices for Delta, 202, 207; surveys of, 187–88, 190, 196, 221; value of, 196, 198; wasting of, 188. *See also individual logging companies*

Hardwoods Manufacturers Institute, 218

Harper, Lewis, 178

Harris, Marvin, 8

Harrison, Robert W., 4, 253

Helena, Ark., 27, 189, 200; first sawmill in, 169

Helm, George M., 234

Helme, Charles, 170

herbicides, 237

Hesse-Wartegg, Ernst von, 123, 130–31, 135

Hickman, Nollie W., 297 (n. 5)

hickories, 41, 68, 95, 96, 198; shagbark, 42; wasting of, 188; water, 42, 225

"High Water Everywhere," 158, 159

Hilgard, Eugene, 178

Hindman, William I., 115, 181–83

*histoire de la longue durée*, 7, 24; as expounded by Braudel, 15–16

Holland Landing, 183
Holly Springs, 104
Holmes County, 28, 43, 244;
    hardwoods in, 187; sawmills in, 187
Homochitto River, 167
honeylocust, 41, 43, 95
Hoover, Herbert, 158
Houma, 72, 73
Howard, Stephen, 169
Hudson Hardwood Flooring
    Company, 214
Humphreys, Benjamin Grubb, 98,
    115, 232
Humphreys County, 28, 238, 241;
    creation of, 292 (n. 28)
hunting, 231–36; enhanced by
    flooding, 143; by Native Americans,
    55, 57, 60–61, 64–68; by planters,
    112, 231–34; by poor whites, 231.
    See also *individual game species*
Hurt, A. B., 178, 216
Hutchins, Thomas, 96

d'Iberville, Pierre Le Moyne, Sieur, 73
Ibitoupa, 72
Illinois Central Railroad, 124, 125–26,
    172, 233
Indian Bayou, 188
Indians. See Native Americans
Innis, Harold, 18
Irving Geo. S., 192–93
Issaquena County, 28, 99, 106, 222,
    241, 243; attempts at flood control
    in, 148; bear hunts in, 112; cypress
    in, 179; land forfeitures in, 239
Itta Bena, 98

Jackson, Andrew, 98
Jackson Point, 223
Jacobs, Annie, 109

Jadwin, Edgar, 159
James, Francis, 205
James, John, 205
James E. Stark Lumber Company, 201
Jefferson, "Blind Lemon," 158
Jefferson, Thomas, 75–76, 84–85, 89,
    93
Jefferson County, 29, 191
John O'Brien Land & Lumber
    Company, 194
Johnson, Fisher, 213
Johnson, J. W., 201
Jolliet, Louis, 72
Jones, Sam, 228
J. R. Gilmore & Co., 190

Kennedy, "Uncle" Hemp, 101–2
Ker, Henry, 40, 92–93, 161, 191, 222
Korn-Conkling Company, 209
Koroa, 72, 73
Küchler, A. W., 30
kudzu, 239
Kurz Brothers Company, 214

Lake George (archaeological site), 69
Lamb-Fish Lumber Company, 194,
    197
land surveys, 93–94; in the Delta, 44,
    94–96; in Louisiana, 96
La Salle, René-Robert Cavelier, Sieur
    de, 64, 72, 77
La Tour, Vitrac de, 142
Lawson, John, 70
Leaf Wilderness Area, 244
Learned, Rufus, 183–85, 192
Learned Lumber Company, logging
    operations by, 183–85; Andrew
    Brown Lumber Company becomes,
    183
Leavenworth, J. H., 200

Leflore County, 28, 134, 193, 222, 244; hardwoods in, 187; land values in, 124; sawmills in, 187

LeFlore family, 84

legislation: on flood control, 144, 146, 148–50, 150–51, 153–54, 159–60; on forest conservation, 214–15, 216, 217

Leland, 158

Lentz, G. H., 218

Leroy Percy State Park, 246

Levee Board of 1858, 146

levees, 142; construction of and standards for, 134, 144, 145, 149, 151, 152, 154; effects of log transportation on, 181; natural, 37; private, 142, 146, 153. *See also* flood control; *and individual levee boards and districts*

"levees only" policy, 151–52, 154, 156, 253

Levins, Richard, 16, 19–20, 22

Lincoln, Abraham, 144

Liquidating Levee Board, 147–48

Little, R. C., 212

Liverpool, England, 134, 188

loggers, working conditions of, 181, 184–85, 188–89, 195, 302 (n. 99)

Louisville, New Orleans & Texas Railroad, 124, 191

Lowe, Ephraim, 178

Lower Mississippi Valley, 27, 28, 31, 35; European and American conquest of, 72–76; seismic activity in, 32

Lucas, James, 116

lumber: cut for fuel, 167–68; production of, in U.S., 166, 171; stealing of, 183, 195; transportation of, 166–67, 169, 194–95. *See also* lumber industry

lumber industry: acquisition of lands by, 173–74, 194; clearing of forests by, 220–21, 223, 225–26; general development of, in Mississippi, 166–67, 174–77, 239–40; general development of, in U.S., 165–66, 171, 177–78, 186–87; organization of, 198; sales of lands by, 124–25, 216–17; technical developments in, 165–66, 174, 175, 177–78, 197; use of pine forests in, 167, 168, 173–78. *See also* lumber; staves and stave making; *and individual companies*

Lumbermen's Club of Memphis, 198

MacArthur, R. H., 226

Madison Parish, La., 214, 227

magnolia, 43

maize. *See* corn

malaria, 104–5, 119, 125, 282 (n. 62)

Malthus, Thomas, 21

mammals, 47, 307 (n. 12); hunting of, by Native Americans, 64–65. *See also individual species*

Mangum, E. C., 234

maples, 42

Marquette, Jacques, 72

Marx, Karl, 9

Massa, Ilmo, 8

McCaslin, Ike (literary character), 164, 255, 302 (n. 96). *See also* Faulkner, William

McClelland, R. L., 201, 205–6

McCoy, "Kansas" Joe, 149

McGee, B. O., 202, 204, 206–7

McIlhenny, John, 233–34

McMahon, R. B., 201

McNeill, A. J., 102–3

McNutt, A. G., 96
Memphis, Tenn., 64, 124, 161, 165, 169, 194, 201, 202; as center of hardwood industry, 188, 189, 197–98, 199, 200, 211, 214; decline of hardwood industry in, 240; as extent of Delta, 26, 27; yellow fever epidemic in, 104
Memphis and Vicksburg Railroad Company, 150
Merchant, Carolyn, 248, 252
Metcalfe, Clive, 233
Minot, 129
Mississippi, creation of, 76,
Mississippi Agricultural Experiment Station, 133
Mississippi Alluvial Plain, 28, 35
Mississippian cultures, 59–60; agriculture of, 60–61, 62–64, 68–70; belief systems of, 61; decline of, 68–69; effects of, on bottomland hardwood forests, 68–71; population dynamics of, 60, 70; use of wild plants by, 64. See also Native Americans
Mississippi Department of Forestry, 216
Mississippi floods: of 1828, 44, 141, 142–43; of 1844, 141, 144; of 1858, 140–41, 145–46; of 1867, 112, 141; of 1882, 141, 150; of 1897, 152, 195; of 1903, 141, 153; of 1912, 141, 153; of 1916, 153; of 1922, 141, 154; of 1927, 141, 156–59, 218, 253; of 1973, 241. See also flood control; flooding
Mississippi Land and Development Association, 216
Mississippi Levee District, 148–49, 150, 151, 154

Mississippi River: drainage basin of, 27–28, 31; formation of, 34–35. See also flood control; flooding; Mississippi floods
Mississippi River Commission, 149–50, 154
Missouri, bottomland hardwood forests in, 311 (n. 62)
Missouri River, 27
mode of production, 9; types of, 16
Mohr, Charles, 178–80, 187, 188
molluscs, 66
Montgomery, Benjamin, 112–14
Montgomery, Isaiah, 114
Morgan, W. A., 203
mosquitoes, 104
moss, Spanish, 43; gathering and ginning of, 190–91
Mound Bayou, 284 (n. 90)
"Mound Builders." See Mississippian cultures
Mound Landing crevasse, 157, 159
mulberry, 41, 95; red, 43
Myllyntaus, Timo, 7, 8

Natchez, 97, 115; as cotton producer, 110, 137, 145; lumber industry in, 167, 169, 180
Natchez (Native American nation), 72, 74
National Audubon Society, 227–28
National Wildlife Refuges (NWR) of Delta, 244–46; Coldwater, 245; Dahomey, 245; Hillside, 244, 245; Mathews Brake, 244; Morgan Brake, 244, 245; Panther Swamp, 244; Tallahatchie, 245; Yazoo, 244, 246
Native Americans: agriculture practiced by, 57–59, 60–61, 62–64,

68–70, 79, 82–84; burning of
forests for hunting by, 65; cultural
chronologies of, in the Southeast,
54–59; and Euro-American trade,
73–74, 80–81, 84–85; and imported
disease, 77–79; land-use practices
of, 52–53; origin of, 54; population
of, 70–71, 77–78; southeastern
tribes of, 71–72; treaties of, with
U.S. government, 75, 84–85; use
of wild plants by, 55, 58; warfare
among, 68, 73–74, 78, 81–82. *See
also* Mississippian cultures; *and
individual tribes and nations*
Neches River, 31
Neuse River, 31
New Orleans, La., 73, 165, 169, 198,
222; flood control in, 142, 159;
lumber industry in, 167, 172, 180;
moss industry in, 191
"New South" ideology, 172–73
Nitta Yuma, 95, 204
Norris, George, 160
Nutt, Rushworth, 80–81, 96
Nuttall, Thomas, 38, 43, 86, 96, 222

Oak Manufacturers Association, 198
oaks, 41, 95, 96, 167, 169, 187,
246; depletion of, 199; laurel, 42;
overcup, 42, 197, 225; swamp
chestnut, 43; uses for, 167, 197;
wasting of, 188; water, 43; white,
42, 198; willow, 41
ocelots, 307 (n. 12)
Odum, Eugene P., 21
Ofo, 78
Ogeechee River, 31
Ohio River, 27
Okefenokee Swamp, 31
Olmstead, Frederick Law, 99–100

opossums, 47, 232
otter, river, 47, 245
Otto, John Solomon, 4

Paepcke, Hermann, 200, 201–2, 204,
214, 238
Paepcke, Walter, 214
Paepcke-Leicht Lumber Company,
199–214, 227, 245; attempts
of, to buy Panther Burn timber,
200–207; competition of, with other
operators, 200–202, 206–7, 208–9;
contracts of, with loggers, 203–4,
213–14; estimates of, on cypress
and hardwood timber, 201, 203,
206, 207, 210, 212; founding and
growth of, 200; merges with other
companies, 214; operations of,
on Bolivar County school section,
212–13; operations of, on Hampton
tract, 207–9; operations of, on
Sheldon tract, 209–12; problems
with landowner of, 212–13;
settlement of accident by, 204–5;
shipments by, 204; timber purchase
contracts by, 207–8, 209–11, 212.
*See also* Chicago Mill and Lumber
Company
Paleo-Indians, 55
Pamlico Swamp, 31
Panola County, 28, 194; sawmills in,
187
Panther Burn, 95. *See also* plantations
parakeet, Carolina, 43, 49–50, 226,
231, 252
Parchman (penal farm), 127–28, 236
Parker, John M., 234
Pascagoula River, 31, 167
Patton, Charlie, 158, 159
Paul, William, 189

Paulding, James Kirke, 82, 106, 108
pawpaw, 41, 43
Payne, Dr., 205
pear, prickly, 43
Pearl River, 31, 167, 178
pecans, 42, 190; commercial growing
    of, 238; uses for, 197
Pee Dee River, 31
pellagra, 61, 119
Penrod-Jurden Company, 214
Percy, LeRoy, 121, 202, 204, 209, 234
Perry County, 244
persimmon, 43, 95, 197
pesticides, 132, 237, 246
Philipp, 193
Phillips and Company, 174
pigeon, passenger, 67; hunting of, 235
pigs, 77; feral, 103
Pinchot, Gifford, 160, 216
pines: loblolly, 174; longleaf, 65, 174;
    shortleaf, 174; slash, 174; white,
    174, 176
planertree, 42
plantations, 88–89, 90; Abydon,
    102; Ashland, 210; Belle Air, 105;
    Brierfield, 112–14; Doro, 98, 233;
    Hurricane, 112–14; Loudon, 210;
    Nanachehaw, 198; Panther Burn,
    102–3, 171, 201–7; post–Civil War
    changes to, 119; run by freedmen,
    112–14; Sunnyside 121; Tokeba,
    112; twentieth-century changes
    to, 236–37; Whitehall, 210; Wild
    Woods, 99; Woodstock, 152
Pollan, Michael, 14
poplar, 198; yellow, 41
possumhaw, 43
Poverty Point culture, 56, 57, 69
Powell, John Wesley, 160
privet, swamp, 43

"Project Flood," 159–60
pulp and paper industry, 217–18,
    219
Putnam, John A., 218–19, 240

Quachita River, 31
Quapaw, 74, 79
Quitman County, 28, 245
Quiver River, 183

raccoons, 47, 232
railroads, 120, 124, 165; acquisition
    of Delta timberlands by, 191, 192–
    93; boosterism by, 125–26, 155;
    and levee construction, 150–51;
    and logging, 174, 175, 176, 185,
    186, 194; sales of Delta lands by,
    156, 191–92. *See also individual
    railroads*
Randolph, W. F., 153
Rappaport, Roy, 8
*Raubwirtschaft*, 21
Red Gum Association, 198
Red River, 31, 96, 163, 174
reptiles, 47; alligators, 235, 245, 267
    (n. 32)
rice, 238, 242, 246
Richardson, Edmund, 120
"Rising High Water Blues," 158
River Bottom Forest, 30. *See
    also* bottomland hardwood
    forests
R. J. Darnell, Inc., 214
Roanoke River, 31
Rodney, 191
Roosevelt, Franklin D., 228
Roosevelt, Theodore, 160, 217,
    233–34, 243
Rosedale, 50, 158, 228
Ruffin, James H., 104

Sabine River, 31
Santee-Cooper River, 31
Santee River swamp, 31
Sardis, 194
Sardis Reservoir, 164
Sargent, Charles Sprague, 178, 187
sassafras, 41, 43, 66
Satartia, 232
Satterfield, W. R., 204–6, 211–12
sawmills, in Mississippi, 167, 169,
    175–77. *See also* lumber; lumber
    industry; *and individual lumber
    companies*
Schenck, Carl A., 216
Schultz, Christian, 48, 191, 235
Sclater, P. L., 40
Shantz, H. L., 30
sharecroppers, 110, 117–20, 236;
    living conditions of, 119. *See also*
    tenancy
Sharkey County, 28, 203, 222, 233,
    241, 243; cotton production of,
    126; creation of, 292 (n. 28); land
    surveys in, 95
Sheldon, George L., 209–12
Sheldon, Rose, 209
Sillers, Walter, Sr., 147
Singer Sewing Machine Company,
    227–28, 243
Singer Tract, La., 227–28
slaves: beliefs of, 101–2; demand in the
    Delta for, 99–100, 110; legislation
    concerning, 215; in levee work, 142,
    147; mortality of, 104–5; in Native
    American communities, 61, 83; in
    plantation economy, 88–89, 90;
    as trade items, 73–74; use of wild
    plants by, 101
Smith, Adam, 19
Smith, Berry, 116, 121, 153

Smith, Bruce D., 54
Smith, L. E., 203–4
Society of American Foresters, 44–45
Sonderegger, V. H., 218
Soper, J. P., 201
Southeastern Bottomland Hardwood
    Forest, 30. *See also* bottomland
    hardwood forests
Southern Alluvial Land Association,
    126
Southern Floodplain Forest, 27,
    29, 30; climate of, 29. *See also*
    bottomland hardwood forests
Southern Forest Experiment Station,
    218, 222
Southern Forest Survey, 196, 218,
    221–26
Southern Hardwoods Laboratory, 240
Southern Pine Association, 217
soybeans, 238–39, 242, 246
Spain & Co., 202, 206
spiders, 46
squirrels, 47, 232
staves and stave making, 167, 188–89,
    193–94, 195–96
steamboats, 145; lumber consumption
    by, 168
Steele Bayou, 28, 179
Steward, Julian, 8
St. Francis River, 31, 163
Stier, Isaac, 116
Stoner, Mr. (owner of Hampton
    timber), 208
Stoneville, 240
stork, wood, 246–47
Sunflower County, 28, 44, 98;
    agriculture in, 97, 121, 127, 129;
    flood of 1882 in, 150; labor in,
    121, 127; logging in, 188, 195–96;
    population of, in 1860, 106

Sunflower River, 43, 100, 141, 150, 179, 183, 189, 192; Big, 28, 129; Little, 179, 233
Suwannee River, 31
Swamp Land Acts, 144, 145, 168, 192
sweetgum (redgum), 41, 42, 95, 96, 223, 246; depletion of, 199, 225–26; promotion of, 190; uses for, 197
sycamore, 40, 41, 43, 68, 199

Tallahatchie County, 28, 195, 196, 245; cypress in, 179–80; lumber industry in, 193, 197; sawmills in, 187
Tallahatchie River, 28, 141, 161, 163, 193; Little, 28, 164
Tallulah, La., 214
Tanner, James T., 227
Taposa, 72
Tate County, 28
"Teddy Bear," 234
tenancy, 117–20; decline of, 236–37; increase of, 129–30; types of, 117–18
Tennessee Valley Authority (TVA), 160
Tensas River, 31, 228
tern, least, 246–47
Thistlethwaite Lumber Company, 218
Thorne, Robert M., 94, 95
Thos. Wellford & Sons, 209
Tiou, 72
tobacco, 63; consumption by loggers, 184–85
Tombigbee River, 31
Trager, H. S., 198
trees. See individual species
Trinity River, 31
Tunica, 78, 79
Tunica County, 28, 99, 103, 106; cotton production in, 110; cypress and hardwoods in, 187; lack of flood control in, 149
tupelos, 199; black (blackgum), 41, 42, 95; uses for, 197; water, 42
turkey, 49, 67, 232, 233, 244
Turner, Frederick Jackson, 276 (n. 51)
Tutwiler, 195
Twain, Mark, 150

Urania Lumber Company, 217
U.S. Army Corps of Engineers, 149, 156, 159, 162–63, 241, 243, 245
U.S. Congress, 77, 85, 126, 144, 149, 159, 163, 221
U.S. Department of Agriculture, 117, 122, 133, 155, 162, 185, 216, 239
U.S. Department of Commerce and Labor, 221
U.S. Fish and Wildlife Service, 44, 228, 241, 244
U.S. Forest Service, 27, 30, 44, 214, 221
U.S. Supreme Court, 77

Vaiden, M. G., 228
Vicksburg, 59, 145, 152, 156, 158, 182; in Civil War, 147; as extent of Delta, 26, 27, 28; lumber industry in, 167, 180, 194; reestablished as port city, 154
Vileisis, Ann, 4
vines, 43

Wall, E. G., 178
walnut, black, 41, 42, 189; high price of, 190; wasting of, 188
Warren County, 28, 29, 108, 112, 198, 222, 241; cypress in, 179; levees in, 290 (n. 8)
Washington, Lake, 99

Washington County, 28, 98;
agriculture in, 99, 100, 102, 103,
105, 108, 127, 241; attempts at
flood control in, 148, 152; cotton
production of, 126; cypress in, 179;
forestry in, 240; hardwoods in,
187; logging in, 210; population
of, 106, 126, 127; sawmills in,
187; settlement of, 97; wildlife in,
232–33, 234; wildlife refuges in,
244, 246
Watson (lawyer), 205
Webb, 196
White, Richard, 80, 85
White River, 31, 163
wilderness: concept of, 14; Delta as, 2,
86, 135, 249
Wilderness Areas: Black Creek, 244;
Leaf, 244
Wildlife Management Areas, in Delta,
246
Wilkinson County, 29, 198
Willis, John C., 4
willow, black, 42, 225; use of, for
revetment work, 293 (n. 39), 295
(n. 61)
Wilson, Edward O., 226
Wilson, Henry, 183–85
Winterville Mounds State Park,
246
Wissler, Clark, 8
Wittfogel, Karl A., 139–40
Wolf, Eric, 16
wolf, red, 47, 226, 233
Wolverine Lumber Company, 200

woodpecker, ivory-billed, 49–50, 226,
252; extinction of, 227–29, 236;
hunting of, 67–68, 234
Woods, Clyde, 4
Worster, Donald, 8–9, 12, 19, 139
Worthington, Amanda Dougherty,
103–4
Worthington, William, 232

Yalobusha River, 28, 124, 163–64
Yazoo (Native American nation), 72,
74, 78
Yazoo & Mississippi Valley Railroad,
124, 191
Yazoo Backwater Project, 162
Yazoo City, 97, 112
Yazoo County, 28, 32, 102, 232, 241,
244; cypress in, 179
Yazoo Headwater Project, 163–64
Yazoo land companies, 76–77, 98
Yazoo-Mississippi Delta. *See* Delta,
the
Yazoo-Mississippi Delta Levee
District, 150, 151–52
Yazoo Pass, 141, 147
Yazoo Pumps Project, 241
Yazoo River: 28, 31; development of
flood control on, 160–64; drainage
basin of, 28, 141, 161–64; flooding
of, 141, 161–62; relocation of
mouth of, 154
yellow fever, 104–5, 125
Yocona River, 28, 163–64

Zon, R., 30